MW00576318

Current Concepts in Forensic Entomology

Jens Amendt · M. Lee Goff
Carlo P. Campobasso · Martin Grassberger
Editors

Current Concepts in Forensic Entomology

Editors
Jens Amendt
Institute of Forensic Medicine
University of Frankfurt am Main
Germany
amendt@em.uni-frankfurt.de

M. Lee Goff
Forensic Sciences Program
Chaminade University of Honolulu
Hawaii
USA

Carlo P. Campobasso
Department of Health Sciences
University of Molise
Campobasso
Italy
carlo.campobasso@unimol.it

Martin Grassberger
Institute of Pathology
Hospital Rudolfstiftung
Vienna
Austria
martin.grassberger@wienkav.at

ISBN 978-1-4020-9683-9 e-ISBN 978-1-4020-9684-6
DOI 10.1007/978-1-4020-9684-6
Springer Dordrecht Heidelberg London New York

Library of Congress Control Number: 2009943054

Cover illustration: One adult *Chrysomya marginalis* and three adult *Chrysomya albiceps* feeding on a White Rhinoceros (*Ceratotherium simum*) carcass in Thomas Baines Nature Reserve, South Africa by Cameron Richards (Natural History Museum London).

Printed on acid-free paper

Springer is part of Springer Science+Business Media (www.springer.com)

Preface

Forensic Entomology deals with the use of insects and other arthropods in medico legal investigations. We are sure that many people know this or a similar definition, maybe even already read a scientific or popular book dealing with this topic. So, do we really need another book on Forensic Entomology? The answer is 13, 29, 31, 38, and 61. These are not some golden bingo numbers, but an excerpt of the increasing amount of annual publications in the current decade dealing with Forensic Entomology. Comparing them with 89 articles which were published during the 1990s it illustrates the growing interest in this very special intersection of Forensic Science and Entomology and clearly underlines the statement: Yes, we need this book because Forensic Entomology is on the move with so many new things happening every year.

One of the most attractive features of Forensic Entomology is that it is multidisciplinary. There is almost no branch in natural science which cannot find its field of activity here. The chapters included in this book highlight this variety of researches and would like to give the impetus for future work, improving the development of Forensic Entomology, which is clearly needed by the scientific community. On its way to the courtrooms of the world this discipline needs a sound and serious scientific background to receive the acceptance it deserves.

This book does not ignore the forensic and entomological basics of the discipline, and gives an update in entomotoxicology, offering a survey about the decomposition of a cadaver (including a protocol for decomposing studies) and keys for identifying the difficult stages of immature insects. Especially the latter topic is an important one, as we believe that, despite the enormous progress made in bar-coding and identification of many taxa via DNA-analysis in recent years, one should not neglect the very basic skills - particularly because using these "easy lab-tools" could give you a speciously feeling of certainty.

Forensic Entomology and Blowflies are very often named in the same breath. We would like to attract the readers to some groups of animals which are neglected or even ignored such as, beetles and mites. Blowflies are much easier to handle in the lab than beetles, which could be the major reason why the majority of developmental studies are dealing with Diptera. If you have ever seen a cadaver infested by thousands of Silphidae or Dermestidae you soon realise that you must know more about them. Mites are not insects, nevertheless they belong traditionally to medical entomology since its early beginnings. So we should recognize them as a part of forensic entomology as well, keeping in mind that the great Mégnin includes them in his famous *Faune des Cadavres* in the late nineteenth century. These arthropods are especially abundant in buildings, which leads to another gap in our knowledge: Indoor scenarios. Interestingly the majority of experiments analysing the insect succession on cadavers take place outside in the field. However we should not ignore that vast amount of corpses found every year indoors. No doubt, it's much easier to conduct experiments out in nature, but we need indoor data sets as well for a better understanding of crime scenes which are located in a building.

Working as a forensic entomologist means mainly working with terrestrial ecosystems, but people die in the sea as well, or their dead bodies are dropped there after a homicide. What happens to those corpses? How do the bodies decompose? And are any arthropods or insects involved in this process? You will know this soon. From deep in the sea to down in the ground: It is surprising that our knowledge of forensic entomology of the soil is so incomplete. Dealing with cases where the bodies were buried always creates a lot of difficulties. Is there a succession in the soil as well as on the surface? Are the species found on the body able to colonize the buried cadaver or did they colonize him before?

Despite all of the scientific possibilities to improve the quality of entomological reports for the court, there are always pitfalls which cannot always be avoided. This book highlights certain caveats, bearing in mind that we are dealing with biological systems which do not always work in the same predictable manner. Due to the variability, we need statistics and probabilities in our expertises, which information is also covered in this book.

A topic such as climate change would not be expected in a book about Forensic Entomology, but the truth is simple: Climate change is everywhere and it will also influence a topic like the use of insects in forensic investigations. Last but not least we dedicate an own chapter to the field of myiasis, which is a well known subject for a veterinary. Insects also infest living humans and feed on them. A forensic entomologist should understand this process because it could bias his work, and at the same time he might be asked to estimate the time of negligence.

Curious? Then join us on our journey through the world of Forensic Entomology, but take care: after reading this book you may find you like this subject so much that perhaps you can find your own field of activity there: It is an exciting field of research.

The editors, 2009

Jens Amendt
M. Lee Goff
Carlo P. Campobasso
Martin Grassberger

Contents

Chapter 1
Early Postmortem Changes and Stages of Decomposition

M. Lee Goff

1.1 Introduction

When faced with the task of estimating a period of time since death, there are generally two known points existing for the worker: the time at which the body was discovered and the last time the individual was reliably known to be alive. The death occurred between these two points and the aim is to estimate when it most probably took place. This will be an estimate since, it is generally accepted that there is actually no scientific way to precisely determine the exact period of time since death. What is done in the case of entomology is an estimation of the period of insect activity on the body. This period of insect activity will reflect the minimum period of time since death or postmortem interval (PMI) but will not precisely determine the time of death. In most cases, the later point is more accurately known than the former. Individuals tend to recall when they first encountered the dead body with considerable precision. This is typically not in their normal daily routine and it makes an impression, even on those accustomed to dealing with the dead.

Once the body is discovered, those processing the scene make meticulous (at least we hope meticulous) notes including times of arrival, departure, movement of the body and, finally, when the body is placed into the morgue. By contrast, the time at which the individual was last reliably known to be alive is often less precise. This is possibly due to the fact that those having the last contact most probably did not anticipate that this would be their last encounter with the individual and nothing of significance took place at the time. In the absence of something unusual, one rarely notes the time one said "good morning" to a neighbor or passed an acquaintance on the street. The last time the individual was reliably known to be alive may involve statements concerning the last time the individual was seen alive. It may involve hearing the individual or a telephone communication. Some instances may involve the touch or smell of the individual. Obviously there is some latitude possible in this determination and the time frame is often incorrect. For this reason, the precision

M.L. Goff
Chaminade University, Forensic Sciences Program, Hawaii, USA

J. Amendt et al. (eds.), *Current Concepts in Forensic Entomology*,
DOI 10.1007/978-1-4020-9684-6_1, © Springer Science+Business Media B.V. 2010

of the time of discovery and collection of specimens become of major significance, as they are the anchor for the estimates. Estimation begins when the insects are collected and preserved, stopping the biological clock.

As the process of estimating the period of insect activity takes place, it must be kept in mind that the parameters of the estimate become progressively wider as the period of time since death increases. The changes to a body that take place immediately following death are often more rapid than those occurring later during the decomposition process. Thus the estimate begins, potentially, with a range of plus or minus minutes, goes to hours, days, weeks, months and finally "its been there a long time." The last is not the most popular with law enforcement agencies as they had already guessed that. It should also be kept in mind that the estimates presented, by their very nature, are not precise. I have found in my experience that it is typically the more inexperienced investigator who provides the most precise and unchanging estimates of the PMI.

Decomposition is a continuous process, beginning at the point of death and ending when the body has been reduced to a skeleton. Although this process is a continuum, virtually every study presented has divided this process into a series of stages. The number of stages has varied from one to as many as nine, depending on author and geographic region (Goff 1993) (Table 1.1). While the number of stages considered has varied, there does not appear to be a firm relationship between these and the total number of species observed in each study. For example, Cornaby (1974) working in Costa Rica using lizards and toads as animal models noted only 1 stage for

Table 1.1 Summary of selected decomposition studies giving numbers of recognized stages and taxa listed

Author and Ref.	Date	Locality	Animal model	# Stages	Total # of arthropod taxa
Avila and Goff	2000	Hawaii	Pigs(burnt)	5	68 species
Blacklith and Blacklith	1990	Ireland	Birds, mice	1	27 species
Bornemissza	1957	Australia	Guinea pig	5	45 groups listed
Braack	1986	Africa	Impala	4	227 species
Coe	1978	Africa	Elephants	3	No totals given
Cornaby	Cornaby, 1974	Costa Rica	Lizards, toads	1	172 species
Davisand Goff	2000	Hawaii	Pigs(intertidal)	5	85 species
Earlyand Goff	1986	Hawaii	Cats	5	133 species
Hewadikaram and Goff	1991	Hawaii	Pigs	5	46 species
Megnin	1894	France	Humans	9	No totals given
Payne	1965	South Carolina	Pigs(surface)	6	522 species
Reed	1958	Tennessee	Dogs	4	240 species
Rodriguez and Bass	1983	Tennessee	Humans	4	10 families listed
Shalaby et al.	2000	Hawaii	Pig(hanging)	5	35 species
Shean et al.	1993	Washington	Pig	–	48 species
Tullisand Goff	1987	Hawaii	Pig	5	45 species

decomposition but recorded 172 different species. By contrast, work in Hawaii by Early and Goff (1986), using domestic cats as the animal model, recognized five stages of decomposition but recorded 133 species. Other studies have recognized other numbers but with no real correlation between stages observed and numbers of taxa reported. To a certain extent, these differences may be related to sampling methods and taxonomic interests of those involved.

1.2 Early Postmortem Changes

As death proceeds, there are a series of early changes to the body that result in a definite change in the physical nature and/or appearance of the body prior to the onset of gross, recognizable decompositional changes. These changes have traditionally been used in estimations of the PMI and may be a source of confusion if not recognized. For that reason, they are described here.

1.2.1 Livor Mortis

One of the early changes observable is livor mortis, also referred to as lividity, postmortem hypostasis, vibices and suggilations. This is a physical process. While the individual is alive, the heart is functioning and circulating the blood. When death occurs, circulation stops and the blood begins to settle, by gravity, to the lowest portions of the body. This results in a discoloration of those lower, dependent parts of the body (Fig. 1.1). Although beginning immediately, the first signs of livor

Fig. 1.1 Livor

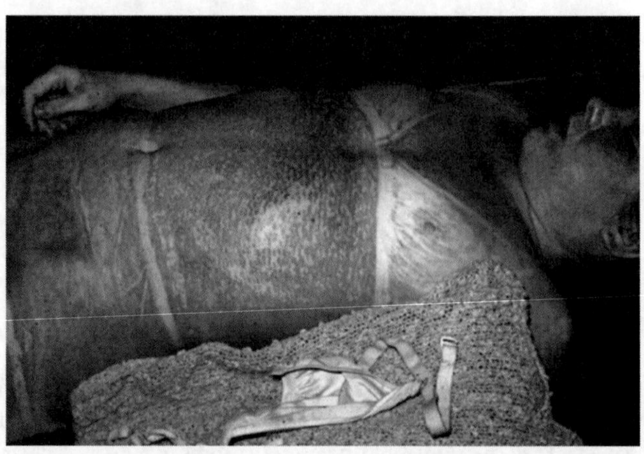

Fig. 1.2 Blanching

mortis are typically observed after a period of approximately 1 h following death with full development being observed 3–4 h following death. At this time, the blood is still liquid and pressing on the skin will result in the blood being squeezed out of the area (blanching), only to return once pressure is removed. This situation continues until 9–12 h following death, at which time the pattern will not change and the livor mortis is said to be "fixed." Any areas of pressure resulting from clothing or continued pressure during this period will not show discoloration (Fig. 1.2).

1.2.2 Rigor Mortis

This is a chemical change resulting in a stiffening of the body muscles following death due to changes in the myofribrils of the muscle tissues. Immediately following death, the body becomes limp and is easily flexed. As ATP is converted to ADP and lactic acid is produced lowering the cellular pH, locking chemical bridges are formed between actin and myosin resulting in formation of rigor. Typically, the onset of rigor is first observed 2–6 h following death and develops over the first 12 h. The onset begins with the muscles of the face and then spreads to all of the muscles of the body over a period of the next 4–6 h (Gill-King 1996). Rigor typically lasts from 24 to 84 h, after which the muscles begin to once again relax. The onset and duration of rigor mortis is governed by two primary factors: temperature and the metabolic state of the body. Lower ambient temperatures tend to accelerate the onset of rigor and prolong its duration while the opposite is found in warmer temperatures. If the individual has been involved in vigorous activity immediately prior to death, the onset of rigor is more rapid. Body mass and rates of cooling following

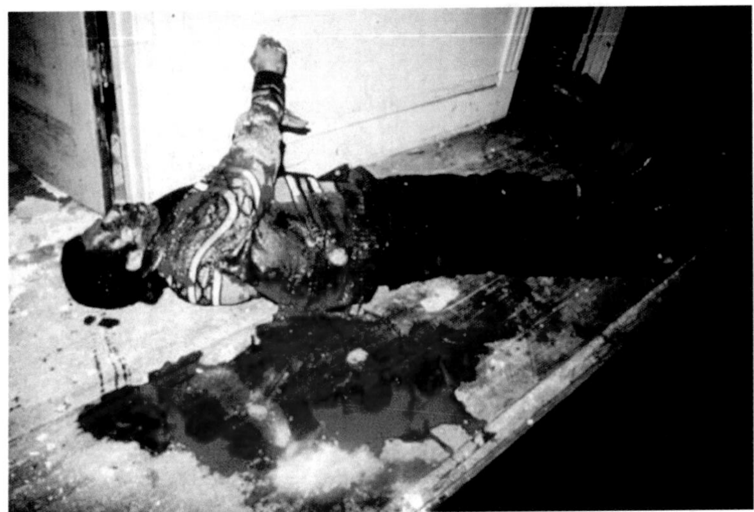

Fig. 1.3 Rigor

death also influence the onset and duration of rigor mortis. As rigor disappears from the body, the pattern is similar to that seen during the onset, with the muscles of the face relaxing first (see Fig. 1.3).

1.2.3 Algor Mortis

Once death has occurred, the body ceases to regulate its internal temperature and the internal temperature begins to approximate the ambient temperature. In most instances this involves a cooling of the body until ambient temperature is reached, most often in a period of 18–20 h (Fisher 2000). Although there are several different approaches, the rate of cooling is most often expressed by the equation:

$$PMI\,(hours) = \frac{98.6\ Body\ Temperature\ (^{\circ}\,F)}{1.5} \tag{1}$$

Any estimate of the postmortem interval obtained using this technique should be limited to the very early stages of death (18 h or less) and treated with care. There are several obvious factors involved in the cooling of the body that may easily influence the rate at which this occurs. The size of the individual is a major factor. A smaller individual will cool more rapidly than a larger individual in the same set of conditions. Exposure to sunlight or heating may also influence the rate of cooling as may clothing and a number of other factors. The most commonly used temperature in these calculations are from the liver although rectal temperature may also be employed.

Fig. 1.4 Glove

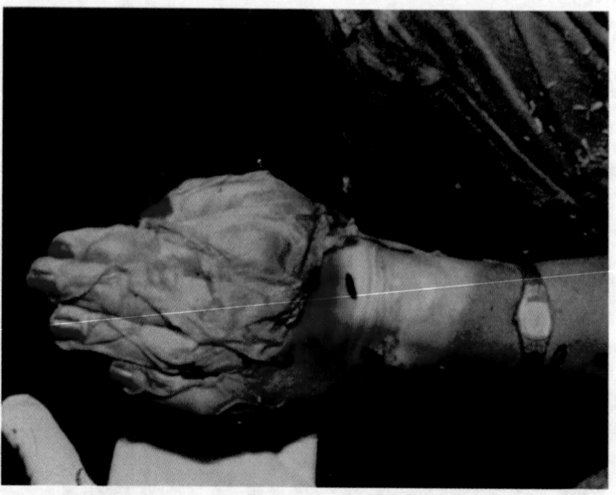

1.2.4 *Tache Noir*

Following death, the eyes may remain open and the exposed part of the cornea will dry, leaving a re-orange to black discoloration. This is termed tache noire (French for "black line") and may be misinterpreted as hemorrhage. Unlike hemorrhage, this will have symmetrical distribution, corresponding to the position of the eyelids (see Fig. 1.5).

1.2.5 *Greenish Discoloration*

As the body decomposes, gasses are produced in the abdomen and other parts of the body. While the exact composition of the gasses may vary from body to body, a significant component of these gasses is hydrogen sulfide (H_2S). This gas is a small molecule and readily diffuses through the body. Hydrogen sulfide will react with the hemoglobin in blood to form sulfhemoglobin. This pigment is greenish and may be seen in blood vessels and in other areas of the body, particularly where livor mortis has formed.

1.2.6 *Marbling*

As the anaerobic bacteria from the abdomen spread via the blood vessels, the subcutaneous vessels take on a purple to greenish discoloration, presenting a mosaic appearance, similar to what is seen in cracking of old marble statuary. Typically this is seen on the trunk and extremities (see Fig. 1.6).

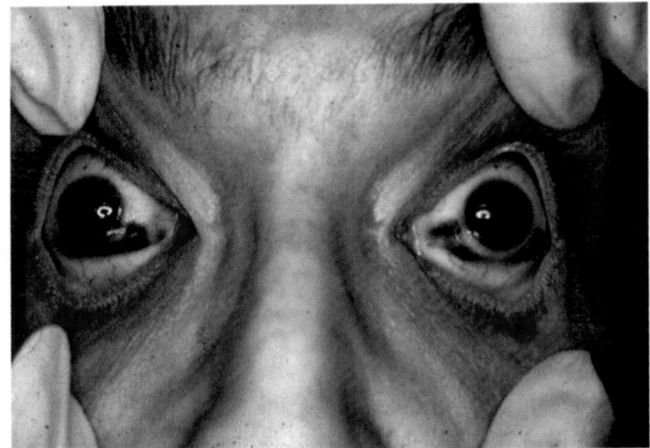

Fig. 1.5 Tache noir

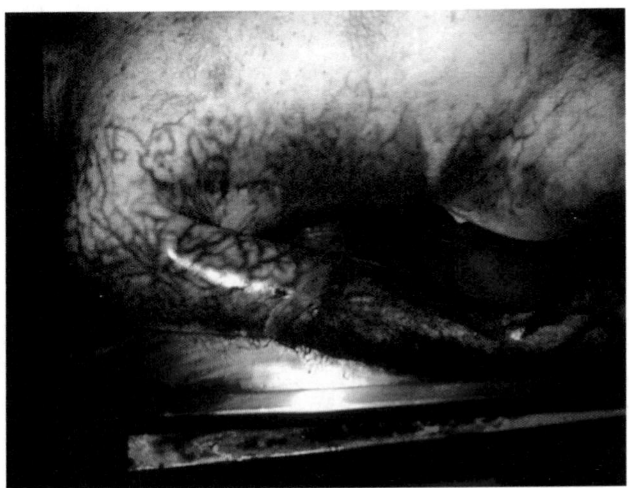

Fig. 1.6 Marbling

1.2.7 Skin Slippage

Upon death, in moist or wet habitats, epidermis begins to separate from the underlying dermis due to production of hydrolytic enzymes from cells at the junction between the epidermis and the underlying dermis. This results in the separation of the epidermis which can be easily removed from the body. Slippage may first be observed as the formation of vesicle formation in dependent portions of the body. In some instances, the skin from the hand may separate from the underlying dermis as a complete or relatively complete unit. This is termed glove formation and can be removed from the hand s an intact unit. This skin can be used for finger printing, often with better results than if the skin remains on the hand (see Fig. 1.4).

1.2.8 Mummification

In a dry climate, a body will dessicate. The low level of humidity will serve to inhibit bacterial action and typically there will be some exclusion of insects and other scavengers from the body. The temperatures will be either very hot or very cold in this type of situation. The dessicated tissues and skin will have a leathery appearance and will survive for long periods of time with minimal change. In hot, dry climates, mummification can occur within a period of several weeks (see Fig. 1.7).

1.2.9 Saponification

This is the process of hydrolysis of fatty tissues in wet, anaerobic situations, such as submersion or in flooded burials. The tissues take on a waxy appearance and consistency. This process requires a period of several months to complete (see Fig. 1.8).

1.2.10 Putrefaction

Putrefaction is nature's recycling process. It is the result of the combined activities of all organisms involved in decomposition, reducing the body to a skeletal state.

Fig. 1.7 Mummification

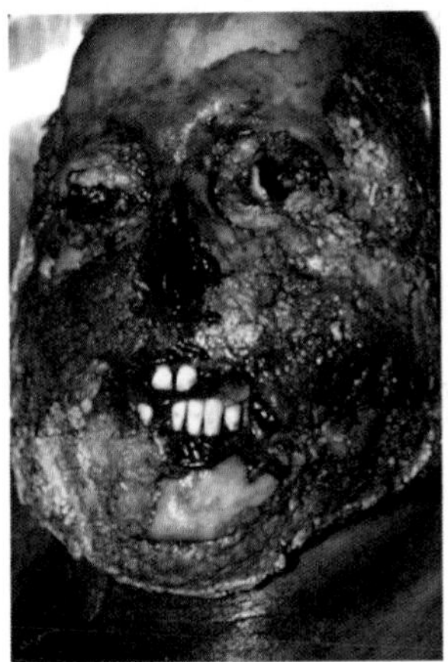

Fig. 1.8 Saponification

1.3 Decomposers

In order to consider the process of decomposition and the stages involved, it helps to have some understanding of what organisms will be involved in the process. Some of these have already been mentioned with respect to the early changes to the body mentioned above. There are four primary categories of organisms involved in decomposition.

1.3.1 Bacteria

There are bacteria associated with both the external and internal aspects of the human body. While alive, the body defends against these organisms and, in fact, many are actually beneficial. There is a large component of anaerobic bacteria associated with the human digestive system. Some of these exist normally in our intestines, such as *E. coli*, and, as long as they remain in place do no damage and may assist in breakdown of food and materials. By contrast, the same organism in the wrong place, such as the kidneys, etc., will result in a serious disease condition. Once the individual dies, there are few barriers to keep them in any particular place and human tissues are excellent growth media. Shortly after death, these bacteria begin to digest the body from the inside out. This activity is particularly evident in the areas of the head and abdomen. The metabolic activities of these bacteria are major components of the decomposition processes.

1.3.2 Fungi/Molds

As noted earlier, the outer surface of the human body is comprised of dead material. This dead outer layer is necessary to assist in the survival of the human body. As a normal process, as new tissues are produced below, the outer *stratum corneum* is shed as dander. As it is shed, any attached spores of molds or fungi are also shed from the body. Following death, the outer layer is no longer shed and the mold and fungi spores with begin to colonize the external surface of the body, often forming significant mats on the body.

1.3.3 Insects

Insects and other arthropods are the primary organisms involved in the major decomposition of the body. They arrive at exposed remains shortly after death, often in less than 10 min, and quickly begin their activities. As the rest of this work is devoted to their activities, no more needs to be said here.

1.3.4 Vertebrate Scavengers

A dead human is a potential food resource for a number of vertebrate scavengers. Carnivores of all sizes can rapidly alter a dead body. Even small rodents can cause significant damage to a body in a relatively short period of time. In the wild it may take less than a week for scavengers to completely skeletonize a body. In addition to non-domesticated animals, common domestic animals and rodents will also feed on the body in the absence of their normal food. Pet dogs and particularly cats will feed on their deceased owners, most often attacking the face and exposed limbs first.

1.4 Factors Delaying Decomposition

While there are a number of different organisms involved in the process of decomposition, there are also several types of factors that serve to stop of retard the rate at which the process continues. These barriers to decomposition fall into three broad categories.

1.4.1 Physical Barriers

Physical barriers to decomposition are those that prevent access of the body by physical means. A body buried in the soil does not decompose as quickly as one exposed on the surface.

In a similar manner, a body enclosed in a sealed casket or placed into some form of sealed container will also exhibit a delayed decomposition.

1.4.2 Chemical Barriers

The embalming process is specifically designed to prevent the decomposition of the body, with natural body fluids being drained and replaced with various preservative fluids. As the body is then typically placed into a casket, the process should, if done properly, delay decomposition for an extended period of time. The presence of insecticides on, in or in the vicinity of the body may also serve to delay the onset of insect activity for a period of time. It should be noted that insecticides will not permanently delay the colonization of the body by insects. In many cases, immature insects are able to survive on a body with concentrations of an insecticide that would prove fatal to the adults of the same species (Gunatilake and Goff 1989)

1.4.3 Climatic Factors

Temperature can serve as a major factor delaying decomposition. At lower temperatures, bacterial growth and insect activity can be retarded or even arrested. At temperatures below 6°C most insect activity ceases but may resume once temperatures rise above this threshold. In a similar manner, high temperatures will also result in cessation of inset activity, and, if in a dry habitat, result in mummification of the body. Wind also serves to inhibit insect flight and thus colonization of the body. Many texts will indicate a wind speed in excess of 16 km/h will inhibit insect flight. This should not be accepted as a firm wind speed as in many tropical and island areas, tradewinds typically blow at a speed greater than this and there is significant insect activity. Rainfall may also serve as a temporary barrier but, once the rain ceases, the insects again become quite active.

1.5 Relationships of Insects to a Body

There are several distinct relationships between an insect and a decomposing body. The population of insects and other arthropods encountered in any given habitat will contain elements unique to that habitat and components having a wider distribution. Within this population there will be species having some type of relationship to the decomposing body. This relationship will vary with taxon and not all relationships will be of equal value to the investigation. All must be considered as, under different circumstances, there will be different values for the relationship. There have been four basic relationships between a decomposing body and insects and other arthropods (after Goff 1993).

1.5.1 Necrophagous Species

Those taxa actually feeding on the corpse. This group includes many of the true flies (Diptera) particularly the blow flies (Calliphoridae) and flesh flies (Sarcophagidae) who are early invaders and Beetles (Coleoptera: Silphidae and Dermestidae) . This group includes species that may be the most significant isolatable taxa for use in estimating a minimum period of insect activity on the body during the early stages of decomposition (days 1–14) (see Figs. 1.9–1.11).

Fig. 1.9 Chrysomya Lv II

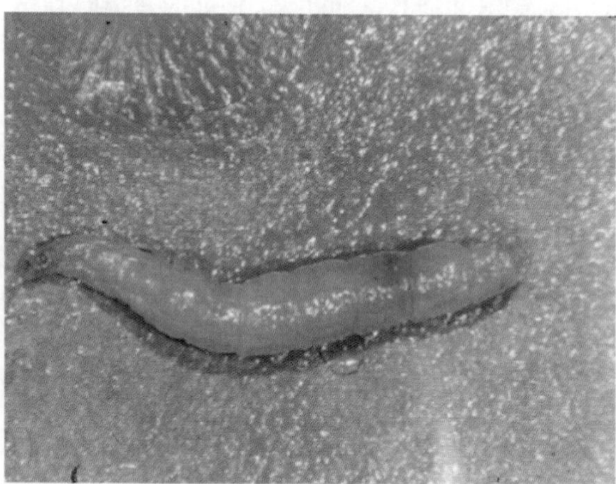

Fig. 1.10 Piophilidae

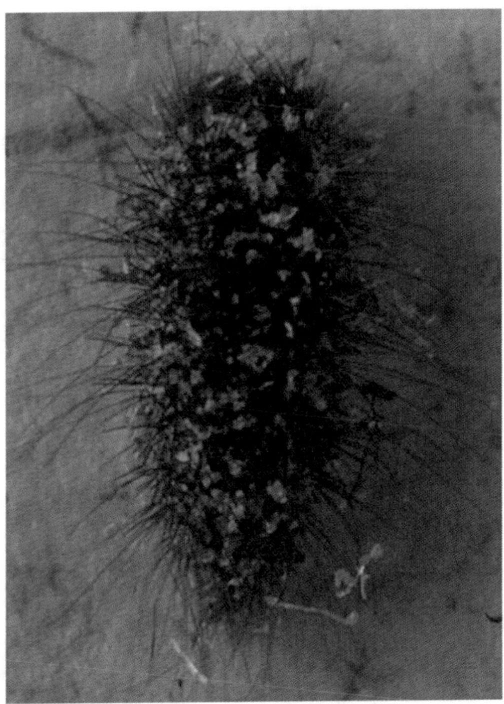

Fig. 1.11 Dermestes

1.5.2 Predators and Parasites of Necrophagous Species

The predators and parasites of the necrophagous species comprise the second most significant group of carrion-freqenting taxa. Many of the beetles (Coleoptera: Silphidae, Staphylinidae, and Histeridae), true flies (Diptera: Calliphoridae, Muscidae and Stratiomyidae), and Wasps (Hymenoptera) parasitic on fly larvae and puparia are included (Figs. 1.12 and 1.13). In some species, fly larvae (maggots) that are necrophages during the early portions of their development become predators on other larvae during the later states of their development, as is the case for *Chrysomya rufufacies and Hydrotaea aenescens.*

1.5.3 Omnivorous Species

Included in this category are the taxa such as wasps, ants and some beetles, that feed on both the corpse and associated arthropods. Early and Goff (1986) observed that large populations of these species may actually retard the rate of carcass removal by depleting the populations of necrophagous species (see Fig. 1.14).

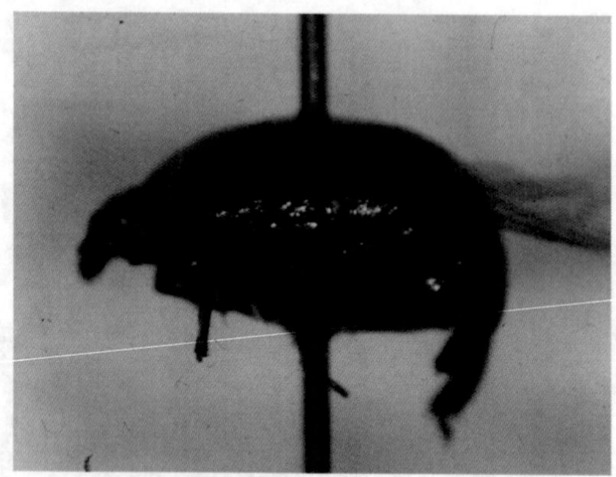

Fig. 1.12 Histeridae

Fig. 1.13 Philonthus

Fig. 1.14 Anopolepsis

1.5.4 *Adventive Species*

This category includes those taxa that simply use the corpse as an extension of their own normal habitat, as is the case for the springtails (Collembola), spiders, centipedes, and millipedes.

1.5.5 *Accidentals*

Another category that is not always recognized but may still be of significance is what might be termed "accidentals." These are species that have no real relationship to the corpse but still are found on the body. These insects may have fallen onto the body from surrounding vegetation, thus possibly supplying some information on postmortem movement of a body. On the other hand, when an insect stops flying, it has to land on something and that "something" might happen to be the body. This is a fact all too often ignored, even by entomologists.

1.6 Stages of Decomposition

1.6.1 *Numbers of Stages*

As noted earlier, there have been a number of different stages proposed for the decomposition process. Keep in mind that the process is a continuum and discrete stages, characterized by physical features and distinctive assemblages of insects, do not exist in nature. Regardless, virtually every study conducted has attempted to divide the process into stages. While artificial, these stages have definite utility. First, they allow for easy organization of research reports and discussion. There is also a utility in court proceedings. Typically in the United States, juries are composed of individuals with little if any background in the biological sciences. They often confused and repulsed by the process of decomposition they are being asked to consider. Under these circumstances, use of stages gives them something to use for reference and makes their task somewhat easier, if not more pleasant.

In studies conducted in Hawaii, five stages have been recognized and these appear to be easily applied to studies conducted in temperate areas (Lord and Goff 2003). These stages are: Fresh, Bloated, Decay, Postdecay and Skeletal or Remains. The most common modification of this set is to subdivide the Decay Stage into Active Decay and Advanced Decay stages. Given the subjective nature of these stages, the Decay Stage is treated here as a single stage. As detailed discussions of insect and arthropod succession are presented elsewhere in this book, the treatment here will be primarily an overview and details of specific insect activity left to the more detailed discussions.

1.6.2 Fresh Stage

The Fresh Stage begins at the moment of death and continues until bloating of the body becomes evident. There are few distinctive, gross decompositional changes associated with the body during this stage although greenish discoloration of the abdomen, livor, skin cracking, tache noir may be observed. The insect invasion of the body generally begins with the natural body openings of the head (eyes, nose, mouth and ears), anus and genitals, and wounds present on the body. The first insects to arrive under most circumstances are the Calliphoridae (blow flies) and Sarcophagidae (flesh flies). Female files will arrive and begin to explore the potential sites for oviposition or larviposition. These flies will often crawl deep into the openings and either deposit eggs of first instar larvae or maggots (Fig. 1.10). While the openings associated with the head are uniformly attractive to flies, the attractiveness of the anus and genital areas may depend on their being exposed or clothed. Wounds inflicted prior to death have been observed to be more attractive to flies for colonization if inflicted prior to death, when blood is flowing, than wounds inflicted postmortem and lacking a blood flow. During this stage, the eggs laid in the body begin to hatch and there is internal feeding activity, although there may be little evidence of this on the surface (see Fig. 1.15).

Fig. 1.15 Fresh stage

1.6.3 Bloated Stage

The principal component of decomposition, putrefaction, begins during the Bloated Stage. The anaerobic bacteria present in the gut and other parts of the body begin to digest the tissues. Their metabolic processes result in the production of gasses that first cause a slight inflation of the abdomen. When this is noted, the Bloated Stage is considered to begin. As this progresses, the body may assume a fully inflated, balloon-like appearance. The combined processes of putrefaction and the metabolic activities of the maggots begin to cause an increase in the internal temperatures of the body. These temperatures can be significantly above ambient temperature (50ºC+) and the body becomes a distinct habitat, in many ways independent of the surrounding environment. The adult Calliphoridae are strongly attracted to the body during this stage in decomposition and significant masses of maggots are observed associated with the head and other primary invasion sites. While these populations are visible externally, there are larger populations present internally. Internal pressures caused by production of gasses result in the seeping of fluids from the natural body openings during this stage and the strong smell of ammonia is noted. These fluids seep into the substrate beneath the body and this becomes alkaline. The normal soil fauna will leave the area under the body as a result in this change in the pH and the invasion of a set of organisms more closely associated with decomposition begins (see Figs. 1.16 and 1.17).

1.6.4 Decay Stage

While the start and termination points for the stages of decomposition are largely subjective, there is a definite physical event marking the start of the Decay Stage.

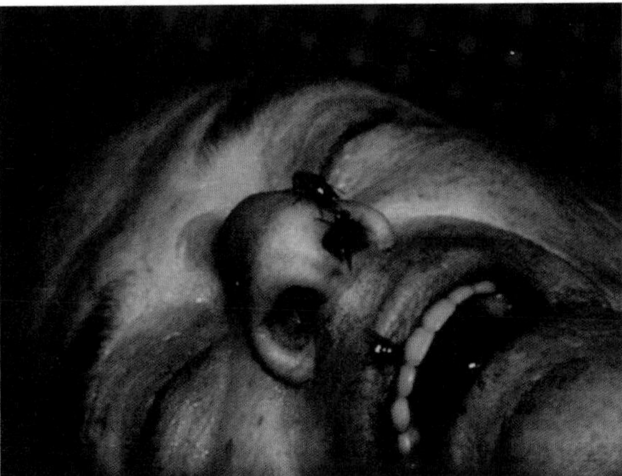

Fig. 1.16 Nose

Fig. 1.17 Bloated stage

This is when the combined activities of the maggot feeding and bacterial putrefaction result in the breaking of the outer layer of the skin and the escape of the gasses from the abdomen. At this point, the body deflates and the Decay Stage is considered to begin. During this stage, strong odors of decomposition are present. The predominant feature of this stage is the presence of large feeding masses of Diptera larvae. These are present internally, externally and often spilling onto the ground beside the body. While some Coleoptera have been arriving during earlier stages of decomposition, they increase in numbers during the Decay Stage and are often quite evident. Some predators, such as the Staphylinidae, are seen during the Bloated Stage and they become more evident now, along with others, such as the Histeridae. In addition to the predators, necrophages are also evident, increasing in numbers as the process continues. By the end of this stage, most of the Calliphoridae and Sarcophagidae will have completed their development and left the remains to pupariate in the surrounding soil. By the end of the Decay Stage, Diptera larvae will have removed most of the flesh from the body, leaving only skin and cartilage (see Fig. 1.18).

1.6.5 Postdecay Stage

As the body is reduced to skin, cartilage and bone, the Diptera cease to be the predominant feature. In xerophytic and mesophytic habitats, various groups of Coleoptera will replace them, with the most commonly seen being the species in

Fig. 1.18 Decay stage

the family Dermestidae. These arrive as adults during the later stages of the Decay Stage but become predominant as adults and larvae during the Postdecay Stage. Their feeding activities remove the remaining dried flesh and cartilage from the bones and the scraping of their mandibles leave the bones with a cleaned, polished appearance. In wet habitats (swamps, rainforests, etc.), the Coleoptera typically are not successful. They are replaced in the process by other groups, including several Diptera families, such as the Psychodidae, along with their respective predator/parasite complexes. Associated with this stage in both types of habitats is an increase in the numbers and diversity of the predators and parasites present. The soil-dwelling taxa increase in number and diversity during this stage (see Fig. 1.19).

1.6.6 Skeletal/Remains Stage

This stage is reached when only bones and hair remain. Typically, there are no obviously carrion-frequenting taxa seen during this stage. During the earlier portions of the Skeletal Stage, there are a number of soil-dwelling taxa, including mites and Collembola, that can be used in estimating the period of time since death. As time passes, the pH of the soil begins to return to the original level and there is a gradual return of components of the normal soil fauna during this stage. There is no definite

Fig. 1.19 Postdecay

Fig. 1.20 Skeletal

end point to this stage and there may be differences in the soil fauna detectable for a period of months or sometimes years, indicating that a body was there at some point in time (see Fig. 1.20).

1.7 Protocol for Decomposition Studies

In order to have adequate data for use in estimations of the period of insect activity, it is necessary to conduct baseline studies of the decomposition process. While there has been a general similarity among studies, there have also been significant differences at the species level in taxa. This is particularly true for those species arriving later in the decomposition process. Many of the Diptera and Coleoptera species involved have somewhat cosmopolitan distributions but other groups tend to be localized. In order to assure the most accurate estimates, it is essential that studies used for estimations be from similar habitats and geographic regions to those in which the body is discovered. For example, work by Early and Goff (1986) and Tullis and Goff (1987) were both conducted on the island of Oahu, Hawaii, and separated by a distance of only 5 miles. The Diptera species were basically the same for both studies but the Coleoptera were markedly different. This was due to the study by Early and Goff being conducted in a xerophytic habitat while that by Tullis and Goff was located in a rainforest. Subsequent studies conducted on a beach some 11 miles away by Davis and Goff 2000) yielded still other species. Other perceived differences may be related to the aims of the study or the taxonomic interests of the investigator. For these reasons, it is important to have data available from baseline studies conducted following a standardized protocol. The protocol presented below is what has been used in studies conducted in the Hawaiian Islands. While some modifications may be needed for particular habitats, it can serve as a general model for conducting decomposition studies.

1.7.1 Animal Model

Animal model should be a domestic pig, ca. 20–30 kg in weight. For each study, three animals will be required. Animals will be dispatched with a single shot from a 38 cal. firearm at 0600 on day 1 of the study. The bullet is to traverse the head laterally, ear to ear. Alternately, the animal will be obtained from a commercial piggery and killed by commercial methods. In this case, the animal will be double bagged and transported to the site immediately following death.

1.7.2 Arrangement of the Animals at the Site

Placement of the animals for the study should be no less than 50 m apart. One animal will be placed directly on the ground and a thermocouple probe inserted into the anus; one animal placed on a wire mesh weight platform (constructed of 2.5 cm^2 welded wire mesh, reinforced with 2.5 cm diameter wooden dowels. Nylon rope should be attached to each corner of the weight platform to allow for weighing with the hand-held scale); and one animal placed on a welded wire mesh screen. There should not be any significant elevation from the ground for either of the animals placed on wire mesh. Each animal should be protected by an exclosure cage constructed from 2.5 cm^2 welded wire mesh. Dimensions of this exclosure cage should be 1 m × 1 m × 0.5 m. Each corner of the exclosure cage should be secured using plastic tent pegs and nylon ropes.

1.7.3 Climatic Data

A hygrothermograph or equivalent instrument along with a high/low thermometer will be placed at the site to record temperature and relative humidity. These will be placed inside a weather station to prevent direct sunlight from altering the temperature data. A rain gauge should also be placed at the site.

1.7.4 Sampling

Sites should be visited at least twice daily for the first 14 days of the study. One visit should be at 1 h past solar zenith. The other visit will be at a time determined by the conditions of the site, but should be constant. Additional visits to the site are desirable, as time allows. Following the first period, the site should be visited on a daily basis for the next 14–21 days. There may be some variations to this schedule, based on differences between sites.

During each visit, weight of the animal on the weight platform will be recorded using a scale. It may prove valuable to construct a tripod over the exclosure cage to hold the scale during the weighing process. This may easily be assembled using 2 × 4's. Rainfall, maximum–minimum temperatures noted. The internal temperature of the animal with the thermocouple probe should be recorded. Temperatures should be recorded for the upper surface of the animal, areas of obvious arthropod activity (maggot masses, etc.) and the soil immediately adjacent to the animal. Observations should be made of the physical condition of each animal. Photographs will be taken of each animal on a daily basis. Observations of arthropod activity should be made for the animals on the weight platform and directly on the ground. No sampling should occur from either of these animals. The animal on the wire mesh screen will be sampled at each visit for arthropods.

Representative samples of immature arthropods collected from the animal on the wire mesh screen should be split into two portions. One part to be fixed in KAA for 1–3 h (depending on the taxon involved) and then transferred into ETOH for identification, and the other placed into rearing chambers to obtain adults. Adult insects should be preserved appropriately.

Soil samples should be taken every 3 days from beneath the sample animal. These samples should be 10 cm in diameter and ca. 0.5 cm deep. The samples should be taken from soil under areas of obvious arthropod activity. Samples should be processed using a Berlese funnel.

1.7.5 *Identifications*

Identifications should be as detailed as possible and confirmed by systematists familiar with the groups. This should be accomplished for all taxa, even those considered to be "common". Voucher specimens should be deposited in an appropriate institution for future reference.

Each study site will be slightly different, and the above steps may be modified for each situation. The aim is to collect as much data from each study as possible. It is better to have a surplus of data as opposed to missing recording of data.

References

Avila FW, Goff ML (2000) Arthropod succession patterns onto burnt carrion in two contrasting habitats in the Hawaiian Islands. J Forensic Sci 43:581–586

Blacklith RE, Blacklith RM (1990) Insect infestations of small corpses. J Nat Hist 24:699–709

Braack LEO (1986) Arthropods associatd with carcasses in the northern Kruger National park. S Aft J Wildl Res 16:91–98

Bornemissza GF (1957) An analysis of arthropod succession in carrion and the effect of its decomposition on the soil fauna. Aust J Zool 5:1–12

Coe M (1978) The decomposition of elephant carcasses in the Tsavo (East) National Park, Kenya. J Arid Environ 1:71–86

Cornaby BW (1974) Carrion reduction by animals in contrasting tropical habitats. Biotropica 6:51–63

Davis JB, Goff ML (2000) Decomposition patterns in terrestrial and intertidal habitats on O'ahu Island and Coconut Island Hawai'i. J Forensic Sci 45:824–830

Early M, Goff ML (1986) Arthropod succession patterns in exposed carrion on the island of O"ahu, Hawaiian Islands, USA. J Med Entomol 23:520–531

Fisher BAJ (2000) Techniques of Crime Scene Investigation, 6th Edition, CRC Press, New York

Gill-King H (1996) Chemical and ultrastructural aspects of decomposition. In: Haglund WD, Sorg MH (eds) Forensic taphonomy: the postmortem fate of human remains. CRC, New York

Goff ML (1993) Estimation of postmortem interval using arthropod development and successional patterns. Forensic Sci Rev 5:81–94

Gunatilake K, Goff ML (1989) Detection of organophosphate poisoning in a putrefying body by analyzing arthropod larvae. J Forensic Sci 34:714–716

Hewadikaram KA, Goff ML (1991) Effect of carcass size on rate of decomposition and arthropod succession patterns. Am J Forensic Med Pathol 12:235–240

Lord WD, Goff ML (2003) Forensic entomology: application of entomological methods to the investigation of death. In: Froede RC (ed) Handbook of Forensic Pathology, 2nd Edn, College of American Pathologists, Illinois.

Megnin P (1894) La faune des cadavers: application de l'entomologie a la medecine legale. Encyclopedia Scientifique des Aide-Memoire. Masson et Gauthier-Villars, Paris, pp 214

Payne JA (1965) A summer carrion study of the baby pig *Sus scrofa* Linnaeus. Ecology 46:592–602

Reed HB (1958) A study of dog carcass communities in Tennessee, with special reference to the insects. Amer Midl Nat 59: 213–245

Rodriguez WC, Bass WM (1983) Insec t activity and its relationship to decay rates of human cadavers in East Tennessee. J Forensic Sci 30:836–852

Shalaby OA, de Carvalho LML, Goff ML (2000) Comparison of patterns of decomposition in a hanging carcass and a carcass in contact with the soil in a xerophytic habitat on the island of O'ahu, Hawai'i. J Forensic Sci 45:1267–1273
Shean BS, Messinger L, Papworth M (1993) Observations of differential decomposition on sun exposed vs. shaded pig carrion in costal Washington State. J Forensic Sci 38:938–949
Tullis K, Goff ML (1987) Arthropod succession in exposed carrion in a tropical rainforest on O'ahu Island, Hawai'i. J Med Entomol 24:332–339

Chapter 2
Keys for Identification of Immature Insects

Patricia J. Thyssen

2.1 Introduction

The study of immature insects is important in forensic entomology, because the identification of the involved species is a crucial step in calculating the post-mortem interval (PMI) and because it is the insect life stage most frequently collected from corpses. The immature stage consists of the egg, nymph or larva with its average of three or four development instars, and the pupa.

Decomposition of a dead body starts through the action of bacteria and fungi, followed by the action of a series of arthropods with a predominance of the dipteran insects (e.g., Carvalho et al. 2000, 2004). Therefore, the anatomical features of the immature stages of these insects are described and taxonomic keys utilized to identify order, family, and species of most of Brazil forensic relevant insects, with emphasis on larvae of the major Diptera families, are presented.

2.2 Anatomical Features of Dipteran Immature Stages

The most recent classification recognizes two suborders, Nematocera and Brachycera (the latter suborder include dipterans that are known as Cyclorrhapha – for classification consult McAlpine et al. 1981). The descriptions below apply to dipterans of the suborder Brachycera and the infraorder Muscomorpha.

P.J. Thyssen
Department of Parasitology, Bioscience Institute, Universidade Estadual Paulista "Júlio de Mesquita Filho", Distrito de Rubião Junior, Botucatu, São Paulo, Brazil, 18618-000

J. Amendt et al. (eds.), *Current Concepts in Forensic Entomology*,
DOI 10.1007/978-1-4020-9684-6_2, © Springer Science+Business Media B.V. 2010

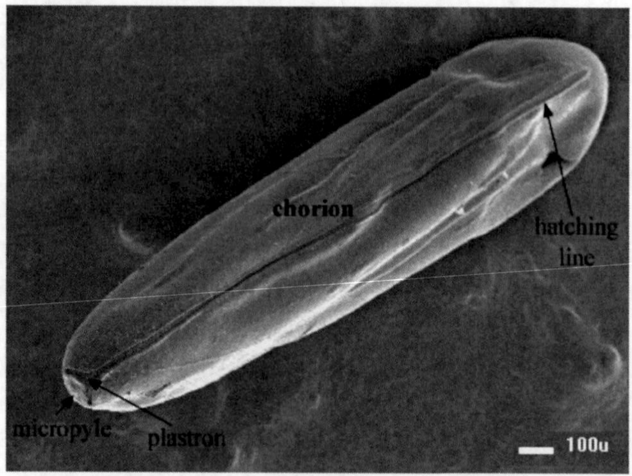

Fig. 2.1 Scanning electron micrograph of the egg of a calliphorid. Scale = 100 μm

2.2.1 Egg Morphology

The egg is the first stage of development in which a series of changes occurs before the hatching of the larvae, since the embryos contain cells and developmental programs for larval structures. A typical egg has the following external characteristics (Fig. 2.1):

- Chorion: outer covering of the egg;
- Micropyle: a pore at the anterior end of the egg that permits entrance of the spermatozoa;
- Plastron: a cell membrane inside and next to the chorion and surrounding the cytoplasm;
- Hatching line: a longitudinal strip that splits off to let the larva emerge.

2.2.2 Larval Morphology

During growth the number of molts varies among insect groups, but in some insect orders this number is rather constant (e.g., two in the muscomorph diptera). The interval between molts is known as stadium, and the form assumed by an insect during a particular stadium is termed an instar (e.g., when an insect hatches from the egg, it is said to be in its first instar) (Fig. 2.2).

The commonest shape of the larvae of muscomorph flies is basically cylindrical, with the anterior end tapering gradually to a slender, pointed head and the posterior end bluntly rounded or somewhat truncated (Fig. 2.3). The body comprises 12 segments: a head (segment I), followed by a prothoracic (segment II), a mesothoracic segment (segment III), a metathoracic segment (segment IV), followed by eight

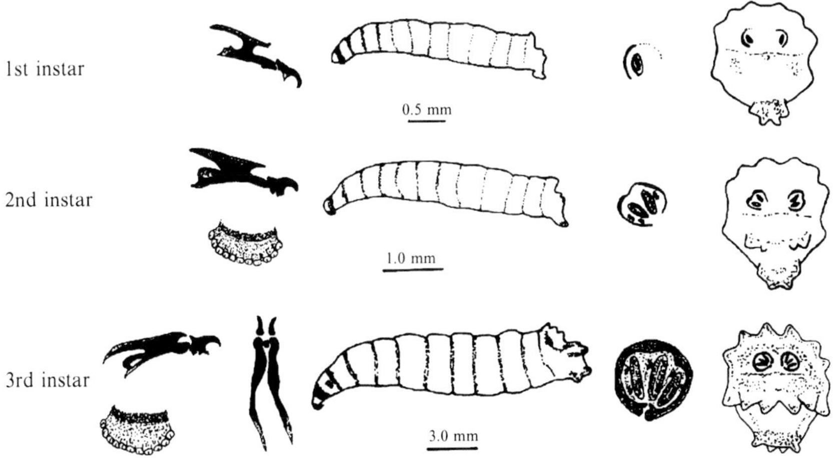

Fig. 2.2 Development of a generalized larvae

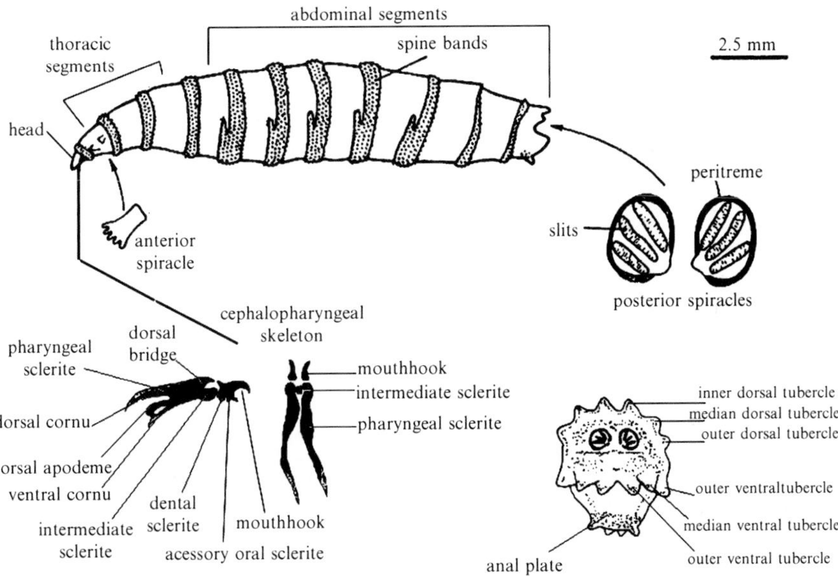

Fig. 2.3 Generalized larval structures

abdominal segments (V-XII). Posterior spiracles are found on the last abdominal segment; and in each spiracle there is a number of slits (according to larval stage) surrounded by a structure called a peritreme. Inside the larval head and extending into the thorax is a chitinous cephalopharyngeal skeleton, which consists of a number of distinct sclerites. Anterior spiracles (when present – generally appear from the second instar) are located on each side of the prothoracic segment and protrude from the body wall.

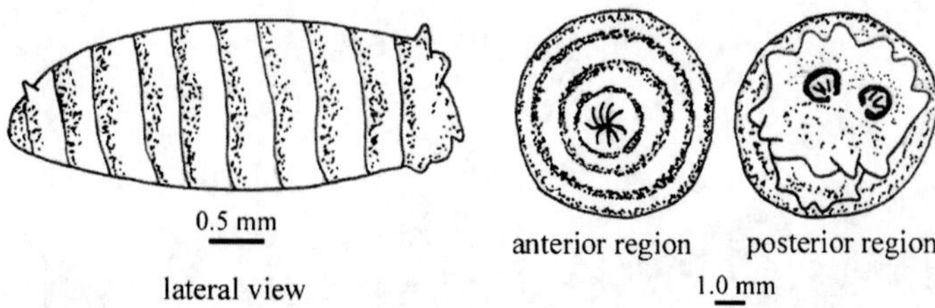

Fig. 2.4 Puparium of a calliphorid

2.2.3 Puparium Morphology

The puparium of the Muscomorpha is most commonly barrel-shaped, heavily sclerotized (formed by hardening of the third instar larval cuticle), with the morphology similar to the previous instar, but smaller in length due to a retraction in the body segments (Fig. 2.4).

2.3 Taxonomic Keys

2.3.1 Key to Larvae and/or Nymphs of Hexapod Orders

This key was modified and adapted from Chu and Cutkomp (1992) (Fig. 2.5a, b and e).

1. Wing pads usually external, nymphs or naiads 2
1′. Wing pads usually internal, larvae 21
2. Chewing mouthparts 3
2′. Sucking mouthparts 18
3. Abdomen 6-segmented, a spring-like organ on the fourth ....Collembola
3′. Abdomen at least 9-segmented, no spring-like structure 4
4. Terrestrial 5
4′. Aquatic 13
5. Cerci absent 6
5′. Cerci present 10
6. Parasitic on birds and some mammals; one tarsal claw .... Mallophaga
6′. Not parasites; two tarsal claws 7
7. Antennae long, longer than body ....Psocoptera
7′. Antennae often inconspicuous 8
8. Cerci minute; social insects ....Isoptera

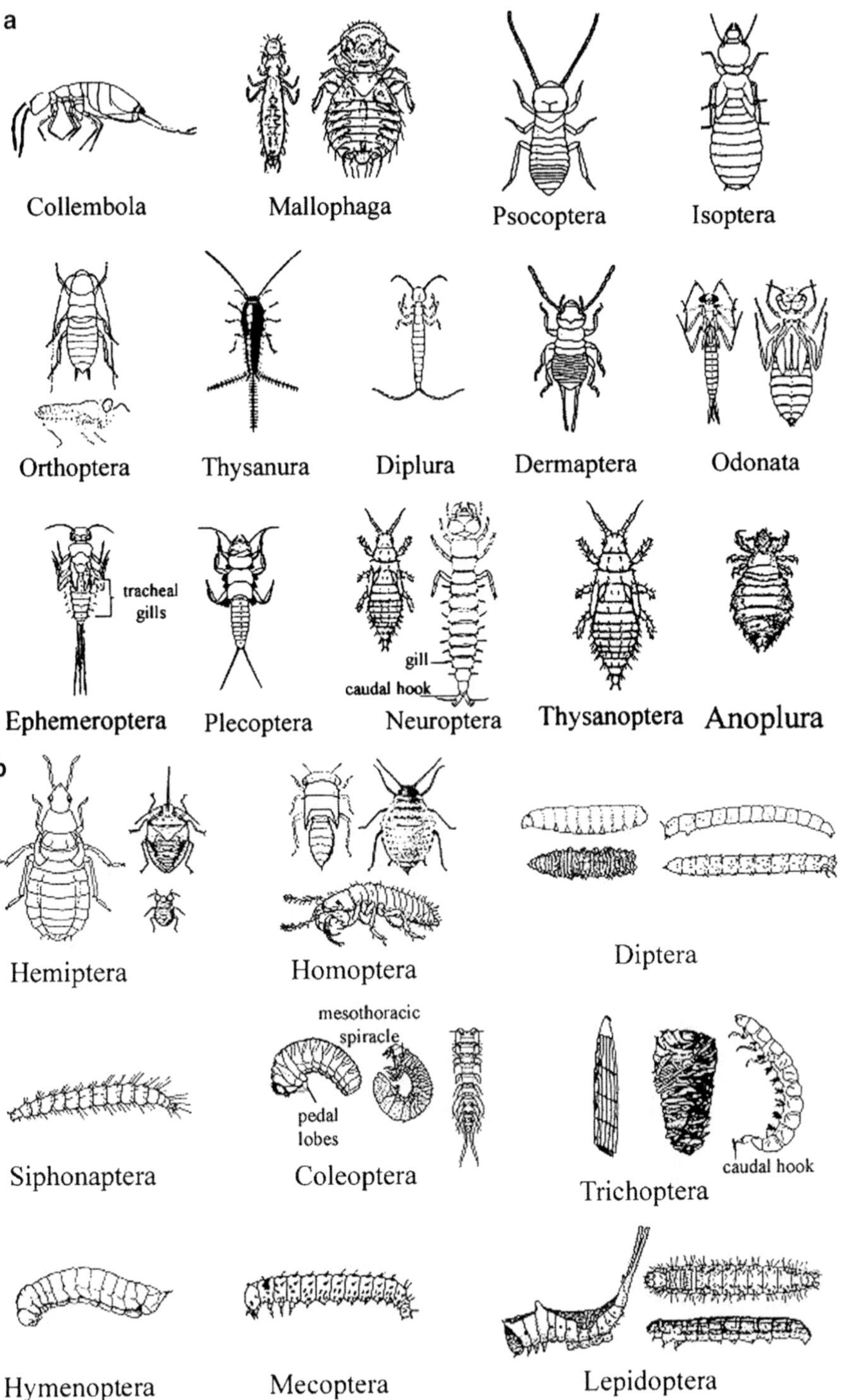

Fig. 2.5 (**a**) Larval morphology of the different Insecta orders (adapted from Chu and Cutkomp, 1992). (**b**) Larval morphology of the different Insecta orders (Adapted from Chu and Cutkomp 1992)

8′. Cerci absent; solitary insects 9
9. Mandibles not sickle-shaped 17
9′. Mandibles sickle-shaped 17
10 Cerci short; with compound eyes .... Orthoptera
10′. Cerci long without compound eyes 11
11. Three-filamented cerci on the end of the abdomen ....Thysanura
11′. Two appendages on the end of the abdomen 12
12. Body not distinctly sclerotized; no wing pads .... Diplura
12′. Body distinctly sclerotized; with wing pads ....Dermaptera
13. (4′) Labium much elongated as a spoon .... Odonata
13′. Labium normal 14
14. Cerci conspicuous 15
14′. Cerci inconspicuous 16
15. Seven or eight abdominal tracheal gills; one tarsal claw ....Ephemeroptera
15′. With thoracic gills; two tarsal claws ....Plecoptera
16. With anal hooks; mandible normal; case bearing 26
16′. Without anal hooks; mandibles sickle-like 17
17. (9) Labial palpi 2-segmented; if gills present, on the sides of the abdomen 24
17′. (9′) Labial palpi if present, more than two segments; if gills present, on the ventral side of abdomen..Neuroptera
18. (2′) Tarsi without claws ....Thysanoptera
18′. Tarsi with claws 19
19. Single tarsal claw .... Anoplura
19′. Two tarsal claws 20
20. Proboscis arising from frontal margin of head ....Hemiptera
20′. Proboscis arising from hind margin of head ....Homoptera
21. (1′) Legless 22
21′. With thoracic legs 25
22. Head capsule not well developed; maggots ....Diptera
22′. Head capsule developed 23
23. Without eyes .... Siphonaptera
23′. Eyes present 24
24. (17) Prothorax without spiracles..25
24′. Prothorax with spiracles 27
25. (24) No distinct prolegs ....Coleoptera
25′. With distinct prolegs 26

26. (16) Only one pair of prolegs on abdomen, located on last abdominal segment .... Trichoptera
26′. Two or more pairs of prolegs on abdomen 27
27. (24′) One large ocellus on each side of head .... Hymenoptera
27′. Two or more small ocelli on each side of head 28
28. Six to eight pairs of prolegs, without crochets .... Mecoptera
28′. Two to five pairs of prolegs, with crochets .... Lepidoptera

2.3.2 Key to larvae of major Diptera families

This key was modified and adapted from Chu and Cutkomp (1992) (Fig. 2.6a–c).

1. Mandibles move horizontally; head complete or, if not, the posterior portion with deep longitudinal incisions, or the thorax and abdomen together consisting of 13 segments (Suborder Nematocera)........ 3
1′. Mandibles move vertically; head incomplete, without a strongly developed upper arched (arcuate) plate 2
2. Maxillae well developed, palpi distinct; mandibles normally sickle-like; antennae well developed on the upper surface of a slightly arcuate, sclerotized dorsal plate (Suborder Brachycera) 17
2′. Maxillae poorly developed, palpi visible only in a few larvae; mandibles short and hook-like; antennae poorly developed or absent, when present situated upon a membranous surface (Suborder Brachycera, old Suborder Cyclorrhapha) 28
3. Head incomplete; thorax and abdomen combined consist of 13 segments; fleshy, pointed paired prolegs on 7 or more abdominal segments; usually with a sclerotized plate on ventral surface of mesothorax .... Cecidomyiidae
3′. Less than 13 segments; other characters differ 4
4. Head and thorax and first and second abdominal segments fused; larvae with minute abdominal spiracles; abdomen with a ventral longitudinal series of sucker-like discs .... Blephariceridae
4′. Head free, or if retracted within or fused with prothorax the other thoracic segments are distinct 5
5. Head complete; mandibles opposed 6
5′. Head incomplete posteriorly, either with three deep wedge-shaped slits (two on dorsum and one on venter), or ventral surface very poorly sclerotized and the dorsal surface posteriorly in the form of four slender heavily sclerotized rods, with a weakly sclerotized divided plate on anterior half of the dorsum (in part, see also 15) .... Tipulidae
6. Thoracic segments fused, flattened, and wider than abdominal segments, forming a complex mass .... Culicidae
6′. Thoracic segments distinct and not wider than abdomen 7

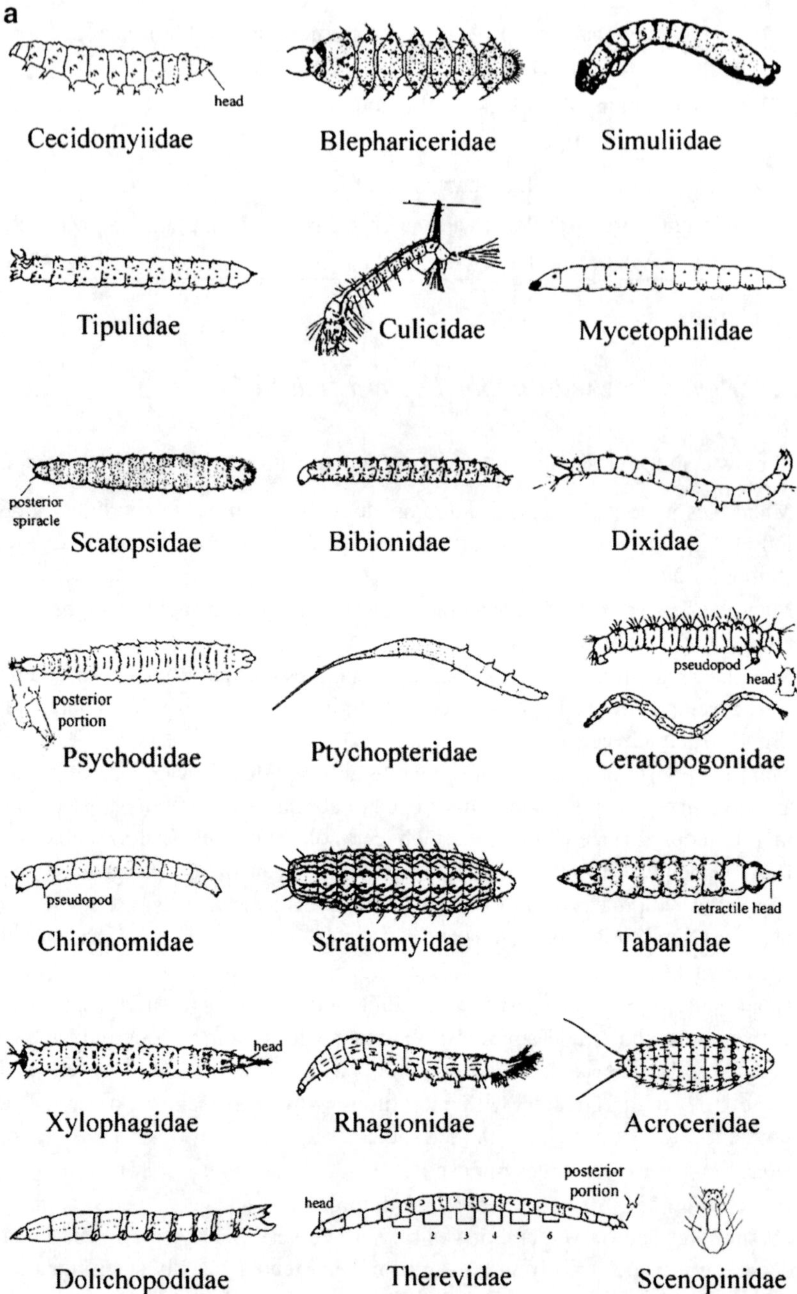

Fig. 2.6 (**a**) Larval morphology of the different Diptera families (adapted from Chu and Cutkomp, 1992). (**b**) Larval morphology of the different Diptera families (adapted from Chu and Cutkomp, 1992). (**c**) Larval morphology of the different Diptera families (Adapted from Chu and Cutkomp 1992)

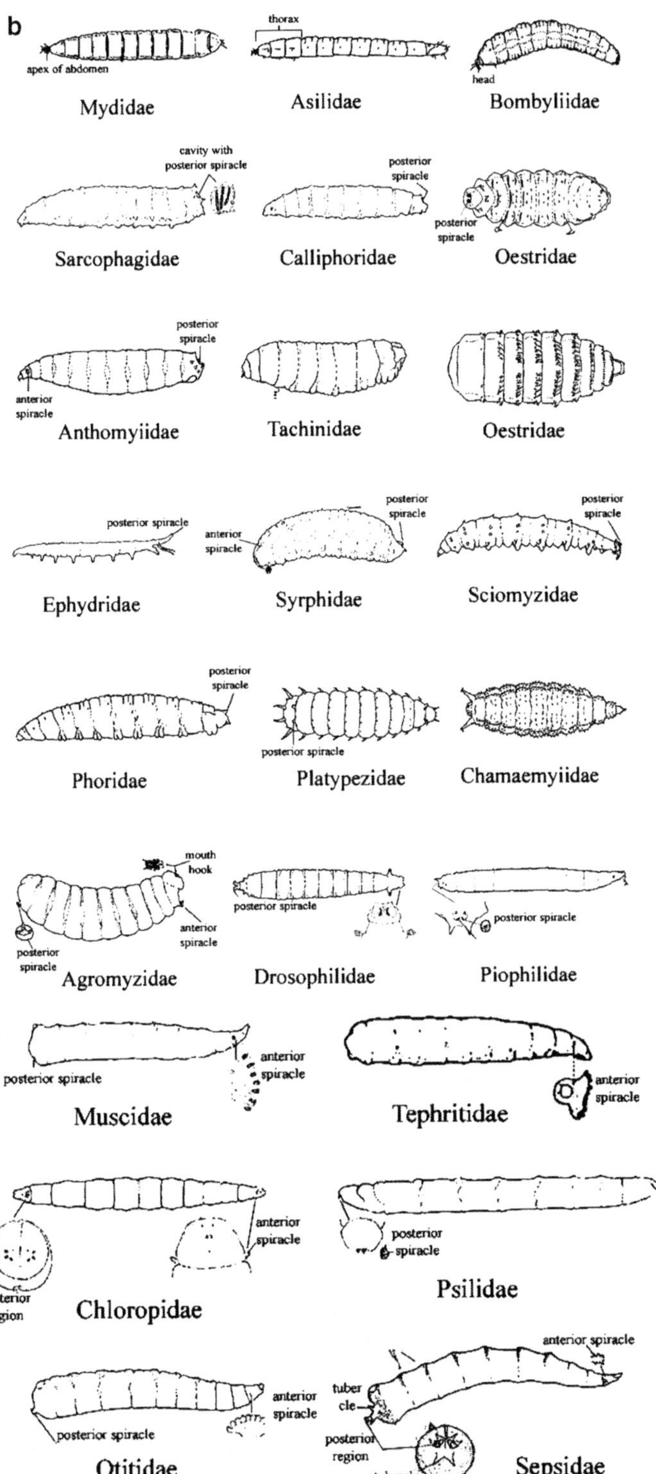

Fig. 2.6 (continued)

7. Larvae with a row of spiracles on each side of body (peripneustic), or with at least rudimentary abdominal spiracles 8
7′. Larvae with spiracles on the prothorax and terminal abdominal segment (amphipneustic) or only on the terminal segment (metapneustic) 11
8. Larvae with rudimentary abdominal spiracles; mouth with a large articulated process on each side that bears a number of long hairs and closes fan-like when at rest; posterior abdominal segments dilated, the last one armed on venter with a sucker-like disc that bears concentric series of bristles........................Simulidae
8′. Larvae with distinct though sometimes small abdominal spiracles; mouth without fan-like processes; posterior abdominal segments not noticeably dilated, the last one without sucker-like disc 9
9. Antennae elongate; body armed with conspicuous bristles or hairs 10
9′. Antennae usually short and inconspicuous, sometimes apparently absent; body without conspicuous bristles 11
10. Posterior spiracles at the apices of a pair of long stalk-like processes .... Scatopsidae
10′. Posterior spiracles not noticeably elevated, situated near base of dorsal surface of posterior segment .... Bibionidae
11. One pair of prolegs on each of first and second abdominal segments; body U-shaped when at rest .... Dixidae
11′. No prolegs on abdominal segments, with possible exception on posterior segment 12
12. All or some of the dorsal segments with narrow, sclerotized strap-like transverse bands; or the apical posterior segment in the form of a short, sclerotized tube; rarely, the ventral abdominal segments bear a central series of sucker-like discs ....Psychodidae
12′. Dorsum without narrow; sclerotized, strap-like bands; posterior segment not in the form of a short sclerotized tube; ventral abdominal segments with sucker-like discs 13
13. Antennae undeveloped, appearing as pale round spots on side of head with sclerites contiguous anteriorly, widely separated posteriorly .... Mycetophilidae
13′. Not as described above 14
14. Antennae stalked (pedunculate) usually well developed; ventral surface of head with sclerites contiguous for entire length, not separated widely posteriorly 15
14′. Antennae greatly reduced; abdomen elongated and inconspicuously segmented; three pairs of prolegs with hooks present .... Ptychopteridae
15. Abdominal segments not subdivided 16
15′. Abdominal segments subdivided by means of transverse constrictions .... Tipulidae
16. Aquatic larvae very slender, tapering toward both ends; without thoracic or anal foot-like appendages (pseudopods) or surface hairs (except about eight at apex of abdomen) or terrestrial larvae stout, with well-defined segments that are armed with strong, bristles, some of which are lanceolate; pseudopods present ....Ceratopogonidae

16′. Larvae rarely very slender, generally of an almost uniform thickness, rarely with the thoracic segments appreciably swollen but not fused; abdominal and thoracic segments frequently with rather noticeable soft hairs, the last segment almost invariably with a conspicuous tuft of hairs on dorsum near the apes; pseudopods almost always present, sometimes (very rare) only the thoracic one is distinguishable in terrestrial forms................Chironomidae

17. Posterior spiracles close together, situated in a terminal or subterminal cleft or chamber, usually concealed; body surface roughened and leathery, or wholly or in part longitudinally striated 18

17′. Posterior spiracles rather widely separated, visible, situated on apical segment, which may be truncated, sclerotized, or armed with apical processes; or spiracles on second to last (penultimate) or third to last (antepenultimate) segment; body surface not roughened or visibly striated 19

18. Head not retractile; body flattened, surface finely roughened, sometimes with lateral abdominal spiracles, without vestigial pseudopods; spiracular fissure transverse, sometimes rather small; pupae enclosed in larval skin ....Stratiomyiidae

18′. Head retractile; body cylindrical, surface not roughened, usually longitudinally striated; abdomen with a girdle of pseudopods on each segment; spiracular fissure vertical ....Tabanidae

19. Posterior spiracles situated on apical segment 20

19′. Posterior spiracles situated on penultimate or antepenultimate segment 24

20. Projecting portion of head and flattened apical plate of terminal abdominal segment heavily sclerotized, the former cone-shaped, entirely closed except at extreme apex and not retractile; the apical plate obliquely truncate and with projecting processes...Xilophagidae

20′. Projecting portion of head more or less retractile, not cone-shaped, the movable portion not enclosed; apical abdominal segment without a heavily sclerotized flattened terminal plate 21

21. Apical abdominal segment ending in two long processes that are fringed with long soft hairs; abdomen with paired pseudopods and fleshy dorsal and lateral appendages (in part, see also 22) ....Rhagionidae

21′. Apical abdominal segment not as described above; paired abdominal pseudopods usually absent; other abdominal appendages always absent 22

22. Apical abdominal segment ending in four short pointed processes or two fleshy lips; internal portion of head with a large, arched, sclerotized upper plate, the longitudinal rods and other head (cephalic) parts are on a horizontal plane ....Rhagionidae

22′. Apical abdominal segment not as described above, or the internal portion of head without an arched upper plate, and the longitudinal cephalic rods and other cephalic parts meet at right angles 23

23. Apical abdominal segment without projecting processes, spiracles very small; first instar with distinct segments and two long apical bristles on abdomen; parasites of spiders ....Acroceridae

23′. Apical abdominal segment frequently with projecting processes; spiracles large; species live in water, mud, earth, or decaying vegetable matter ....Empididae or Dolichopodidae

24. Posterior spiracles situated on the antepenultimate segment; abdominal segments 1–6 subdivided, the body apparently consisting of 20 segments exclusive of the head 25

24′. Posterior spiracles situated on penultimate segment; abdominal segments simple, the body apparently consisting of 11 or 12 segments exclusive of the head 26

25. Posterior dorsal internal extension of head spatulate at apex; ventral posterior projections in the form of two short sclerotized rods ....Therevidae

25′. Posterior dorsal extension of head not spatulate at apex; ventral posterior projections absent ....Scenopinidae

26. Penultimate abdominal segment longer than ultimate, with a deep transverse depression near its apex giving it the appearance of two distinct segments; ultimate segment terminating in a sharp ridge with a median sharp point, on either side, of which dorsally and ventrally four very closely approximated hairs are situated ....Mydidae

26′. Penultimate abdominal segment shorter than ultimate, or if longer then without a deep transverse depression; apical segment not as describe above, the hairs not closely approximated 27

27. Thoracic segments each with two long hairs, one on each side on ventrolateral margin; apical segment with six or eight long hairs; head well developed, forward protruded, and more or less cone-shaped when viewed from above, appearing flattened when viewed from side; penultimate abdominal segment usually shorter than ultimate or not much longer; body straight when alive ....Asilidae

27′. Thoracic segments without hairs or, if present, they are very weak; apical segment without distinguishable hairs; head slightly protruded, directed downward, not cone-shaped, with a dorsal protuberance when viewed from side; penultimate segment distinctly longer than ultimate; body usually curved in a half circle when alive ....Bombyliidae

28. Parasitic 29

28′. Nonparasitic or predaceous 33

29. Parasitic on insects and other arthropods 32

29′. Parasitic on other animals 30

30. Midle portion of body enlarged with strong spinous girdles 31

30′. Body tapering; spines minute 54

31. Parasitic under animal skin, in nasal sinuses, or in throat; large and stout larvae that parasitize rabbits and rodents ....Cuterebridae

31′. Parasitic as describe above, and some species can be found in animal digestive system; larvae is known as bot or warble flies ....Oestridae

32. Endoparasitic; portion end of body truncate with a button on posterior spiracles ....Tachinidae

32′. Some species are endoparasitic 33
33. Aquatic 34
33′. Terrestrial 38
34. Body smooth, without transverse folds (plicae) or tubercles 35
34′. Body rough, with transverse plicae or tubercles 37
35. Prolegs distinct and with hooks 36
35′. Prolegs not distinct; posterior spiracles in a cavity (see also 54) .... Sarcophagidae
36. Posterior spiracles knob-like on conical projections; posterior prolegs larger (see also 42′) .... Anthomyiidae
36′. Posterior spiracles on a prolongation of posterior segment; 6–8 pairs of prolegs, but ventral prolegs absent .... Ephydridae
37. Posterior spiracles on end disks of upturned posterior segment; ventral prolegs absent .... Sciomyzidae
37′. Posterior spiracles contiguous; ventral prolegs present (see also 39′) .... Syrphidae
38. Posterior spiracles contiguous or fused 39
38′. Posterior spiracles not as described above 40
39. Saprophagous; body segments with processes .... Phoridae
39′. Predaceous or phytophagous; body with transverse plicae .... Syrphidae
40. Not associated with fungi 41
40′. Associated with fungi .... Platypezidae
41. Body with end (caudal), lateral, and dorsal processes 42
41′. Not as described above 43
42. Predators; body processes short .... Chamaemyiidae
42′. Scavengers; body with spiny or setiferous processes 43
43. Anterior (prothoracic) spiracles not adjacent 44
43′. Anterior (prothoracic) spiracles adjacent .... Agromyzidae
44. Posterior portion truncate or rounded 46
44′. Posterior portion somewhat elongated or terminating in two subspiracular processes; if inconspicuous processes occur, the posterior spiracles are inconspicuous and usually light-colored 45
45. Posterior spiracles on two short processes .... Drosophilidae
45′. Posterior spiracles not on two short processes .... Piophilidae
46. Two or more mouthhooks; posterior spiracles not sinuous 47
46′. One mouthhook; posterior spiracles sinuous (Fig.2.7) .... Muscidae
47. Anterior (prothoracic) spiracles arranged in a circle or ellipse 48
47′. Anterior (prothoracic) spiracles never so arranged .... Tephritidae
48. End portion with four or more processes 52
48′. Not described above 49
49. Posterior spiracles conspicuous, deeply pigmented 51

49′. Not described above 50
50. Posterior spiracular slits oval .... Chloropidae
50′. Posterior spiracular slits elongate .... Tephritidae
51. Posterior spiracles with a distinct pointed process on the dorsal margin of ring .... Psilidae
51′. Not as described above .... Otitidae
52. Posterior spiracles on elevated projections .... Sepsidae
52′. Posterior spiracles sessile 53
53. Slits of posterior spiracles slender and long 54
53′. Slits of posterior spiracles short and radial .... Anthomyiidae
54. Slits of posterior spiracles nearly transverse, button almost always distinct (Fig.2.7); posterior spiracles not in a cavity (Fig.2.8) .... Calliphoridae
54′. Slits of posterior spiracles nearly vertical, button indistinct (Fig.2.7); posterior spiracles are enclosed within a deep cavity (Fig.2.8) .... Sarcophagidae

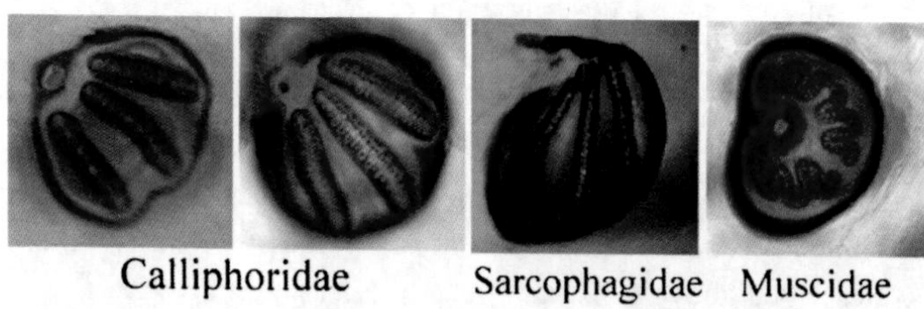

Fig. 2.7 Posterior spiracles of major Diptera families found on corpses

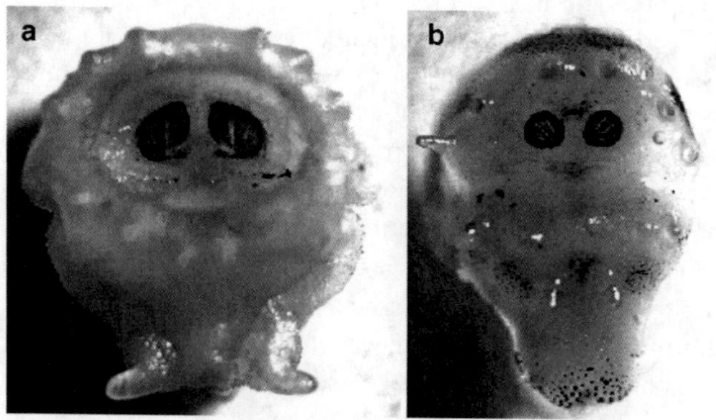

Fig. 2.8 Posterior aspect of a sarcophagid larvae (**a**) and a calliphorid larvae (**b**). The posterior spiracles of sarcophagid larvae are enclosed within a deep stigmal cavity

2.3.3 Key to Third Instar Larvae of the Most Carrion Breeding and Feeding Dipteran Species from Brazil

This key was produced by reference to specimens in the author's own collection. Other keys and descriptions by the following authors were also consulted: Zumpt (1965), Ishijima (1967), Prins (1982), Greenberg and Szyska (1984), Smith (1986), Erzinclioglu (1987), Queiroz and Carvalho (1987), Liu and Greenberg (1989), Guimarães and Papavero (1999), Wells et al. (1999), Greenberg and Kunich (2002), and Thyssen and Linhares (2007).

2.3.3.1 Nomenclature according to Fig. 2.3

1. Larvae with an obvious head that is sclerotized and clearly differentiated from the rest of the body; body flattened; mouthhook moving vertically, parallel to each other like a pair of hooks..............................*Hermetia illucens*(Stratiomyiidae)

1′. Larvae without an obvious head, which merges with the rest of the body although the mouthparts may be obvious 2

2. Surface of the body segments with obvious fleshy or spinous processes 3

2′. Surface of the body segments without processes, although the integument itself may have strong spines 5

3. Body flattened dorsoventrally and with numerous projections 4

3′. Body cylindrical 5

4. Each projections with short and broad spines on the basal portion .... *Fannia canicularis*(Fanniidae)

4′. Each projections with large branches; lateral part of segment 12 with three feather-like large projections, with more than five forked branches ....*Fannia scalaris*(Fanniidae)

5. Larger longer pointed fleshy processes laterally and dorsally ending up in a tuff of hairs or crown of spines strongly pigmented on the apex; posterior spiracles in a cleft on posterior face of anal segment and consisting of flattened plates with three slits ....*Chrysomya albiceps*(Calliphoridae)

5′. Small white, slightly flattened larvae up to 4 mm long with short processes on the dorsal and lateral surfaces; posterior spiracles on brown, sclerotized tubercles, each with a narrow opening ....*Megaselia scalaris*(Phoridae)

6. Posterior spiracles are enclosedin a deep cavity .... *Peckia*(*Pattonella*) *intermutans*(Sarcophagidae)

6′. Posterior spiracles not enclosed in a deep cavity 7

7. Peritreme of posterior spiracle complete, although sometimes weaker in the region of button 8

7′. Peritreme of posterior spiracle incomplete and not enclosing button, or it is sometimes indistinct 13

8. Slits of posterior spiracles strongly sinuous; inner surface of anal plate covered with small spines; anal plate surrounded by small spines ....*Musca domestica*(Muscidae)

8′. Slits of posterior spiracles straight or at most arcuate 9

9. Slits of posterior spiracles almost parallels; anterior spiracles with 10 lobes; anal segment with ventral lobes much longer than the dorsal lobes ....*Piophila casei*(Piophilidae)

9′. Peritreme with button distinct 10

10. Slits of posterior spiracles bent distally (3 o'clock position); posterior margin of ventral surface of segment 10 with a row of numerous small spines; anterior margin of anal plate convex and posterior margin concave, lateral part of anal plate bent posteriorly ....*Ophyra chalcogaster*(Muscidae)

10′. Accessory oral sclerite absent 11

11. Posterior dorsal margin of segment 11 withoutrows of spines .... *Lucilia eximia*(Calliphoridae)

11′. Posterior dorsal margin of segment 11 with rows of spines 12

12. Inner dorsal tubercles of the posterior region separated from each other by a distance approximately equal to the distance between the inner and median dorsal tubercles; very few spines present dorsal to anus ....*Lucilia sericata*(Calliphoridae)

12′. Inner dorsal tubercles of the posterior region separated from each other by a distance approximately equal to the distance between the inner and outer dorsal tubercles; strong patch of spines present dorsal to anus ....*Lucilia cuprina*(Calliphoridae)

13. Posterior margin of segment 11 without dorsal spines; in other segments of the body, spines multipointed, each with at least two teeth (sometimes up to four); the teeth are well separated from one another and having rounded tips; main tracheal trunks leaving anterior spiracles are not black-pigmented ....*Cochliomyia macellaria*(Calliphoridae)

13′. Posterior margin of segment 11 with dorsal spines; button distinct or indistinct 14

14. All spines single-pointed and small ...*Sarconesia chlorogaster*(Calliphoridae)

14′. Spines single or multipointed 15

15. Lateral fusiform areas absent; with accessory dental sclerite 16

15′. Lateral fusiform areas present; with or without accessory dental sclerite 17

16. Spines single and multipointed, especially large and numerous ventrad on segment 12; peritreme is moderately sclerotized with delicate peritreme; anal plate well extended..................................*Hemilucilia semidiaphana*(Calliphoridae)

16′. Spine pattern follows the described above;peritremeis strongly sclerotized; a littledistinguished button; anal plate is short ....*Hemilucilia segmentaria*(Calliphoridae)

17. With accessory dental sclerite ....Compsomyiops spp. (Calliphoridae)

17′. With or without accessory dental sclerite 18

18. Spine pattern on anal protuberance convex or bell-shaped; mouthhook devoid of accessory oral sclerite; anterior spiracleswith 10–12 branches .... *Chrysomya putoria*(Calliphoridae)

18′. Spine pattern on anal protuberance V- or U-shaped 19

19. Mature third instar blue-gray; devoid of accessory oral sclerite .... *Paralucilia fulvinota*(Calliphoridae)

19′. Mature third instar non-pigmented and with accessory dental sclerite......... 20

20. Mouthhook with the tooth-like apical portion longer than greatest depth of the basal portion; accessory oral sclerite present between the mouthhook; peritreme complete and posterior spiracles smaller and separated by a distance equal to or greater than the width of a single spiracle ....*Calliphora vicina*(Calliphoridae)

20′. Mouthhook with small accessory oral sclerite and comma-shaped; peritremes of the posterior spiracles separated one from the other by a distance equal to approximately one-third to one-half the diameter of one of the peritremes; anterior spiracles with 8–10 short branches .... Chrysomya megacephala(Calliphoridae)

References

Chu HF, Cutkomp LK (1992) How to know the immature insects. WCB, USA, p 346

Carvalho LML, Thyssen PJ, Linhares AX, Palhares FB (2000) A checklist of arthropods associated with carrion and human corpses in southeastern Brazil. Mem Inst Oswaldo Cruz 95(1):135–138

Carvalho LML, Thyssen PJ, Goff ML, Linhares AX (2004) Observations on the succession patterns of necrophagous insects onto a pig carcass in an urban area of Southeastern Brazil. Aggrawal's Int J For Med Toxicol 5:33–39

Erzinclioglu YZ (1987) The larvae of some blowflies of medical and veterinary importance. Med Vet Entomol 1:121–125

Greenberg B, Szyska M (1984) Immature stages and biology of fifteen species of Peruvian Calliphoridae (Diptera). Ann Entomol Soc Am 77:488–517

Greenberg B, Kunich JC (2002) Entomology and the law: flies as forensic indicators. Cambridge University Press, USA, p 356

Guimarães JH, Papavero N (1999) Myiasis in man and animals in the Neotropical region. Plêiade/FAPESP, São Paulo

Ishijima H (1967) Revision of the third stage larvae of synanthropic flies of Japan (Diptera: Anthomyiidae, Muscidae, Calliphoridae and Sarcophagidae). Japanese J San Zool 18:48–100

Liu D, Greenberg B (1989) Immature stages of some flies of forensic importance. Ann Entomol Soc Am 82:80–93

Mc Alpine, Peterson BV, Shewell GE, Teskey HJ, Vockeroth JR, Wood DM (eds) (1981) Manual of Neartic Diptera, vol 1. Research Branch Agriculture Canada, Ottawa

Prins AJ (1982) Morphological and biological notes on six South African blow-flies (Diptera, Calliphoridae) and their immature stages. Ann S Afr Mus 90:201–217

Queiroz SMP, Carvalho CJB (1987) Chave pictórica e descrições de larvas de 3º instar de Díptera (Calliphoridae, Muscidae e Fanniidae) em vazadouros de resíduos sólidos domésticos em Curitiba, Paraná. An Soc Entomol Brasil 16:265–288

Smith KGV (1986) A manual of forensic entomology. Cornell University Press, Ithaca, NY 205

Thyssen PJ, Linhares AX (2007) First description of the immature stages of *Hemilucilia segmentaria* (Diptera: Calliphoridae). Biol Res 40:271–280

Wells JD, Byrd JH, Tantawi TI (1999) Key to third-instar Chrysomyinae (Diptera: Calliphoridae) from carrion in the continental United States. J Med Entomol 36:638–641

Zumpt F (1965) Myiasis in man and animals in the old world. Butherworths, London 267

Chapter 3
Key for the Identification of Third Instars of European Blowflies (Diptera: Calliphoridae) of Forensic Importance

Krzysztof Szpila

3.1 Introduction

In Europe larvae of blowflies are the main group of insects responsible for decomposition of exposed vertebrate remains, including the human body. This determines their high forensic importance and frequent application for estimation of PMI. The importance of proper identification of insects collected in forensic cases and experiments to the species level is underlined by all manuals of forensic entomology (e.g. Smith 1986; Byrd and Castner 2001; Greenberg and Kunich 2002). Especially difficult is the identification of the larval stages, where breeding to the adult stage or DNA-based methods are recommended. Fortunately, the available knowledge of the morphology of third instars of Calliphoridae is sufficiently good to allow the preparation of a complete identification key for at least all European species of forensic importance. Eleven species are included in the key. Most of them are widespread through Europe (Rognes 2004) and have been frequently reported from both real cases and carrion experiments, and the necessity of their inclusion into the key cannot be questioned. There are: *Calliphora vicina*, *C. vomitoria*, *Chrysomya albiceps*, *Phormia regina*, *Protophormia terraenovae*, *Lucilia caesar*, *L. illustris*, *L. sericata*. The author has also decided to add three additional species to the key: *Cynomya mortuorum*, *Chrysomya megacephala*, and *Lucilia ampullacea*. Recently, the larvae of *Cynomya mortuorum* were recorded from human corpses at least twice (Stærkeby 2001; Benecke 2002). Smith (1986) points out this species as rather a late newcomer in comparison to other blowflies, but recent research on succession shows that in spring conditions *Cynomya mortuorum* may be among the first colonizers of pig carrion (Szpila et al. 2008). *Chrysomya megacephala* is the newly discovered species in Europe, with distribution still restricted to the Spanish mainland, Canary Is. (Rognes 2004), Malta (Ebejer 2007), and Madeira (Martínez and Rognes 2008). In the tropical regions,

K. Szpila
Institute of Ecology and Environmental Protection,
Nicolaus Copernicus University, Toruń, Poland

J. Amendt et al. (eds.), *Current Concepts in Forensic Entomology*,
DOI 10.1007/978-1-4020-9684-6_3, © Springer Science+Business Media B.V. 2010

this species is very common and abundant on human corpses; in continental Spain, the first record of the development of *Chrysomya megacephala* on pig carrion was recently reported (Velásquez et al. 2008). *L. ampullacea* may develop on large vertebrate carrion at least in Central European conditions (Grunwald et al. 2009) and has been reported from human corpses (Benecke 1998). A few species of *Calliphora* Robineau-Desvoidy other than *C. vicina* or *C. vomitoria* were also recorded in pig carrion experiments, but only as single adult flies and they are hence not included in this key. Also, *Lucilia cuprina* (Wiedemann, 1830) known from the Iberian Peninsula (Rognes 2004), is not included there. This facultative parasite may also develop in carrion but has not been recorded so far in any real case or carrion experiment in European conditions.

Original information concerning the morphology of third instars of European, forensically important blowflies is scattered in many papers (Knipling 1936; Hall 1948; Zimin 1948; Kano and Sato 1952; Schumann 1954, 1965, 1971; Ishijima 1967; Kitching 1976; Teskey 1981; Prins 1982; Holloway 1985, 1991; Erzinçlioğlu 1985, 1987a, b, 1988, 1990; Smith 1986, 1989; Shewell 1987; Liu and Greenberg 1989; Carvalho Queiroz et al. 1997; Fan et al. 1997; Wells et al. 1999; Wallman 2001; Povolný 2002; Grassberger et al. 2003; Sukontason et al. 2003, 2008; Zumpt 1965). Several keys have been compiled so far, but none of them cover the current complete list of species. The most comprehensive, according to this list, is the key of Ishijima (1967) where only *C. mortuorum* and *Ch. albiceps* are absent. Other important keys have been provided by Schumann (1954, 1971); especially valuable is his contribution to the knowledge of the larval stages of European species of *Lucilia* Robineau-Desvoidy. Erzinçlioğlu (1985, 1987a, b, 1988, 1990) published a series of excellent papers with descriptions and keys covering all third instars of the European, forensically important Calliphorinae and Chrysomyinae. Data about third instar morphology of Central European Calliphoridae were summarized by Draber-Mońko (2004). In her monograph (written in Polish) are accumulated all the figures useful for identification of the larvae of Central European blowflies, published before year 2003.

A critical review of the morphological characters of third instars of Calliphoridae was provided by Erzinçlioğlu (1985). Experience from the continuous work of the author on larval morphology of blowflies makes it possible to point a few morphological details that should be used for taxonomic purposes with special caution. They are the small sclerotised spot below the posterior tip of ventral cornua (used in *Lucilia* identification), the direction of the process on the postero-dorsal angle of the basal part of mouthhooks (*Lucilia*), spinulation on the last abdominal segments (*Lucilia*), and the shape of the peritreme (Chrysomyinae vs other blowflies). The most difficult part of the presented key is in distinguishing the larvae of the two closely related species *L. caesar* and *L. illustris*. It may be done only on the basis of two characters (from those listed above, see also key), both of which are difficult to see and need preparation of light microscope slides. The presence of an interrupted peritreme of posterior spiracles cannot be used as feature is characteristic only for larvae of subfamily Chrysomyinae (Erzinçlioğlu 1985; Wallman 2001). Such form of peritreme is present at least in some specimens of *C. vicina* (Erzinçlioğlu 1985; see also Fig. 3.4i). Also the value of spiracular distance factor

(SDF) may vary according to the size of the maggots and the techniques of preparation. This measure should be used only for the fully grown third instars. Moreover, Wallman (2001) recommends using this measure only for freshly killed larvae. It is also important to mention significant doubts, reported recently (Hale et al. 2008) concerning the reliability of some important morphological characters used in the present key (presence of sclerotised oral sclerite). However, the continuing extensive work of the author on larval material has not confirmed this observation so far.

The present key for the identification of third instar larvae is the first to cover all European species of forensic importance. Thanks to the opportunity provided by editors, all significant characters are illustrated in the form of color pictures, taken using a digital camera mounted on the microscopes. The black and white figures are not included here but are easily available in the references listed below. This key has been seriously tested before publication and seems to work well; in doubtful cases, however, the author recommends that the identifications should be checked against the keys and descriptions published in the listed references.

3.2 Material and Methods

Third instars of *C. vicina*, *C. vomitoria*, *P. regina*, *P. terraenovae*, *L. ampullacea*, *L. illustris*, and *L. sericata* were bred from eggs deposited by females collected in the city of Toruń in Northern Poland (53°00′N, 18°35′E). At least a part of the larvae in all the cases were bred to adult form for unquestionable species identification. Third instars of *C. mortuorum*, *Ch. albiceps*, and *L. caesar* were collected during research on the insect succession on pig carrion conducted on Biedrusko Military Range in Western Poland (52°31′N, 16°54′E) and identified as larvae using suitable references. Specimens of *C. megacephala* were available for investigation thanks to Professor Kabkaev Sukontason and Professor Kom Sukontason (Chiang Mai University) who provided larval material from Thailand.

All larvae were killed by soaking in hot water (about 95°C), and then stored in 80% ethanol. This technique of preservation is often recommended for the forensic entomologist as it is very convenient and can be used even in poorly equipped laboratories.

For preparation of slide, larvae were macerated for 24 h in a cold solution of 5% KOH. Next, the particular fragments of the body were mounted in Hoyer's medium or dehydrated through 80%, 90%, and 99.5% ethanol and mounted in Euparal. For the cephaloskeletons, concave slides, and for other morphological details, flat slides were used.

A digital Nikon 8400 camera mounted on a Nikon Eclipse E200 microscope and a Nikon SMZ1500 stereomicroscope were used for photomicrography.

Larval terminology follows Courtney et al. (2000) and Szpila and Pape (2007). The spiracular distance factor was calculated according to Erzinçlioğlu (1985) (SDF = a/b, see Fig. 3.4m).

In the references, species originally figured in a particular paper are listed in square brackets after the reference.

3.3 General Morphology

The body of larvae in necrophagous Calliphoridae follows the general pattern for Calyptrata in being divided into a bilobed pseudocephalon, three thoracic segments (termed TI–TIII below), seven abdominal segments (AI–AVII), and the anal division (AD) (Fig. 3.1a). Third instar is easily distinguishable from the other instars by the

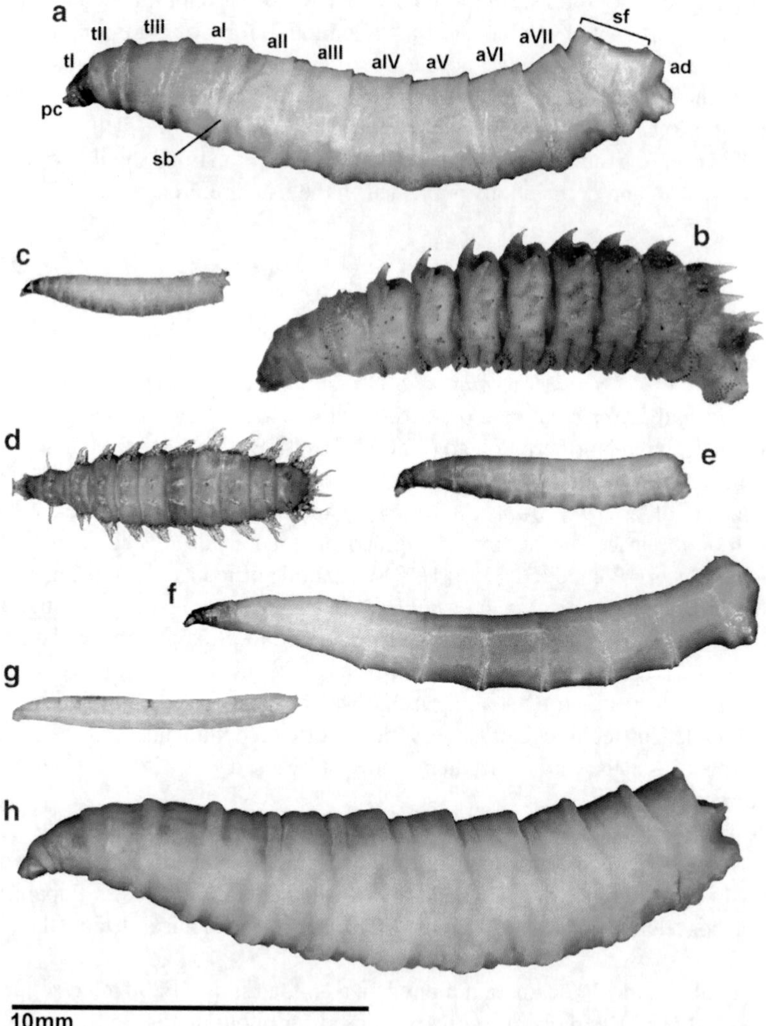

Fig. 3.1 Third instars of necrophagous flies: **a** – Calliphoridae, *Calliphora vomitoria*; **b** – Calliphoridae, *Chrysomya albiceps*; **c** – Sepsidae, *Nemopoda nitidula*; **d** – Fannidae, *Fannia coracina*, dorsal view; **e** – Heleomyzidae, unidentified species; **f** – Muscidae, *Hydrotaea dentipes*; **g** – Piophilidae, *Stearibia foveolata*; **h** – Sarcophagidae, *Sarcophaga caerulescens*. Abbreviations: ad – anal division, aI-VII – abdominal segments, pc – pseudocephalon, sb – spinose band, sf – spiracular field, tI-III – thoracic segments

presence of anterior spiracles (vs first instar) and the number of slits of posterior spiracles (vs second instar) (Fig. 3.2a–f). Each of the pseudocephalic lobes of a larva has an antennal complex with the antennal dome situated on a basal ring (Fig. 3.2g). The maxillary palpus is located on the anterior surface of the pseudocephalic lobe and has the form of a flattened protuberance with numerous sensilla. Above and lateral to the mouth opening is the ventral organ. The functional mouth opening is closed from below by a triangular labial lobe with two sensilla of labial organ. Numerous oral ridges are present on ventro-lateral surfaces of the pseudocephalon. The internal cephaloskeleton consists of massive paired mouthhooks, a small oral sclerite, small paired dental sclerites, unpaired intermediate and labial sclerites and paired basal sclerites with parastomal bar, dorsal bridge, vertical plate, and dorsal and ventral cornua (Fig. 3.4a). The

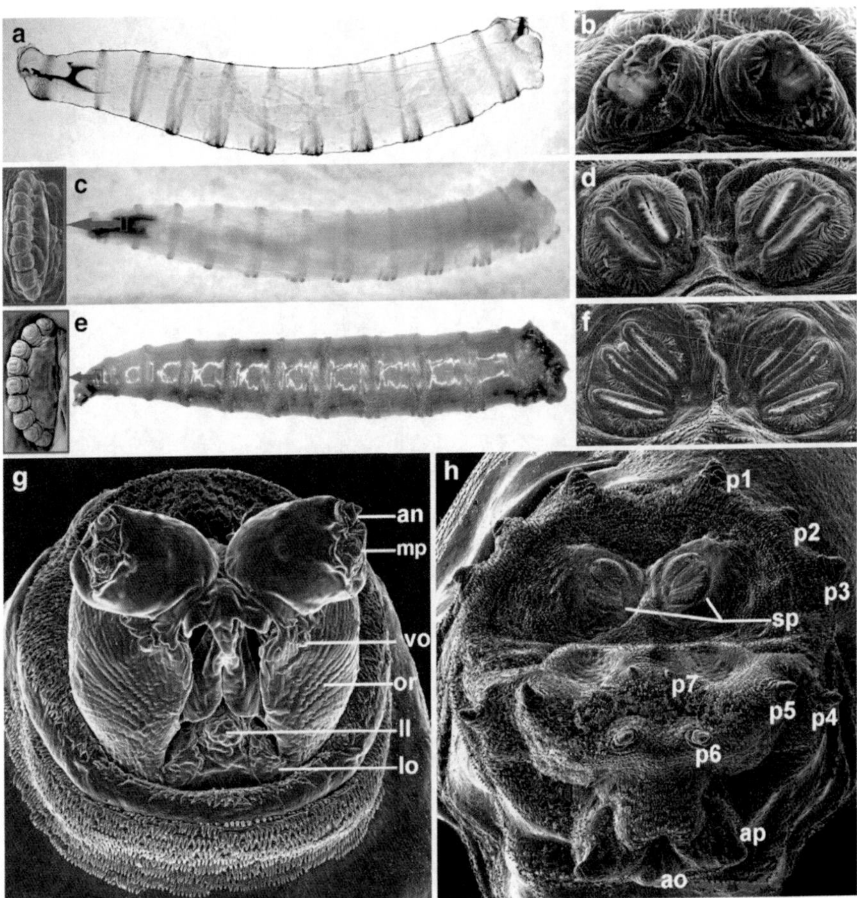

Fig. 3.2 Larvae of necrophagous Calliphoridae: **a** – first instar, habitus; **b** – first instar, posterior spiracles; **c** – second instar, habitus and anterior spiracles; **d** – second instar, posterior spiracles; **e** – third instar, habitus and anterior spiracles; **f** – third instar, posterior spiracles; **g** – third instar, pseudocephalon; **h** – third instar, anal division. Abbreviations: an – antenna, ao – anal opening, ap – anal pad, ll – labial lobe, lo – labial organ, mp – maxillary palpus, or – oral ridges, p1-7 – papillae 1-7, sp – posterior spiracles, vo – ventral organ

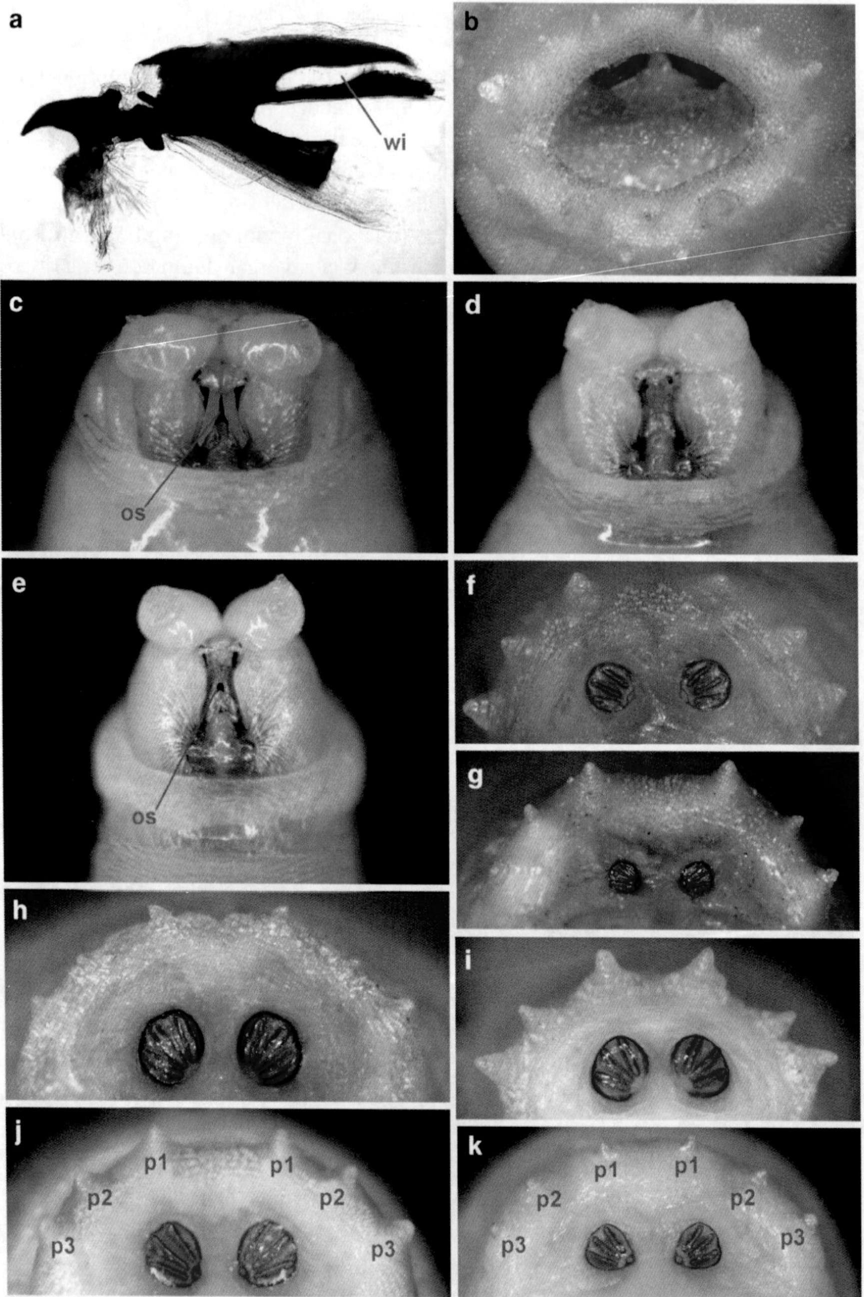

Fig. 3.3 Third instars of necrophagous Sarcophagidae and Calliphoridae: **a** – *Sarcophaga caerulescens*, cephaloskeleton, lateral view; **b** – *Sarcophaga caerulescens*, anal division, spiracular cavity; **c** – *Calliphora vomitoria*, pseudocephalon, ventral view; **d** – *Lucilia sericata*, pseudocephalon, ventral view; **e** – *L. ampullacea*, pseudocephalon, ventral view; **f** – *Calliphora vicina*, anal division, upper half of spiracular field; **g** – *Cynomya mortuorum*, anal division, upper half of spiracular field; **h** – *Phormia regina*, anal division, upper half of spiracular field; **i** – *Protophormia terraenovae*, anal division, upper half of spiracular field; **j** – *Lucilia illustris*, anal division, upper half of spiracular field; **k** – *Lucilia sericata*, anal division, upper half of spiracular field. Abbreviations: os – oral sclerite, p1-3 – papillae 1-3, wi – window

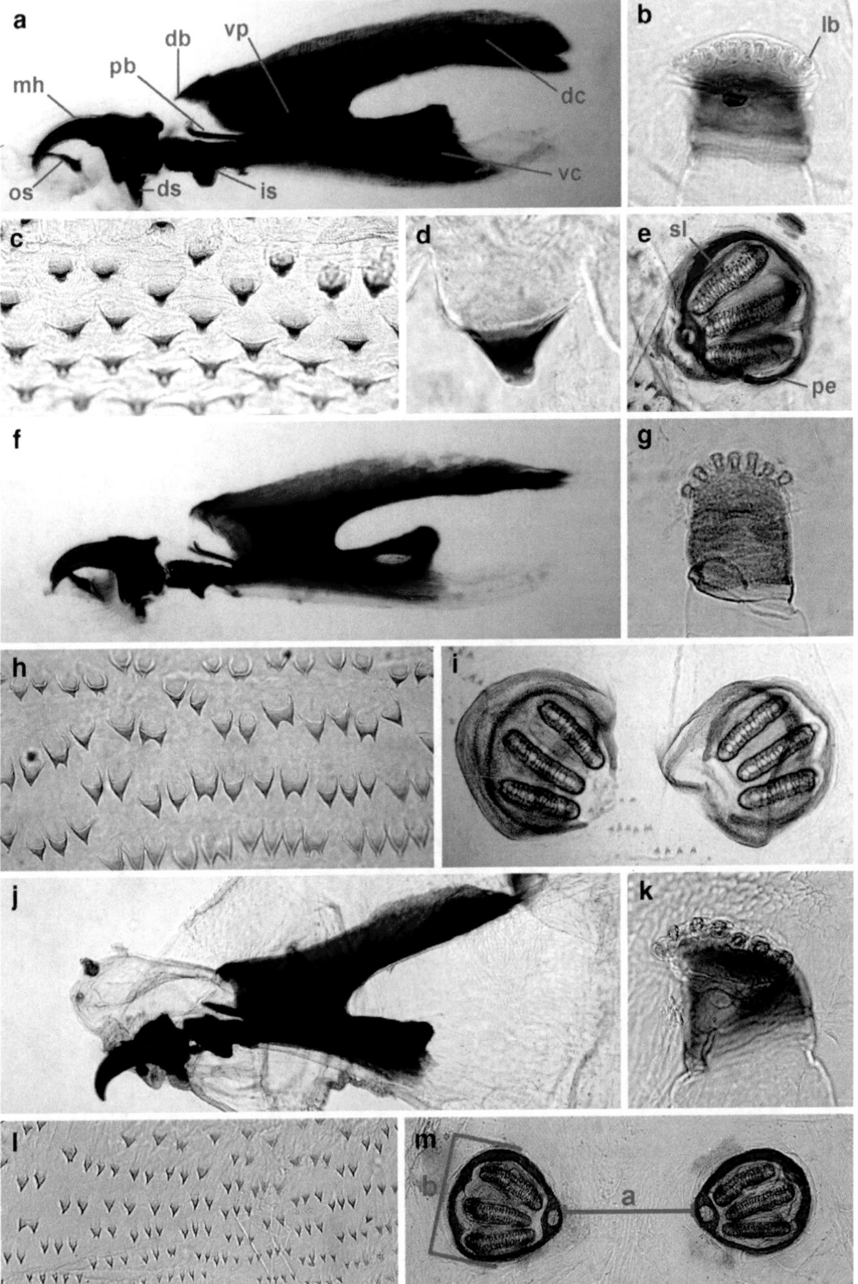

Fig. 3.4 Third instars of *Calliphora* and *Cynomya*: **a** – *Calliphora vomitoria*, cephaloskeleton, lateral view; **b** – *C. vomitoria*, anterior spiracle; **c** – *C. vomitoria*, thoracic segment III, spines; **d** – *C. vomitoria*, thoracic segment III, spine; **e** – *C. vomitoria*, posterior spiracle, **f** – *Calliphora vicina*, cephaloskeleton, lateral view; **g** – *C. vicina*, anterior spiracle; **h** – *C. vicina*, thoracic segment III, spines; **i** – *C. vicina*, posterior spiracles; **j** – *Cynomya mortuorum*, cephaloskeleton, lateral view; **k** – *C. mortuorum*, anterior spiracle; **l** – *C. mortuorum*, thoracic segment III, spines; **m** – *C. mortuorum*, posterior spiracles. Abbreviations: a – distance between posterior spiracles, b – diameter of posterior spiracle, db – dorsal bridge, dc – dorsal cornu, ds – dental sclerite, is – intermediate sclerite, lb – lobe of anterior spiracle, mh – mouthhook, os – oral sclerite, pb – parastomal bar, pe – peritreme, sl – slit of posterior spiracle, vc – ventral cornu, vp – vertical plate

Fig. 3.5 Third instars of Chrysomyinae: **a** – *Chrysomya megacephala*, cephaloskeleton, lateral view; **b** – *C. megacephala*, anterior spiracle; **c** – *C. megacephala*, thoracic segment III, spines; **d** – *C. megacephala*, thoracic segment III, spines; **e** – *C. megacephala*, posterior spiracles, **f** – *Phormia regina*, cephaloskeleton, lateral view; **g** – *P. regina*, anterior spiracle; **h** – *P. regina*, thoracic segment III, spines; **i** – *P. regina*, posterior spiracles; **j** – *Protophormia terraenovae*, cephaloskeleton, lateral view; **k** – *P. terraenovae*, anterior spiracle; **l** – *P. terraenovae*, thoracic segment III, spines; **m** – *P. terraenovae*, posterior spiracles

mouthhooks are strongly sclerotised and divided into sharp curved anterior part and broad basal part. The intermediate sclerite is located between the mouthhooks and the basal sclerite (Fig. 3.4a). The basal sclerite is the most posterior part of the cephaloskeleton. Both parts of the basal sclerite are connected dorso-anteriorly by a dorsal bridge. The parastomal bar has the form of a thin rod directed anteriorly. The vertical plate is broad. The dorsal cornu is longer than the ventral cornu. The postero-dorsal part of the ventral cornu is equipped in a small window (Fig. 3.4f). Segments TI–TIII are equipped with spinose bands only anteriorly (Fig. 3.1a). On each lateral surface of TI is the anterior spiracle with the number of lobes varying according to particular species (Figs. 3.4b, g, k, 3.5b, g, k, and 3.6g, j, m). The number of lobes also shows some intraspecific variation. Segments AI–AVII are armed with both anterior and posterior spinose bands. The width of the anterior bands decreases toward the posterior end of the body, whereas the width of the posterior spinose bands increases in this same direction. Spinose bands are often incomplete, especially on terminal segments (AV–AVII). The shape of the spines shows infraspecific variation in size, shape, and arrangement. The tip of the spines may be single (Figs. 3.4c, l and 3.6c) or serrated (Fig. 3.5d, h, l). Particular spines are arranged separately (Figs. 3.4c and 3.5h) or in irregular rows (Figs. 3.4h, l and 3.6c). Abdominal segments are followed by the terminal region of the larval body, the anal division. The most conspicuous parts of the anal division of the blowfly's third instar are the spiracular field (Fig. 3.1a) and the anal area. The spiracular field consists of seven pairs of papillae situated marginally along its outer surface and posterior spiracles situated centrally (Figs. 3.2h and 3.3f–k). The size and position of the papillae are characteristic for particular species and the dorsalmost pairs of papillae have special taxonomic importance (Figs. 3.3f–k). Each posterior spiracle possesses three linear slits (Fig. 3.4e). The peritreme of the posterior spiracle may completely surround the spiracle (Figs. 3.4a, m and 3.6f, i, l) or be interrupted at some distance (Figs. 3.4i and 3.5e, i, m). The ratio of the distance between the posterior spiracles and their diameter is also of taxonomic value (Fig. 3.4 m). In the anal area, two conical and fleshy anal pads flank the slit-like anal opening (Fig. 3.2h). The shape of the anal pads is rather conservative among the third instars of necrophagous blowflies. The arrangement of spines around the anal area shows intraspecific variation.

3.4 Key

1. – body cylindrical and tapering (Fig. 3.1a, b), cephaloskeleton without long window in dorsal cornua and with developed parastomal bar (Fig. 3.4a), anterior spinose bands on all thoracic and most abdominal segments fully developed (Fig. 3.1a, b), posterior spiracles never in deep spiracular cavity or on long stalks, around spiracular field seven pairs of papillae (Figs. 3.2h and 3.3f–k), slits of posterior spiracles linear (Fig. 3.4e) . . . . 2 (Calliphoridae)
   – other combination of characters (Figs. 3.1c–g and 3.3 a, b) . . . . . other Diptera
2. – abdominal segments of the larva with numerous fleshy protuberances (Fig. 3.1b) . . . . Chrysomya albiceps (Wiedemann, 1819)

– abdominal segments of the larva without such protuberances. 3

3. – oral sclerite at least partly sclerotised (Figs. 3.3c, e, 3.4a, f, j, 3.5a and 3.6a). . . 4
   – oral sclerite unsclerotised (Figs. 3.3d, 3.5f, j and 3.6e, h, k). 8
4. – sclerotised part of the oral sclerite small, almost circular (Figs. 3.3e, 3.5a and 3.6a) 5
   – oral sclerite well sclerotised along the whole length (Figs. 3.3c and 3.4a, f, j) 6
5. – spines big, robust, often with serrated tips (check on the thoracic segments) (Fig. 3.5c, d); posterior spiracles close to each other and with incomplete peritreme (Fig. 3.5e) .... Chrysomya megacephala (Fabricius, 1794)
   – spines small, with single tips, arranged in short rows (check on the thoracic segments) (Fig. 3.6c); posterior spiracles wide apart and with complete peritreme (Fig. 3.6d) . . . . Lucilia ampullacea Villeneuve, 1922
6. – spines big, arranged separately (check on the thoracic segments) (Fig. 3.4c, d) . . . . Calliphora vomitoria (Linnaeus, 1758)
   – spines small, arranged in short rows (check on the thoracic segments) (Fig. 3.4h, l) 7
7. – apical part of mouthhooks gently curved (Fig. 3.4f); posterior spiracles relatively close together (SDF $\approx$ 1.0) (Figs. 3.3f and 3.4i) . . . . .Calliphora vicina (Robineau-Desvoidy, 1830)
   – apical part of mouthhooks abruptly curved (Fig. 3.4j); posterior spiracles very wide apart (SDF> 1.2) (Figs. 3.3g and 3.4m) . . . . Cynomya mortuorum (Linnaeus, 1761)
8. – spines on thoracic segments predominantly with serrated tips (Fig. 3.5h, l), spines of different sizes 9
   – spines with serrated tips on thoracic segments absent (Fig. 3.6c), all spines small of similar size 10
9. – papillae around the spiracular field very big (Fig. 3.3i), all posterior spinose bands incomplete . . . . Protophormia terraenovae (Robineau-Desvoidy, 1830)
   – papillae around the spiracular field small (Fig. 3.3h), at least some posterior spinose bands complete . . . . Phormia regina (Meigen, 1826)
10. – cephaloskeleton without sclerotised area below the posterior tip of ventral cornua (Fig. 3.6k), distance between each P1 similar to distance between P1 and P2 (Fig. 3.3k) . . . . Lucilia sericata (Meigen, 1826)
    – cephaloskeleton with sclerotised area below the posterior tip of ventral cornua (Fig. 3.6e, h), distance between each P1 larger than distance between P1 and P2 (Fig. 3.3j) 11
11. – postero-dorsal angle of the basal part of the mouthhook with process directed postero-dorsally (Fig. 3.6e), posterior spinose band on abdominal segment VI interrupted dorsally (Fig. 3.6n). . . . . Lucilia caesar (Linnaeus, 1758)
    – postero-dorsal angle of the basal part of the mouthhook with process directed posteriorly (Fig. 3.6h), posterior spinose band on abdominal segment VI complete (Fig. 3.6o) . . . . Lucilia illustris (Meigen, 1826)

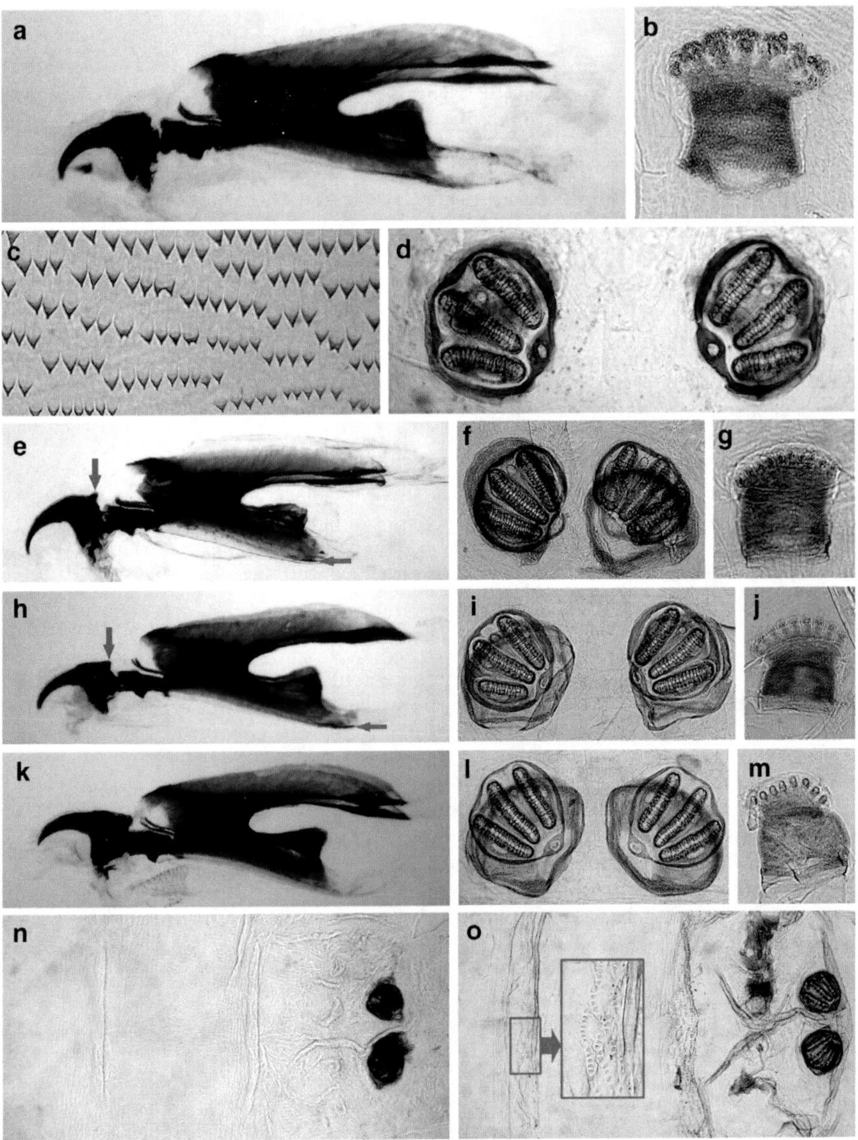

Fig. 3.6 Third instars of *Lucilia*: **a** – *Lucilia ampullacea*, cephaloskeleton, lateral view; **b** – *L. ampullacea*, anterior spiracle; **c** – *L. ampullacea*, thoracic segment III, spines; **d** – *L. ampullacea*, posterior spiracles, **e** – *L. caesar*, cephaloskeleton, lateral view; **f** – *L. caesar*, posterior spiracles; **g** – *L. caesar*, anterior spiracle; **h** – *L. illustris*, cephaloskeleton, lateral view; **i** – *L. illustris*, posterior spiracles; **j** – *L. illustris*, anterior spiracle; **k** – *L. sericata*, cephaloskeleton, lateral view; **l** – *L. sericata*, posterior spiracles; **m** – *L. sericata*, anterior spiracle; **n** – *L. caesar*, abdominal segment VII, dorsal view; **o** – *L. illustris*, abdominal segment VII, dorsal view

References

Benecke M (1998) Six forensic cases: description and commentary. J Forensic Sci 43(4): 797–805

Benecke M (2002) Insects and Corpses. In: Baccino E: 16th Meeting of the International Association of Forensic Sciences, Monpellier, France, Sept. 2–7, 2002. Monduzzi Editore, Bologna, pp 135–140

Byrd JH, Castner JL (2001) Forensic Entomology – The Utility of Arthropods in Legal Investigations. CRC, Boca Raton, FL

Carvalho Queiroz MM, Pinto de Mello R, Lima MM (1997) Morphological aspects of the larval instars of *Chrysomya albiceps* (Diptera, Calliphoridae) reared in the laboratory. Mem Inst Oswaldo Cruz 92:187–196 [*Ch. albiceps*]

Courtney GW, Sinclair BJ, Meier R (2000) Morphology and terminology of Diptera larvae. In: Papp L, Darvas B (eds) Contributions to a Manual of Palaearctic Diptera (with special reference to flies of economic importance). Science Herald, Budapest, pp 85–161

Draber-Mońko A (2004) Calliphoridae, Plujki (Insecta Diptera). Fauna Poloniae 23, Natura Optima Dux Fundation & MIZ PAS, Warsaw [in Polish]

Ebejer MJ (2007) The occurrence of *Chrysomya megacephala* (Fabricius) (Diptera, Brachycera) in Malta and records of other Calliphoridae from the Maltese Islands. Entomol Mon Mag 143:165–170

Erzinçlioğlu YZ (1985) Immature stages of British *Calliphora* and *Cynomya*, with re-evaluation of the taxonomic characters of larval Calliphoridae (Diptera). J Nat Hist 19:69–96 [*C. vicina*, *C. vomitoria*, *C. mortuorum*]

Erzinçlioğlu YZ (1987a) The larvae of some blowflies of medical and veterinary importance. Med Vet Entomol 1:121–125 [*Ch. albiceps*, *L. sericata*]

Erzinçlioğlu YZ (1987b) The larval instars of the African blowfly, Calliphora croceipalpis Jaennicke, with a key to the genera of the third instars of African carrion-breeding Calliphoridae (Diptera). Bull Ent Res 77:575–580 [*L. sericata*]

Erzinçlioğlu YZ (1988) The larvae of species of *Phormia* and *Boreellus*: Northern, cold-adapted blowflies (Diptera: Calliphoridae). J Nat Hist 22:11–16 [*P. regina*, *P. terraenovae*]

Erzinçlioğlu YZ (1990) The larvae of two closely-related blowfly species of the genus *Chrysomya* (Diptera, Calliphoridae). Entomol Fenn 1:151–153 [*Ch. megacephala*]

Fan Z, Zhizi C, Jianming F, Shensheng Z, Zhenliang T (1997) Diptera: Calliphoridae. Fauna Sin, Insecta, 6: x + 1–707 [in Chinese with English summary] [*C. vomitoria*, *C. mortuorum*, *Ch. megacephala*, *P. terraenovae*, *L. caesar*, *L. illustris*, *L. sericata*]

Grassberger M, Friedrich E, Reiter C (2003) The blowfly *Chrysomya albiceps* (Wiedemann) (Diptera, Calliphoridae) as a new forensic indicator in Central Europe. Int J Legal Med 117:75–81 [*Ch. albiceps*]

Greenberg B, Kunich JC (2002) Entomology and the law – flies as forensic indicators. Cambridge University Press, Cambridge

Grunwald J, Swoboda S, Reckel F (2009) A comparative study on the arthropod succession on dressed and undressed pig carrion in the city of Munich (Germany). Programme of the 7th meeting EAFE, Uppsala, p 46

Hale C, Hall M, Wardhana A, Adams Z, Ready P (2008) Molecular identification of the agents of traumatic myiasis of small mammals in UK. Proceedings of the 6th meeting EAFE, Kolymbari, Crete, p 65

Hall DG (1948) The blowflies of North America. The Thomas Say Foundation, Baltimore, MD [*C. vicina*, *C. vomitoria*, *P. regina*, *L. illustris*, *L. sericata*]

Holloway BA (1985) Immature stages of New Zealand Calliphoridae. In: Dear JP (ed) Calliphoridae (Insecta: Diptera). Fauna of New Zealand 8, DSIR, Wellington, pp 12–14, 80–83 [*C. vicina*, *L. sericata*]

Holloway BA (1991) Identification of third-instar larvae of flystrike and carrion associated blowflies in New Zealand (Diptera: Calliphoridae). N Z Entomol 14:24–28 [*Ch. megacephala*, *L. sericata*]

Ishijima H (1967) Revision of the third stage larvae of synanthropic flies of Japan (Diptera: Anthomyiidae, Muscidae, Calliphoridae and Sarcophagidae). Jpn J Sanit Zool 18:47–100 [*C. vicina*, *C. vomitoria*, *Ch. megacephala*, *P. regina*, *P. terraenovae*, *L. ampullacea*, *L. caesar*, *L. illustris*, *L. sericata*]

Kano R, Sato K (1952) Notes on flies of medical importance in Japan. (Part VI) Larvae of Luciliini in Japan. Jpn J Exp Med Tokyo 22:33–42 [*L. ampullacea*, *L. illustris*, *L. sericata*]

Kitching RL (1976) The immature stages of the Old-World screw-worm fly, *Chrysomya bezziana* Villeneuve, with comparative notes on other Australasian species of *Chrysomya* (Diptera, Calliphoridae). Bull Entomol Res 66:195–203 [*Ch. megacephala*]

Knipling EF (1936) Some specific taxonomic characters of common Lucilia larvae-Calliphorinae-Diptera. Iowa State College J Sci 10(3):275–293 [*L. illustris*, *L. sericata*]

Liu D, Greenberg B (1989) Immature stages of some flies of forensic importance. Ann Entomol Soc Am 82:80–93 [*C. vicina*, *P. regina*, *L. illustris*, *L. sericata*]

Martínez AJ, Rognes K (2008) Calliphoridae (Diptera). In: Borges PAV, Abreu C, Aguiar AMF, Carvalho P, Jardim R, Melo I, Oliveira P, Sérgio C, Serrano ARM, Vieira P (eds) A list of terrestrial fungi, flora and fauna of Madeira and Selvanges archipelagos. Direcção Regional do Ambiente da Madeira und Universidade dos Açores, Funchal and Angra do Heroísmo, p 329

Prins AJ (1982) Morphological and biological notes on six south African blow-flies (Diptera, Calliphoridae) and their immature stages. Ann South Afr Mus 90:201–217 [*L. sericata*, *Ch. megacephala*]

Povolný D (2002) *Chrysomya albiceps* (Wiedemann, 1819): the first forensic case in Central Europe involving this blowfly (Diptera, Calliphoridae). Acta univ agric et silvic Mendel Brun, L3:105–112 [*Ch. albiceps*]

Rognes K (2004) Fauna Europaea: Diptera, Calliphoridae. Fauna Europaea version 1.1, http://www.faunaeur.org

Shewell GE (1987) Calliphoridae. In: McAlpine JF, Peterson BV, Shewell GE, Teskey HJ, Vockeroth JR, Wood DM (eds) Manual of Nearctic Diptera, vol 2. Research Branch Agriculture Canada, Ottawa, pp 1133–1145 [*L. sericata*]

Schumann H (1954) Morphologisch-systematische Studien an Larven von hygienisch wichtigen mitteleuropäischen Dipteren der Familien Calliphoridae – Muscidae. Wiss Zeitschr Univ Greifswald, Jahrgang III, 1953/54 Mathematisch-naturwissenschaftliche Reihe Nr. 4/5:245–274 [*C. vicina*, *C. mortuorum*, *P. terraenovae*, *L. ampullacea*, *L. sericata*]

Schumann H (1965) Merkblatter uber angewandte Parasitenkunde und Schadlingsbekampfung. Merkblatt Nr. 11. Die Schmeissfliegengattung Calliphora. Angew Parasitol [Suppl.] 6(3):1–14 [*C. vicina*]

Schumann H (1971) Merkblatter uber angewandte Parasitenkunde und Schadlingsbekampfung. Merkblatt Nr. 18. Die Gattung Lucilia (Goldenfliegen). Angew Parasitol [Suppl.] 12(4):1–20 [*L. sericata*]

Smith KGV (1986) A Manual of Forensic Entomology. British Museum (Natural History), London, and Cornell University Press, Ithaca, NY [*C. vicina*, *C. mortuorum*, *Ch. albiceps*, *P. terraenovae*, *L. ampullacea*, *L. illustris*, *L. sericata*]

Smith KGV (1989) An introduction to the immature stages of British flies. Diptera larvae with notes on eggs, puparia and pupae. Handbooks for the Identification of British Insects, vol 10, part 14 [*C. vicina*, *L. ampullacea*, *L. illustris*]

Stærkeby M (2001) Dead larvae of Cynomya mortuorum (L.) (Diptera, Calliphoridae) as indicators of the post-mortem interval – a case history from Norway. Forensic Sci Int 120:77–78

Sukontason KL, Sukontason K, Piangjai S, Boonchu N, Chaiwong T, Vogtsberger RC, Kuntalue B, Thijuk N, Olson JK (2003) Larval morphology of *Chrysomya megacephala* (Fabricius) (Diptera: Calliphoridae) using scanning electron microscopy. J Vector Ecol 2003:47–52 [*Ch. megacephala*]

Sukontason K, Piangjai S, Siriwattanarungsee S, Sukontason KL (2008) Morphology and developmental rate of blowflies *Chrysomya megacephala* and Chrysomya rufifacies in Thailand: application in forensic entomology. Parasitol Res 102:1207–1216 [*Ch. megacephala*]

Szpila K, Pape T (2007) Rediscovery, redescription and reclassification of Beludzhia phylloteliptera (Diptera: Sarcophagidae, Miltogramminae). Eur J Entomol 104:119–137

Szpila K, Matuszewski S, Bajerlein D, Konwerski S (2008) Blowflies (Diptera: Calliphoridae) visiting pig carcasses in selected forests of Central Europe – preliminary report. Proceedings of the 6th meeting EAFE, Kolymbari, Crete, p 83

Teskey HJ (1981) Morphology and terminology – larvae. In: McAlpine JF, Peterson BV, Shewell GE, Teskey HJ, Vockeroth JR, Wood DM (eds) Manual of Nearctic Diptera. Research Branch, Agriculture Canada, Ottawa, pp 68–88 [*P. regina*]

Velásquez Y, Martínez-Sánchez A, Rojo S. (2008) Autumn colonization of pig carrion by blowflies (Diptera: Calliphoridae) in a Mediterranean urban area (SE, Spain). Proceedings of the 6th meeting EAFE, Kolymbari, Crete, p 84

Wallman JF (2001) Third instar larvae of common carrion-breeding blowflies of the genus *Calliphora* (Diptera: Calliphoridae) in South Australia. Invertebr Taxon 15:37–51 [*C. vicina*]

Wells JD, Byrd JH, Tantawi TI (1999) Key to third-instar Chrysomyinae (Diptera: Calliphoridae) from carrion in the continental United States. J Med Entomol 36(5):638–641 [*Ch. albiceps, Ch. megacephala, P. regina, P. terraenovae*]

Zimin LS (1948) Key to the third instar larvae of synathropic flies of Tadzhikistan. Opred Faune SSSR 28:1–114 [in Russian] [*C. vicina, Ch. albiceps, P. regina, L. sericata*]

Zumpt F. (1965) Myasis in man and animals in the Old World. Butterworths, London [*Ch. albiceps*]

Chapter 4
The Utility of Coleoptera in Forensic Investigations

John M. Midgley, Cameron S. Richards, and Martin H. Villet

4.1 Introduction

Forensic entomology is a developing field of forensic science, so there are many avenues to investigate. These avenues include novel directions that have never been addressed, as well as more critical and rigorous research into areas which have already been explored. Most research in forensic entomology has focused on flies, and beetles (Coleoptera) have been at best under-emphasized. A good example of this is the review by Smith (1986), where 70 pages are dedicated to Diptera and only 12 to Coleoptera; this situation has changed little in the subsequent 20 years. To contextualize the neglect, throughout the world there are at least as many species of Coleoptera that may visit a particular carcass as Diptera (Braack 1986; Louw and van der Linde 1993; Bourel et al. 1999; Lopes de Carvalho et al. 2000; Pérez et al. 2005; Shea 2005; Watson and Carlton 2005a; Salazar 2006; Martinez et al. 2007). A common assumption underlying the neglect of Coleoptera is that Diptera locate corpses faster, and thus give a more accurate estimate of minimum Post Mortem Interval (PMI_{min}). Recent observations (Midgley and Villet 2009b) have shown that *Thanatophilus micans* (Silphidae) can locate corpses and start breeding within 24 h of death, and thus the potential utility of estimates based on this species is equal to that of those based on flies.

Beetles form a taxonomically and ecologically diverse part of the carrion insect community (Smith 1986; Braack 1986; Bourel et al. 1999; Shea 2005; Tabor et al. 2004; Watson and Carlton 2005a; Salazar 2006), thus providing a wide spectrum of sources of potential evidence. They are also integral to postmortem biology. For instance, larder beetles can, given ideal conditions of desiccation or even mummification, accelerate the decomposition of a corpse (Schröder et al. 2002), and produce characteristic postmortem changes (Voight 1965) that should be distinguished from antemortem trauma. They also complicate other forensic analyses (e.g. Offelle et al. 2007). This chapter aims to highlight potential uses of beetles in forensic entomology and review the relevant literature.

J.M. Midgley (✉), C.S. Richards and M.H. Villet
Southern African Forensic Entomology Research Laboratory, Department of Zoology and Entomology, Rhodes University, Grahamstown, 6140, South Africa

J. Amendt et al. (eds.), *Current Concepts in Forensic Entomology*,
DOI 10.1007/978-1-4020-9684-6_4,

4.2 Forensic Applications of Coleoptera

4.2.1 Developmental Biology

Insects are used in forensic investigations primarily to develop an estimate of PMI_{min}. These estimates can be based on the duration of the immature stages of the insects found on a corpse or on the community composition of insects on the corpse (Byrd and Castner 2001). The duration of the immature stages is generally longer in Coleoptera than in Diptera (Fig. 4.1), which means that Coleoptera are useful to estimate PMI_{min} not only during early decomposition, but also in later stages of decomposition. In addition, many beetles utilize corpses in advanced decomposition and can be used to estimate PMI_{min} by analyzing the community present on a corpse (Smith 1986). In these cases many fly larvae have already left the corpse, leaving mostly Coleoptera from which to make estimates.

The most precise method of estimating PMI_{min} using insects is to use models based on development of immature stages (Higley and Haskell 2001). These models can either use size as a surrogate for age or use physiological age by identifying developmental landmarks. The latter models are less biased and more precise (Dadour et al. 2001), as they measure actual age, and not size, which can be affected by many factors other than age (Villet et al. 2009, Chapter 7). Development of flies has been investigated extensively and refined models for various species are available (Grassberger and Reiter 2001, 2002; Higley and Haskell 2001; Villet et al. 2006; Richards et al. 2008). Statistically robust models for coleopteran development are not as common (Midgley and Villet 2009a) and so data for most species should be interpreted with caution. This is not to say that all data should be disregarded, but further study is required to develop statistically robust models. The development of *T. micans* has been thoroughly modelled (Midgley and Villet 2009a)

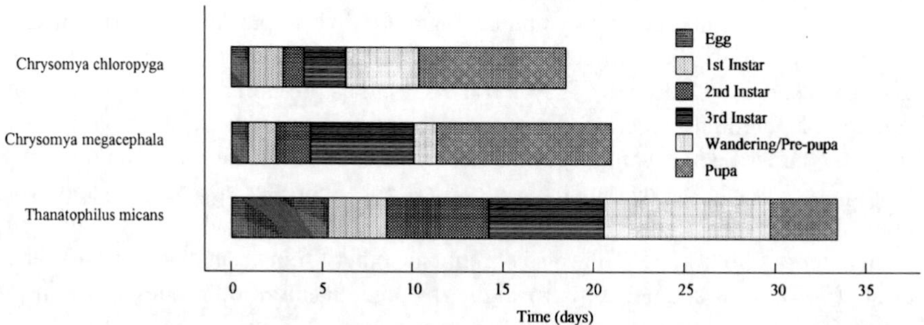

Fig. 4.1 Comparison of the time taken for development of three forensically important insect species at 20°C, showing extended development time typical in Coleoptera, Thanatophilus micans, compared to Diptera, *Chrysomya chloropyga* and *Chrysomya megacephala*, based on data from Midgley and Villet (2009a) (*T. micans*), Richards et al. (2009) (*C. chloropyga*) and Richards and Villet (2009) (*C. megacephala*)

and shows that with more research, development of Coleoptera can be a useful tool for forensic entomologists. The models produced for the developmental landmarks of *T. micans* not only meet the minimum statistical requirements for regression modelling (Richards and Villet 2008), but have coefficients of determination greater than 0.98 for all post-hatching stages (Midgley and Villet 2009a). This shows that beetle development is predictable and, coupled with the rapid location of corpses, shows that at least *T. micans* and probably other sexton beetles (Silphidae) are reliable forensic indicators.

In many cases live insects are not available for PMI_{min} estimation, usually because they are collected by non-specialists (Lord and Burger 1983). In such cases the length, width or mass of the collected specimens is the only reliable measure for estimating PMI_{min}. Size-at-age data is not available for most forensically relevant beetle species, with the exception of *T. micans* (Midgley and Villet 2009a). Change in specimens' sizes during storage is a well known fact in forensic entomology (Lord and Burger 1983; Adams and Hall 2003; Amendt et al. 2007; Midgley and Villet 2009b) and this must be considered when using developmental models based on length. The killing method used to preserve samples has an effect on the change during storage and must also be considered. For fly larvae, killing with ethanol is not recommended, as significant changes in length occur (Tantawi and Greenberg 1993; Adams and Hall 2003). This is not the case with beetle larvae: killing with ethanol causes the least change in length of silphid larvae (Midgley and Villet 2009b) and is therefore the most suitable preservation method. This is because beetle larvae have extensively sclerotized exoskeletons, and so are more rigid that fly larvae. Similarities and differences between Coleoptera and Diptera must be considered when samples are taken at a crime scene to obtain accurate estimates of PMI_{min}.

An advantage of estimating PMI_{max} from beetle larvae is that they are solitary and furtive, while maggots aggregate into maggot balls or maggot masses. This results in beetle larvae experiencing temperatures close to ambient, which simplifies the application of thermal accumulation models of development. Blowfly larvae in maggot masses collectively generate enough heat to warm themselves as much as 25°C above ambient temperatures. Accounting for this while estimating a PMI_{max} is a source of error that can be avoided by using both flies and beetles in a given estimate.

The use of development to estimate PMI_{min} is not limited to the primary consumers of decaying corpses. Parasitoids and predators, such as rove beetles (Staphylinidae) and clown beetles (Histeridae), of these species can also be used, as their larvae are also obligate corpse dwellers. The precision and accuracy of these estimates may be decreased because they are subject to the developmental variability of the necrophilous parasite or predator in addition to that usual in the necrophagous species. The latter may even be modified by parasitoids. Fly pupae of several species and families can be parasitized by species of *Aleochara* (Staphylinidae) (Gauvin 1998; Ferreira de Almeida and Pires do Prado 1999). *Aleochara* is however a large and diverse genus, with between 300 and 400 species (Maus et al. 2001), many of which are geographically localized or do not parasitize necrophagous Diptera. Identification of locally relevant species and the generation of developmental models for these species is critical before *Aleochara* can reach its potential in estimating the PMI_{min}.

4.2.2 Community Ecology

When corpses are in advanced decomposition, it is not always possible to use development to accurately estimate PMI_{min}. Enclosed pupae and puparia can give inaccurate PMI_{min} estimates, as they remain unchanged for long periods. In such cases the community of animals present on a corpse can be used to estimate PMI_{min} (Smith 1986; Schoenley et al. 1996, 2005, 2007). These communities are usually dominated by Coleoptera (Braack 1986; Smith 1986; Catts 1992; Bourel et al. 1999; Shea 2005) and understanding their biology is vital to making accurate PMI_{min} estimates.

Certain species are known to feed on dried animal material such as *Trox* spp., but the adults of these species have been observed on and around corpses at very early stages of decomposition. It is therefore crucial that collections from corpses in advanced stages of decomposition be undertaken by trained professionals who are aware of where to look for entomological evidence. Ecological successions are qualitatively predictable, but quantitatively variable by their nature, and thus an adequate error margin should always be considered in PMI_{min} estimates made with this technique. Good progress has been made in validating this approach (Schoenley et al. 1996, 2005, 2007), but there is still a relative paucity of data on the 'windows of activity' of beetles on carrion and corpses (but see e.g. Bourel et al. 1999; Tabor et al. 2004; Watson and Carlton 2005a; Matuszewski et al. 2008). A way of addressing this problem is outlined in Section 4.3.2.

4.2.3 Toxicology

It has proven possible to extract traces of certain drugs from the larvae (Bourel et al. 2001a), pupae (Bourel et al. 2001b) and exuviae of beetles (Miller et al. 1994). Beetle exuviae are moderately tough, and may remain at a death scene long after the corpse has decomposed beyond toxicological analysis, thus providing a useful source of toxicological samples.

Although the developmental responses of fly larvae to drugs in their food have been well studied, the same is not true of beetles. Studies have been conducted on beetle responses to some plant materials (Egwunyenga et al. 1998; Fasakin and Aberejo 2002), and using this information can assist in some PMI_{min} estimates. More dedicated research in this field is needed to refine PMI_{min} estimates.

4.3 Existing Data with Forensic Relevance

4.3.1 Taxonomy

One of the key needs of a forensic entomologist is a method of identifying insects found on corpses. Necrophagous Coleoptera found on older or dryer corpses can be identified using the keys provided in the stored product literature, such as

Hinton (1945) and Gorham (1987). Works more specifically about stored product pests are also useful. Dermestidae (Fig. 4.2) adults (Mroczkowski 1968; Peacock 1993) and larvae (Rees 1947; Adams 1980; Zhantiev and Volkova 1998, 1999) are

Fig. 4.2 Two common and near-cosmopolitan beetles found on corpses are *Dermestes maculatus* (*above*) and *Necrobia rufipes* (*below*). Both species occur on corpses of all sizes and in stored products

common on corpses and most are cosmopolitan, making identification easier. Cleridae are represented by a few species of *Necrobia*, particularly *N. rufipes* (Fig. 4.2), that are widespread stored product pests, which simplifies their identification (Smith 1986; Gorham 1987; Rajendran and Hajira Parveen 2005). Ptinidae are also found on dessicated bodies, and can be identified using Brown (1940), Harney (1993) or Irish (1999).

For exclusively necrophilous species, such as Silphidae, Staphylinidae and Histeridae, identification is not as easy because many of them are neither cosmopolitan nor pestilent. Their identification therefore depends on the taxonomic advancement of the broad geographic area in which the corpse is located. African and Australian Silphidae can be identified using Schawaller (1981, 1987) and Peck (2001), and the Afrotropical Trogidae using Scholtz (1980, 1982) and van der Merwe and Scholtz (2005); these works allow easy identification of these species. With a little more effort, one can identify many genera of Histeridae using Caterino and Vogler (2002) and all tribes of Staphylinidae using Solodovnikov and Newton (2005); Catalogues of local necrophilous beetles, such as the Turkish species found by Özdemir and Sert (2008), can provide easy identification of beetles in a given area, but should be used with caution outside of the geographic range treated.

4.3.2 Biology

Once the species on a corpse are identified, information will be needed about their biologies. Despite the fact that little research on beetles has been focussed explicitly on forensics (Williams and Villet 2006), notes and data on the development of several coleopteran species can be found in various publications. The more common and widespread species are often stored product pests, and significant research has been carried out in efforts to control these species. A good example of this is *Dermestes maculatus*, a cosmopolitan pest of stored products (Rajendran and Hajira Parveen 2005) that is common on mummified remains (Schröder et al. 2002). There are developmental data for this species (Scoggin and Tauber 1951; Paul et al. 1963; Richardson and Goff 2001) and notes on behaviour (Archer and Elgar 1998), control methods (Egwunyenga et al. 1998; Fasakin and Aberejo 2002) and other topics (Azab et al. 1972; Archer and Elgar 1999). Developmental data are also available for *Dermestes haemorrhoidalis* and *D. peruvianus* (Coombs 1979), but attention needs to be given to the requirements for a robust developmental model (Richards and Villet 2008). The available developmental data are summarised in Table 4.1. By amalgamating the available information on forensically important species, a balanced view of their biology can be obtained. If this is done for a variety of stored product species that also occur on corpses, such as *N. rufipes* and the locally common species of *Dermestes* (Rajendran and Hajira Parveen 2005), the information can be used to predict PMI_{min} by community analysis. This approach essentially requires a hybrid technique that uses the developmental data to produce traditional estimates of when each species is active on a particular carcass (Villet et al. 2009, Chapter 7), and then uses these

Table 4.1 A summary of some developmental parameters already published on necrophilous Coleoptera. In cases where D_0 and K were not supplied in the original publication, they were calculated according to the methods of Ikemoto and Takai (2001)

Species	Temperature (°C)	Development time (days)	D_0	K	Source
Thanatophilus micans	25	15.7	10.3	7.62	Midgley and Villet (2009a)
Dermestes haemorrhoidalis	25	40.6	12.6	26.98	Coombs (1979)
Dermestes peruvianus	25	52.5	11.4	44.5	Coombs (1979)
Dermestes maculatus	25	59.2			Richardson and Goff (2001)
Dermestes maculatus	Unknown	37.5	n/a	n/a	Scoggin and Tauber (1951)

estimates as refined 'windows of activity' that are analysed by the methods more usually used in community analysis (Schoenley and Reid 1987; Schoenley 1992; Schoenley et al. 1992, 1996, 2005, 2007). This incidentally allows succession data for necrophagous species to be refined to account for weather conditions through the well known effects of temperature on growth. Suitable data have been published for three North American silphids (Watson and Carlton 2005b).

It is important to analyse the beetle community as a whole, as not all beetle species will oviposit immediately after death. This delay means that the precision of an analysis using only one species will be reduced. When the biology of *Dermestes* spp. is used to estimate PMI_{min}, it becomes clear that adjustments may need to be made to the estimate because key factors modifying growth in *Dermestes* spp. are moisture content in food (Scoggin and Tauber 1951) and relative humidity (Coombs 1979). It is therefore important to adjust PMI_{min} estimates based on the development of these species to account for relative humidity and dietary moisture. In cases where dietary moisture remains extremely high or low during decomposition, the development of *Dermestes* spp. will not give unbiased PMI_{min} estimates as development will not occur at normal rates (Schröder et al. 2002). In these cases other species found on the corpse should be used in conjunction with *Dermestes* spp. for PMI_{min} estimates, such as *T. micans* and other Silphidae during early decomposition and *N. rufipes* and *Aleochara* spp. in advanced decomposition. By assessing the development of as many species as possible, a crossvalidated view of the community can be obtained for analysis, making oviposition and biological variations less important and giving a more unbiased and precise PMI_{min} estimate.

4.4 Future Research

A key opportunity is the development of sexton beetles (Silphidae), in particular the genera *Silpha* and *Thanatophilus*. These genera are widespread, but none of the species are cosmopolitan (Schawaller 1981; Peck 2001). Because of this, development models must be created for each species. By generating models for more of these species, accurate predictions of extended PMI_{min}s will be possible over a larger geographic area. In addition to this, PMI_{min}s estimated using these species can crossvalidate estimates made using dipteran larvae, as it is preferable to use as much data as possible to predict a PMI_{min}. The same is true for *Aleochara* (Staphylinidae), as none of the species in the genus are cosmopolitan. Such studies should sample according to published standards to produce robust developmental models (Ikemoto and Takai 2001; Richards and Villet 2008).

While the generation of development models should be prioritized to increase the utility of Coleoptera in forensic entomology, community composition studies are also important. It is vital that forensic entomologists adjust PMI_{min} estimates to correct for factors that affect insect development. To understand the effects on development of different species, the biology of these species must be understood (Anderson 2001). Important beetles for community analysis will differ from region to region, but in general the most important families will be the Silphidae, Staphylinidae, Trogidae, Dermestidae and Histeridae (Braack 1986). These families contain several necrophilous or necrophagous species and an understanding of the factors that influence their use of corpses and interactions with each other will allow better understanding of the carcass community during the advanced stages of decay.

Acknowledgements The authors would like to thank Terence Bellingan and Kendall Crous for assistance with laboratory work. Rhodes University funded the work.

References

Adams RG (1980) *Dermestes leechi* Kalik in stored products and new diagnostic characters for *Dermestes* spp. (Coleoptera: Dermestidae). J Stored Prod Res 16:119–122

Adams ZJO, Hall MJR (2003) Methods used for the killing and preservation of blowfly larvae, and their effect on post-mortem larval length. Forensic Sci Int 138:50–61

Amendt J, Campobasso CP, Gaudry E, Reiter C, LeBlanc H, Hall M (2007) Best practice in forensic entomology – standards and guidelines. Int J Leg Med 121:90–104

Anderson G (2001) Insect succession on carrion and its relationship to determining time of death. In: Byrd JH, Castner JL (eds) Forensic entomology: The utility of arthropods in legal investigations. CRC, Boca Raton, pp 143–175

Archer MS, Elgar MA (1998) Cannibalism and delayed pupation in hide beetles, *Dermestes maculatus* DeGeer (Coleoptera: Dermestidae). Aust J Entomol 37:158–161

Archer MS, Elgar MA (1999) Female preference for multiple partners: sperm competition in the hide beetle, *Dermestes maculatus* (DeGeer). Anim. Behav 58:669–675

Azab AK, Tawfik MFS, Abouzeid NA (1972) The biology of *Dermestes maculatus* De Geer (Coleoptera: Dermestidae). Bull Soc Ent Egypt 56:1–14

Bourel B, Martin-Bouyer L, Hedouin V, Cailliez JC, Derout D, Gosset D (1999) Necrophilous insect succession on rabbit carrion in sand dune habitats in northern France. J Med Ent 36:420–425

Bourel B, Tournel G, Hédouin V, Deveaux M, Goff ML, Gosset D (2001a) Determination of drug levels in two species of necrophagous Coleoptera reared on substrates containing morphine. J Forens Sci 46:600–603

Bourel B, Tournel G, Hédouin V, Deveaux M, Goff ML, Gosset D (2001b) Morphine extraction in necrophagous insects remains for determining ante-mortem opiate intoxication. Forens Sci Int 120:127–131

Braack LEO (1986) Arthropods associated with carcasses in the northern Kruger National Park. S Afr J Wildl Res 16:91–96

Brown WJ (1940) A key to the species of Ptinidae occuring in dwellings and warehouses in Canada (Coleoptera). Can Ent 72:115–122

Byrd JH, Castner JL (eds) (2001) Forensic entomology: the utility of arthropods in legal investigations. CRC, Boca Raton, pp 418

Caterino MS, Vogler AP (2002) The phylogeny of the Histeroidea (Coleoptera: Staphyliniformia). Cladistics 18:394–415

Catts EP (1992) Problems in estimating the postmortem interval in death investigations. J Agric Ent 9:245–255

Coombs CW (1979) The effect of temperature and humidity upon the development and fecundity of *Dermestes haemorrhoidalis* Küster and *Dermestes peruvianus* Laporte de Castelnau (Coleoptera: Dermestidae). J Stored Prod Res 15:43–52

Dadour IE, Cook DF, Fissioli JN, Bailey WJ (2001) Forensic entomology: application, education and research in Western Australia. Forensic Sci. Int 120:48–52

Ferreira de Almeida MA, Pires do Prado A (1999) *Aleochara* spp. (Coleoptera: Staphylinidae) and pupal parasitoids (Hymenoptera: Pteromalidae) attacking symbovine fly pupae (Diptera: Muscidae, Sarcophagidae and Otitidae) in southeastern Brazil. Biol Control 14:77–83

Egwunyenga OA, Alo EB, Nmorsi OPG (1998) Laboratory evaluation of the repellency of *Dennettia tripetala* Baker (Anonaceae) to *Dermestes maculatus* (F.) (Coleoptera: Dermestidae). J Stored Prod Res 34:195–199

Fasakin EA, Aberejo BA (2002) Effect of some pulverized plant materials on the developmental stages of fish beetle, *Dermestes maculates* Degeer in smoked catfish (*Clarias gariepinus*) during storage. Bioresour Tech 85:173–177

Gauvin MJ (1998) Reproductive and developmental biology of *Aleochara bilineata* Gyllenhall (Coleoptera: Staphylinidae). Master of Science Thesis, McGill University, Quebec, Canada

Gorham JR (ed) (1987) Insect and mite pests in food; an illustrated key. U.S. Department of Agriculture, Agriculture Handbook Number pp 655, 767

Grassberger M, Rieter C (2001) Effect of temperature on *Lucilia sericata* (Diptera: Calliphoridae) development with special reference to the isomegalen- and isomorphen-diagram. Forensic Sci Intl 120:32–36

Grassberger M, Rieter C (2002) Effect of temperature on development of the forensically important holarctic blow fly *Protophormia terraenovae* (Robineau-Desvoidy) (Diptera: Calliphoridae). Forensic Sci Intl 128:177–182

Harney M (1993) A guide to the insects of stored grain in South Africa. Plant Protection Research Institute, Pretoria

Higley LG, Haskell NH (2001) Insect development and forensic entomology. In: Byrd JH, Castner JL (eds) Forensic entomology: the utility of arthropods in legal investigations. CRC, Boca Raton, pp 287–302

Hinton HE (1945) A monograph of the beetles associated with stored products. British Museum (Natural History), London

Ikemoto T, Takai K (2001) A new linearized formula for the law of total effective temperature and the evaluation of line-fitting methods with both variables subject to error. Environ Ent 29:671–682

Irish J (1999) First records of pest species of Ptinidae (Coleoptera) from southern Africa. Afr Ent 7:149–150

Lopes de Carvalho LM, Thyssen PJ, Linhares AX, Palhares FA (2000) A checklist of arthropods associated with pig carrion and human corpses in southeastern Brazil. Mem Inst Oswaldo Cruz 95:135–138

Lord D, Burger JF (1983) Collection and preservation of forensically important entomological materials. J Forensic Sci 28:936–944

Louw SvdM, van der Linde TC (1993) Insects frequenting decomposing corpses in central South Africa. Afr Ent 1:265–269

Martinez E, Duque P, Wolff M (2007) Succession pattern of carrion-feeding insects in Paramo, Colombia. Forens Sci Int 166:182–189

Matuszewski S, Bajerlein D, Konwerski S, Szpila K (2008) An initial study of insect succession and carrion decomposition in various forest habitats of Central Europe. Forens Sci Int 180:61–69

Maus C, Peschke K, Dobler S (2001) Phylogeny of the genus *Aleochara* inferred from mitochondrial cytochrome oxidase sequences (Coleoptera: Staphylinidae). Mol Phylog Evol 18: 202–216

Midgley JM, Villet MH (2009a) The effect of killing method on post-mortem change in length of larvae of *Thanatophilus micans* (Fabricius, 1794) (Coleoptera: Silphidae) stored in 70% ethanol. Int J Leg Med 123:285–292

Midgley JM, Villet MH (2009b) Development of *Thanatophilus micans* (Fabricius 1794) (Coleoptera: Silphidae) at constant temperatures. Int J Leg Med 123:103–108

Miller ML, Lord WD, Goff ML, Donnelly B, McDonough ET, Alexis JC (1994) Isolation of amitriptyline and nortriptyline from fly puparia (Phoridae) and beetle exuviae (Dermestidae) associated with mummified human remains. J Forensic Sci 38:1305–1313

Mroczkowski M (1968) Distribution of the Dermestidae (Coleoptera) of the world with a catalogue of all known species. Ann Zool 26:15–191

Offelle D, Harbeck M, Dobberstein R, Wurmb-Schwark N, Ritz-Timme S (2007) Soft tissue removal by maceration and feeding of *Dermestes* sp.: impact on morphological and biomolecular analyses of dental tissues in forensic medicine. Int J Legal Med 121:341–348

Özdemir S, Sert O (2008) Systematic studies on male genitalia of Coleoptera species found on decomposing pig (*Sus scrofa* L.) carcasses at Ankara Province. Hacettepe J Biol Chem 36:137–161

Paul CF, Shukla GN, Das SR, Perti SL (1963) A life-history study of the hide beetle *Dermestes vulpinus* Fab. (Coleoptera: Dermestidae). Indian J Ent 24:167–179

Peacock ER (1993) Adults and larvae of hide, larder and carpet beetles and their relatives (Coleoptera: Dermestidae) and of derodontid beetles (Coleoptera: Derodontidae). In: Dolling WR (ed) Handbooks for the identification of British insects, vol 5. London: Royal Entomological Society of London, pp 1–144

Peck SB (2001) Review of the carrion beetles of Australia and New Guinea (Coleoptera: Silphidae). Aust J Ent 40:93–101

Pérez SP, Duque P, Wolff M (2005) Successional behavior and occurrence matrix of carrion-associated arthropods in the urban area of Medellin, Colombia. J Forens Sci 50:448–454

Rajendran S, Hajira Parveen KM (2005) Insect infestation in stored animal products. J Stored Prod Res 41:1–30

Rees BE (1947) Taxonomy of the larvae of some North American species of the genus *Dermestes* (Coleoptera: Dermestidae). Proc Ent Soc Wash 49:1–14

Richards CS, Villet MH (2008) Factors affecting accuracy and precision of thermal summation models of insect development used to estimate postmortem intervals. Int J Leg Med 122:401–408

Richards CS, Villet MH (2009) Data quality in thermal summation development models for forensically important blowflies. Med Vet Ent 23(3):269–276

Richards CS, Paterson ID, and Villet MH (2008) Estimating the age of immature *Chrysomya albiceps* (Diptera: Calliphoridae), correcting for temperature and geographical latitude. Int J Leg Med 122:271–279

Richards CS, Crous KL, Villet MH (2009) Models of development for blowfly sister species Chrysomya chloropyga and Chrysomya putoria. Med Vet Ent 23(1):56–61

Richardson MS, Goff ML (2001) Effects of temperature and intraspecific interaction on the development of *Dermestes maculates* (Coleoptera: Dermestidae). J Med Ent 38:347–351

Salazar JL (2006) Insectos de importancia forense en cadáveres de ratas, Carabobo - Venezuela. Rev Peruana Med Exp Salud Publ 23:33–38

Schawaller W (1981) Taxonomie und Faunistik der Gattung *Thanatophilus* (Coleoptera: Silphidae). Stutt Beitr Naturk 351:1–21

Schawaller W (1987) Faunistische und systematische Daten zur Silphiden-Fauna Sudafrikas (Coleoptera, Silphidae). Entomofauna 8:277–288

Schoenley KG (1992) A statistical analysis of successional patterns in carrion-arthropod assemblages: implications for forensic entomology and determination of the postmortem interval. J Forensic Sci 37:1489–1513

Schoenley KG, Reid W (1987) Dynamics of heterotrophic succession in carrion arthropod assemblages: discrete seres or a continuum of change? Oecologia 73:192–202

Schoenley KG, Goff ML, Early M (1992) A BASIC algorithm for calculating the postmortem interval from arthropod successional data. J Forensic Sci 37:808–823

Schoenley KG, Goff ML, Wells JD, Lord WD (1996) Quantifying statistical uncertainty in succession-based entomological estimates of the postmortem interval in death scene investigations: a simulation study. Am Ent 42:106–112

Schoenley KG, Shahid SA, Haskell NH, Hall RD (2005) Does carcass enrichment alter community structure of predaceous and parasitic arthropods? A second test of the arthropod saturation hypothesis at the Anthropology Research Facility in Knoxville, Tennessee. J Forensic Sci 50:134–141

Schoenley KG, Haskell NH, Hall RD, Gbur JR (2007) Comparative performance and complementarity of four sampling methods and arthropod preference tests from human and porcine remains at the Forensic Anthropology Center in Knoxville, Tennessee. J Med Ent 44:881–894

Scholtz CH (1980) Monograph of the genus *Trox* F. (Coleoptera: Trogidae) of subsaharan Africa. Cimbebasia Mem 4:1–104

Scholtz CH (1982) Catalogue of world Trogidae (Coleoptera: Scarabaeoidea). Ent Mem Dept Agric Rep S Afr 54:1–27

Schröder H, Klotzbach H, Oesterhelweg L, Püschel K (2002) Larder beetles (Coleoptera, Dermestidae) as an accelerating factor for decomposition of a human corpse. Forensic Sci Int 127:231–236

Scoggin JK, Tauber OE (1951) The bionomics of *Dermestes maculatus* Deg II. Larval and pupal development at different moisture levels and on various media. Ann Ent Soc Am 44:544–550

Shea J (2005) A survey of the Coleoptera associated with carrion at sites with varying disturbances in Cuyahoga County, Ohio. Ohio J Sci 105:17–20

Smith KGV (1986) A manual of forensic entomology. British Museum (Natural History), London, pp 205

Solodovnikov AYu, Newton AF (2005) Phylogenetic placement of Arrowinini trib. n. within the subfamily Staphylininae (Coleoptera: Staphylinidae), with revision of the relict South African genus *Arrowinus* and description of its larva. Syst Ent 30:398–441

Tabor KL, Brewster CC, Fell RD (2004) Analysis of the successional patterns of insects on carrion in Southwest Virginia. J Med Ent 41:785–795

Tantawi TI, Greenberg B (1993) The effect of killing and preservative solutions on estimates of maggot age in forensic cases. J Forensic Sci 38:702–707

van der Merwe Y, Scholtz CH (2005) New species of flightless Trogidae from South African relict forest fragments. Afr Ent 13:181–186

Villet MH, MacKenzie B, Muller WJ (2006) Larval development of the carrion-breeding flesh fly, *Sarcophaga* (*Liosarcophaga*) *tibialis* Maquart (Diptera: Sarcophagidae), at constant temperatures. Afr Ent 14:357–366

Villet MH, Richards CS, Midgley JM (2009) Contemporary Precision, Bias and Accuracy of Minimum Post-Mortem Intervals Estimated Using Development of Carrion-Feeding Insects, In: Amendt J, Goff ML, Campobasso CP, Grassberger M (eds) Current Concepts in Forensic Entomology. Springer, Berlin

Voight J (1965) Specific postmortem changes produced by larder beetles. J Forensic Med 12:76–80

Watson EJ, Carlton CE (2005a) Insect succession and decomposition of wildlife carcasses during fall and winter in Louisiana. J Med Ent 42:193–203

Watson EJ, Carlton CE (2005b) Succession of forensically significant carrion beetle larvae on large carcasses (Coleoptera: Silphidae). Southeast Nat 4:335–346

Williams KA, Villet MH (2006) A history of southern African research relevant to forensic entomology. S Afr J Sci 102:59–65

Zhantiev RD, Volkova TG (1998) Larvae of dermestid beetles of the genus *Dermestes* (Coleoptera, Dermestidae) of Russia and adjacent countries: 1 Subgenus *Dermestes*. Ent Rev 78:962–968

Zhantiev RD, Volkova TG (1999) Larvae of dermestid beetles of the genus *Dermestes* (Coleoptera, Dermestidae) of Russia and adjacent countries: 2 Subgenus *Dermestinus*. Ent Rev 79:107–117

Chapter 5
Phoretic Mites and Carcasses: Acari Transported by Organisms Associated with Animal and Human Decomposition

M. Alejandra Perotti, Henk R. Braig, and M. Lee Goff

5.1 Introduction

Ephemeral and fluctuant, the fauna associated with a corpse provides a rich diversity of species. Several groups of arthropods are known to visit a carcass of a vertebrate at its various stages of decay; however, forensic investigations have so far been primarily limited to insects, focussing mainly on flies (Diptera) and beetles (Coleoptera) as often the largest and most persistent representatives. These insects might fly, walk or occasionally swim to reach the corpse and sooner or later abandon it in a similar way. Most will build transitional food webs that will lead to a faunal succession of species that will reflect the degree of decay under given environmental conditions.

From the very beginning, some insect and arthropod species will visit a dead body without laying eggs or showing signs of feeding that are destructive to the integrity of the body. These species might hold important information but do not leave behind any obvious marks of their former presence. Also species that will eventually colonise a body days or weeks later often visit the body early on. However, these insects don't come alone. Practically all of them will carry an extraordinary diversity of mites (Acari) with them, phoretic mites.

The Acari are a group of arthropods within the Subphylum Chelicerata or Cheliceriformes. Chelicerates differ conspicuously in their body plan in respect to the other arthropods. Their bodies are divided in two parts, the prosoma (front or anterior part) and the opistosoma (posterior part). Chelicerates do not have head, thorax and abdomen. The prosoma holds all the appendages; the chelicerae or mouthparts being the first pair, followed by the palps or pedipalps and four pairs of walking legs (in most adult forms).

M.A. Perotti (✉)
University of Reading, School of Biological Sciences, UK

H.R. Braig
Bangor University, School of Biological Sciences, UK

M.L. Goff
Chaminade University, Forensic Sciences Program, Hawaii, USA

J. Amendt et al. (eds.), *Current Concepts in Forensic Entomology*,
DOI 10.1007/978-1-4020-9684-6_5,

According to their origin, the Acari are grouped into three major taxa. The Opilioacariformes is by far the smallest, accounting only for roughly 20 species placed in a single family of a single order, Opilioacarida (Walter and Proctor 1999). Opilioacariformes resemble small harvestmen (opilionids). The Parasitiformes includes ticks, which form the order Ixodida (also known as Metastigmata), and several species of mites parasitic on vertebrates in the orders Mesostigmata and Holothyrida. The Acariformes are all small mites encompassing three orders, the Sarcoptiformes, the Trombidiformes and the Endeostigmata. The Sarcoptiformes combine two well-known orders of an older but still widely used classification, the Astigmata and Cryptostigmata (or Oribatida), which assemble a huge diversity of soil species. The Trombidiformes (or part of the older order Prostigmata) combine many species of economic and medical importance.

Phoretic mites are not an occasional occurrence or an academic curiosity but an ecological consequence of the ephemeral and fluctuant nature of decomposition as a habitat. Regular changes in a habitat lead to an increased synchronisation of life history tactics of arthropods. Irregular, unpredictable disturbances lead to an increased representation of phoresy tactics (Athias-Binche 1994; Bajerlein and Błoszyk 2003; Binns 1982; Elzinga et al. 2006; Farish and Axtell 1971; Krantz and Whitaker 1988; Siepel 1995). In this chapter we would like to sketch some of the biological features of phoretic mites to outline the information these mites might provide for forensic investigations; information additional or complementary to entomological analyses.

5.2 Historic Roots and Importance of Mites

In one of the first reports of modern forensic entomology, Brouardel describes the case of a newborn child that was found as a mummified body in January 1878 (Benecke 2001; Brouardel 1879). The time of death was independently estimated based on caterpillars and mites present on the corpse. The caterpillars were studied by Perier and identified as belonging to the genus Aglossa and might have been *A. caprealis* (Pyralidae, Lepidoptera), also known as murky meal moth, fungus moth, small tabby, or a similar species. The moth would have infested the corpse the summer before. For the mites, Mégnin was consulted. The corpse was covered with large quantities of mite feces and exuviae, the skins that are cast during moulting, which produced a brownish layer on top of the body. Large numbers of a single mite species, *Tyroglyphus longior*, now known as *Tyrophagus longior* (Acaridae, Astigmata), were present inside the cranium. Mégnin estimated that the entire corpse carried about 2.4 million mites, dead or alive. Assuming a generation time of 15 days for the species based on his own observations and life tables for *T. mycophagus* and assuming every female gave birth to ten female and five male offspring, he calculated back the number of generations that would have been required to account for the number present on the corpse. The mites would have arrived by phoresy some two months after death, when the corpse had lost enough humidity to support the development of the mite colony. The total estimate for the time of death was around 7–8 months before the autopsy (Mégnin 1894). This early report links phoresy, mites and the determination of a post-mortem interval. Since the case of the mummy in 1878, mites have been detected regularly on human and animal remains and noted in many reports. However, often these mites have just been listed as Acari and not used for any forensic deduction. Few studies addressed mites and arthropods other than insects directly (Anderson and Vanlaerhoven 1996; Bourel et al. 1999; Braack 1986; Goff et al. 1986; Goff and Odom 1987; Grassberger and Frank 2004; Hewadikaram and Goff 1991; Hunziker 1919; Leclercq and Vaillant 1992; Leclercq and Verstraeten 1988b; Mégnin 1887, 1889, 1892; Pérez et al. 2005; Richards and Goff 1997; Shalaby et al. 2000; Vance et al. 1995; Yovanovitch 1888). Leclercq and Verstraeten recorded mites for decomposing human remains, Goff for carcasses of domestic cats and Centeno and Perotti for carcasses of domestic pigs (Centeno and Perotti 1999; Goff 1989; Leclercq and Verstraeten 1988a). The following table presents a list of the mite genera which are likely to arrive as phoretics in any of the five stages of decomposition according to Goff (2009) (Table 5.1). A complete catalogue compiling all the phoretic species known to arrive at a corpse carried by forensically

Table 5.1 Occurrence of phoretic mites according to the insect carriers associated with the stages of decomposition. The most common carriers are indicated for each wave or stage of decomposition; when only a few genera are known, the predominant genera are presented otherwise the family is indicated.

1. **Initial decay, fresh stage**
   Main carriers are flies (Calliphoridae, Muscidae, Phoridae and Sciaridae). Most Mesostigmata and Prostigmata are specific to the genera of their hosts.

   MESOSTIGMATA
   - Family Macrochelidae
     - *Macrocheles* (phoretic females)
   - Family Uropodidae
     - *Uroboovella*
   - Family Trachytidae
     - *Uroseius*

   PROSTIGMATA
   - Family Pygmephoridae
     - *Pediculaster*

   ASTIGMATA
   - Family Histiostomatidae
     - Mostly *Myianoetus* and *Histiostoma*
   - Family Acaridae
     - *Sancassania* and *Acarus*

2. **Putrefaction, bloated stage**
   Main new arrivals are carrion beetles (Silphidae).

   MESOSTIGMATA
   - Family Macrochelidae
   - Family Eviphididae

(continued)

Table 5.1 (continued)

Family Uropodidae
Uroboovella

ASTIGMATA
Family Histiostomatidae
Spinanoetus

3. **Black putrefaction, active decay, decay stage**
Most representative carriers of this wave are carrion and clown beetles (Silphidae and Histeridae).

MESOSTIGMATA
Family Parasitidae
Parasitus and *Poecilochirus*

Family Macrochelidae
Several species specific to beetles
Family Eviphididae
Family Rhodacaridae
Family Uropodidae
Uroboovella

PROSTIGMATA
Family Pygmephoridae
Bakerdania
ASTIGMATA
Family Histiostomatidae
Spinanoetus
Family Acaridae
Schwiebea and *Sancassania*

4. **Butyric fermentation, advanced decay, post-decay stage**
Expected new carriers arriving at this stage of decomposition are cheese flies (Piophilidae); some vinegar flies (Drosophilidae), dark-winged fungus gnats and hump-backed flies (Sciaridae and Phoridae) depending on the degree of fermentation; hide and carcass beetles (Dermestidae and Trogidae) depending on the level of dryness of tissues.

MESOSTIGMATA
Family Parasitidae
Few *Poecilochirus* are specific on Trogidae
Some species on Drosophilidae
Family Macrochelidae
Macrocheles and *Neopodocinum* carried by Trogidae. *Macrocheles* on Drosophilidae, expected on Piophilidae, Phoridae and Sciaridae as well
Family Halolaelapidae
On Sciaridae and Phoridae flies
Family Digamasellidae
On Sciaridae
Family Ascidae
Hoploseius and *Proctolaelaps* on Drosophilidae
Family Uropodidae
Uroboovella on Dermestidae
Family Trachytidae
Uroseius on Phoridae

(continued)

Table 5.1 (continued)

PROSTIGMATA
Family Pygmephoridae
Pediculaster mostly on Trogidae and Sciaridae

ASTIGMATA
Family Histiostomatidae
Rhopalanoethus and *Histiostoma*, on Trogidae, Sciaridae and Phoridae
Family Lardoglyphidae
Mostly *Lardoglyphus* on Trogidae and Dermestidae
Family Acaridae
On Trogidae
Family Euglycyphagidae
On Trogidae
Family Winterschmidtiidae
On Trogidae

5. Dry decay, dry decomposition, skeletal stage, remains stage
Main carriers at this stage are moths (Tineidae and Pyralidae) and beetles such as hide, larder or carpet beetles (Dermestidae) or carcass beetles (Trogidae) might arrive to the corpse depending of the level of dryness.

MESOSTIGMATA
Family Parasitidae
Few *Poecilochirus* are specific on Trogidae
Some species on Drosophilidae
Family Macrochelidae
Macrocheles and *Neopodocinum* carried by Trogidae.
Family Laelapidae
On other moths (Noctuidae)
Family Ascidae
Blattisocius on Tineidae and other moths
Family Uropodidae
Uroboovella on Dermestidae

PROSTIGMATA
Family Pygmephoridae
Pediculaster mostly on Trogidae
Family Cheyletidae
Cheletomorpha on Tineidae

ASTIGMATA
Family Histiostomatidae
Rhopalanoethus and *Histiostoma* on Trogidae
Family Lardoglyphidae
Mostly *Lardoglyphus* on Trogidae and Dermestidae
Family Acaridae
Various genera on Trogidae and *Sancassania* on moths
Family Euglycyphagidae
On Trogidae
Family Winterschmidtiidae
On Trogidae

important scavengers is presented in Perotti and Braig (2009). This table offers just a general guide, because the insects can be different according to region or circumstances of death and the phoretic fauna will change accordingly.

The human follicle mites *Demodex folliculorum* (Demodicidae, Prostigmata), which are primarily found in the hair follicles of the eyelashes and eyebrows, and *D. brevis*, which live in sebaceous glands connected to hair follicles, have gained the attention of pathologists during autopsies (Ozdemir et al. 2003). More than a 100 years after mites had first been used in France to establish a post-mortem interval for a homicide case, Lee Goff again used acarological data in a case of a burial on the island of Oahu, Hawai'i (Goff 1991).

The localized distribution patterns of mites led ultimately to a conviction in a murder trail in California. A police officer and another 20 members of a search team got bitten by chiggers of the mite *Eutrombicula belkini* (Trombiculidae, Prostigmata) when investigating the body of a 24-old woman at a rural location. This mite species was restricted to a very limited area around the crime scene (Prichard et al. 1986; Webb et al. 1983). It normally feeds on lizards and birds and only occasionally of humans (Bennett and Webb 1985). One of the suspects exhibited a similar characteristic biting pattern to the police officer. The extant of inflammation at the bite site of the suspect was used to estimate the time of the original bites. In this case mites put the suspect both at a specific place as well as at a specific time.

The cover of one of the major monographs, Forensic Entomology: the Utility of Arthropods in Legal Investigations, might serve as a final example of how widespread and easily overlooked phoretic mites are in a forensic setting; it shows a carrion beetle and on its right front leg a tiny phoretic parasitiform mite (Byrd and Castner 2001).

5.3 Phoresy

Phoresy, sometimes called phoresis, describes the transport of one animal through another one. The movement or dispersal is directional and has been compared, especially for mites, to assisted migration (Binns 1982). Many desperate taxa have developed it as a live history trait. Its frequency in a taxon is inversely correlated to the size of the animal. For vertebrate species, it is a rare trait. A recent example is the phoresy of remoras, fish of the family Echeneidae, on marine turtles (Sazima and Grossman 2006). For invertebrate taxa, phoresy becomes a more common strategy such as a sea anemone transported on a predator black snail, pseudoscorpions on flying insects or bromeliad annelids and ostracods using frogs, lizards and snakes for transport (Aguiar and Buhrnheim 1998; Lopez et al. 2005; Luzzatto and Pastorino 2006). Many insect taxa display far reaching adaptations for phoresy. Examples include species of ischnoceran lice (Phthiraptera) phoretic on louseflies (Diptera), simuliid black flies (Diptera) with heptageniid mayflies (Ephemeroptera), mealybugs (Hemiptera) on ant queens (Hymenoptera) or midge larvae (Diptera) on the shells of water snails (de Moor 1999; Gaume et al. 2000; Prat et al. 2004). Eggs of torsalo, the human bot fly *Dermatobia hominis* (Diptera), are transported by mosquitoes and muscoid flies to their human host (Disney 1997). The bot fly itself also carries the phoretic mite, *Macrocheles muscaedomesticae* (Macrochelidae, Mesostigmata) (Moya Borja 1981). Depending on the agricultural or medical importance of the phoretic or phoront, the phoretic might be classified as parasite and the transport host as vector. The sphaerocerid fly *Norrbomia frigipennis* (Diptera) uses the time of transport on scarabid dung beetles to mate (Petersson and Sivinski 2003). Some pseudoscorpions also mate during carriage on their harlequin beetles. Larvae of the blister beetle *Meloe franciscanus* (Coleoptera) attach themselves to males of the solitary bee *Habropoda pailida* (Hymenoptera) for transportation, then transfer to a female bee during mating of the bees to continue transportation to the bees' nests (Saul-Gershenz and Millar 2006). Scarab beetles of the genera *Canthon, Sylvicanthon, Parahyboma and Caththidium* use arboreal mammals to reach carcass-baited traps in suspended at a hight of 10 m in semi-deciduous rainforest in the state of Minas Gerais, Brazil (Vaz-De-Mello and Louzada 1997).

In microarthropods and microinvertebrates like mites and nematodes dispersal by active locomotion becomes of minor importance compared with dispersal by wind (anemochory), by water or by animal carriers (phoresy) (Siepel 1994). The many species of nematodes involved in the decomposition of animal excrements rely for a significant part on phoresy by beetles and flies to reach their feeding grounds. The phoretic nematodes establish a clear succession of species (Sudhaus 1981; Sudhaus et al. 1988). Such a succession of phoretic mites has not yet been shown for animal and human remains.

The major carriers for mites are insects and most insect species carry phoretic mites. Phoretic association of the Astigmata have been reviewed by Houck and OConnor and of the mesostigmata by Hunter and Rosario (Houck and OConnor 1991; Hunter and Rosario 1988; OConnor 1982). While phoretic mites have been well described for fleas (Siphonaptera), they are not known for lice (Phthiraptera) (Fain and Beaucournu 1993; Schwan 1993). Other arthropods such as woodlice (Isopoda), centipedes (Chilopoda) or sand hoppers (Amphipoda) might serve as carriers equally well (Bloszyk et al. 2006; Colloff and Hopkin 1986; Pugh et al. 1997). However, phoretic mites have not (yet) been described for spiders (Araneae). Also vertebrates such as lizards, hummingbirds, small mammals and bats may function as transport hosts (Athias-Binche 1984; Colwell 2000; Domrow 1981; Krantz and Whitaker 1988; Tschapka and Cunningham 2004).

Phoresy in an evolutionary perspective is a transitional stage (Athias-Binche 1990; Athias-Binche and Morand 1993; Houck and OConnor 1991; Sivinski et al. 1999). This means that phoresy spans the whole spectrum from beneficial associations to parasitism. Definitions for phoresy in acarology can therefore be controversial (Walter and Proctor 1999). Some might limit phoresy to the middle ground where the phoretic or phoront has no interaction with its carrier other than transport, especially not feeding on its carrier. Phoresy implies that normal behaviour and physiological processes change. In many species it will lead to vast morphological alterations. In cases such as bless or betsy beetles (Passalidae, Coleoptera) where feeding off the host of numerous mesostigmatic mites has been observed, it does not seem overly detrimental to the carrier (Walter and Proctor 1999). In fly species, the phoretic mites might actually be predators of the newly oviposited eggs of the carrier (Jalil and Rodriguez 1970; Polak 2003). The mite, *Proctolaelaps* sp. (Ascidae, Mesostigmata), transported by the economically important flower weevil of the African oil palm feeds during transfer on phoretic nematodes that share the same carrier (Krantz and Poinar 2004). Thread-footed mites (Tarsonemidae, Prostigmata) phoretic on southern pine beetles carry a fungal species that can outcompete the symbiotic mycangial fungus carried by the beetle itself (Lombardero et al. 2003). Phoretic mites might become the food of phoretic pseudoscorpions riding on the same beetles (Zeh and Zeh 1992).

Phoretic associations of mites can lead to unexpected manifestations. Imagine a forensic investigator approaching a corpse that just has been discovered; he/she might be using protective clothing including disposable boot covers, not only for protection but also to avoid any contamination of the crime scene. Ants searching for a corpse behave as curious and perhaps careful as a forensic scientist, however they do not wear boot covers. At least two species of *Macrocheles* mites have become the boot covers or 'healing foot pads' of the larger workers of army ants (Ecitoninae, Hymenoptera). Females of *Macrocheles* cling to the pulvilli of leg III of the ant attaching themselves with their chelicerae for transportation. The mites have enlarged legs IV, which are now used as the new 'tarsal claws' by the ant and support all the weight. The diversity of associations between army ants and mites is remarkable. A collection on 1,600 army ant colonies yielded over 45,000 mites. Only 3% have been studied but resulted in three new families and 149 new species of mites (Elzinga et al. 2006). Thus, *Macrocheles* species do not travel alone. Phoronts belonging to different families attach to the most disparate areas of the body of ants. *Trichocylliba* species (Uropodidae, Mesostigmata) fasten themselves symmetrically to the insect abdomen; *Planodiscus* species (Uropodoidea incertae sedis, Mesostigmata) probably ride on the underside of the middle or hind tarsi of workers; *Circocylliba crinita* (Circocyllibamidae, Mesostigmata) is not only restricted to the mandibles of soldiers of *Eciton dulcium* but has also been found riding on the inner curve of the mandibles; *Antennophorus* species (Antennophoridae, Mesostigmata) clasp the venter of the head of the worker ant with the forelegs directed towards the mouthparts where they obtain food by trophallaxis from the ants; *Messoracarus* (Messoracaridae Mesostigmata) species sit below the head of the ant, there they are palpated by the ant's antennae and may steal provisions like cleptoparasites; *Urodiscella* species (Oplitidae, Mesostigmata) cling to the antenna cleaners on the forelegs of ants and apparently scavenge on debris combed by the ants; and others, like Laelapidae mites (Mesostigmata) occur on the body and are equally ignored by the insect during their journey (Eickwort 1990; Elzinga et al. 2006).

Looking at a single species of a medically and veterinary important fly, an incredible diversity of mite taxa were discovered. Adult stable flies, *Stomoxys calcitrans* (Diptera), collected in the United Kingdom, carried 12 species of mites from 10 families and three orders (McGarry and Baker 1997). A single insect depending on species can carry anything from a single mite to several hundred individuals of several species.

That phoresy is indeed the most important mechanism or in this case the only mechanism, by which forensically important mite species arrive at a carcass has been shown by Goff in an experiment in which he studied the mites on carcasses of adult cats and in the seapage zone beneath the carcasses (Goff 1989). He found 22 different species belonging to 15 different families. Mites of four of these families, Macrochelidae, Parasitidae, Uropodidae and Pachylaelapidae (all Mesostigmata) might be potenial indicators of post mortem interval. Mites of all these four gamasid families use phoresy to arrive at the carcass.

5.4 Phoretic Adaptations

Phoresy might be facultative for a mite species or obligate. If it is obligate, a particular species cannot develop to adulthood or reproduce without prior engaging in phoretic transport. Examples for obligate phoresy at immature or juvenile stages include *Histiostoma brevimanus* (Histiostomatidae, Astigmata), *Parasitus coleoptratorum* (Parasitidae, Mesostigmata), for obligate phoresy as imagoes (adults), *Macrocheles muscaedomesticae*. Carrier specificity spans the whole range from specific for a host family to a particular body part of a single species. Occasionally, phoretic mites might have an alternative carrier. The phoretic deutonymphs of the uropodine mite, *Fuscuropoda marginata* (Urodinychidae, Mesostigmata) usually attach to dung beetles but can switch to ground skinks, *Scincella lateralis*. This mite and other acarine taxa may be preadapted to shift between arthropod hosts covered with chitinous sclerites and vertebrate hosts covered with keratinous scales (Mertins and Hartdegen 2003). Necrophagous mites might shift carrier as carrion beetle species pass in and out of their reproductive seasons (Brown and Wilson 1992).

Athias-Binche gives a comprehensive overview of the many, often inimitable life history traits that have been realised in mites to accommodate phoresy (Athias-Binche 1994). There are five basic types of phoretic adaptations in the acari: regular nymphs and females, two specialised nymphs and specialised females. A population of mites can be monomorphic or facultative dimorphic at a specific stage of development or gender. A phoront often needs to be modified to resist desiccation during transportation. The adaptations for phoresy often have such an impact on the morphology that phoretic stages have been described as different species even belonging to different families.

The most basic forms are the regular adult females or regular deutonymphs, which have not evolved elaborate morphological modifications but exhibit physiological changes for transportation. In the Mesostigmata, suborder Dermanyssina, and in the Prostigmata, superfamily Cheyletoidea, it is the adult female that makes the move. Most *Macrocheles* mites provide good examples for females adapted for transportation (Fig. 5.1); they use their chelicerae to hold on to the hairs of their beetle or fly carriers. Some *Parasitus* species (Parasitidae, Mesostigmata) start their journey when they reach the stage of deutonymph. The claws of their front legs seem to be used to grasp the setae of their host, however, some of these species are

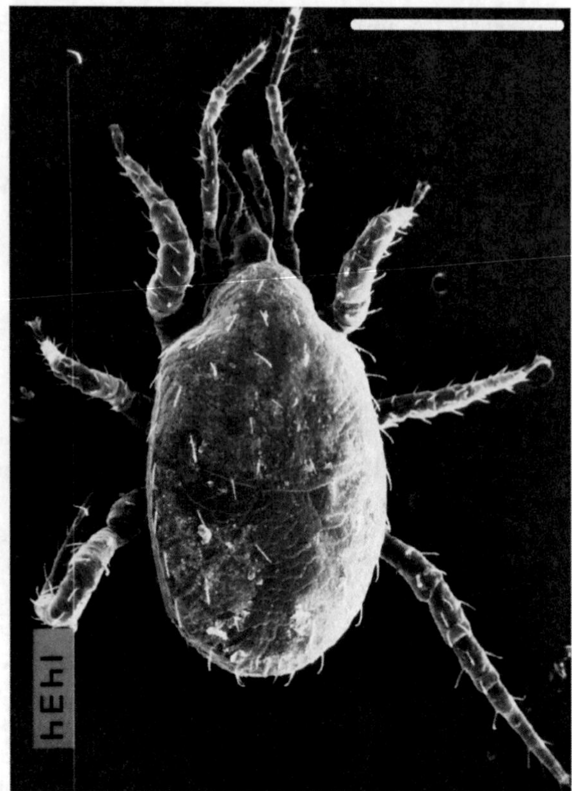

Fig. 5.1 Forensically important phoretic mite. Scanning electronmicrograph of dorsal view of an adult female, phoretic stage, of *Macrocheles* sp. Scale bar = 500 μm

not fixed to a certain spot but likely to be seen moving around the dorsal side of their host during transport.

Drastic modifications in both the morphology and the physiology of adults and nymphs are documented in the orders Mesostigmata, Astigmata and Prostigmata. They include inactive or quiescent stages in which the gut, for example, might not develop completely. These mites might be monomorphic for a given population or engage in a dimorphic population where phoretic and non-phoretic forms can be seen side by side. The term 'hypopus' (hypopi, hypopodes), originally a specific epithet, refers to the altered phoretic deutonymphs existing in several species of Astigmata. Interestingly it was Pierre Mégnin, one of the fathers of modern forensic entomology who introduced the term.

> … une mue bien extraordinaire: dans l'enveloppe des *nymphes* (jeunes sujets octopods non encore sexués), – et des nymphes seulement, – se forme, non pas comme dans les circonstances ordinaries un individu semblable à elles, plus avancé en développement, mais un Acarien cuirassé totalement différent: ses organs buccaux sont atrophiés et il est muni d'un

groupe de ventouses sous et post abdominals qui lui permettront de s'attacher à tout être animé qui passera à sa portée et à fuir un lieu de désolation où il mourrait de faim comme cela arrive forcément à ses parents adultes et aux jeunes larves hexapodes qui n'ont, ni les uns ni les autres, le pouvoir de subir cette métamorphose. Lorsque les hasards du voyage ont fait arriver le nouvel Acarien cuirassé dans un lieu d'abondance, il descend de son omnibus improvisé, se dépouille de son costume de voyage, reprend sa forme ancestrale et se met en devoir de constituer une nouvelle colonie. A la suite de la découverte de ce fair curieux nous avons nommé cette mue particulière, *mue hypopiale*, parce que ces Acariens cuirassés, masqués, en habit de voyage, avaient été pris par les Acariologues pour des espèces définies qu'ils avaient nommées *Hypopus*, … etc., ces noms ont disparu pour faire place à l'adjectif hypopial. (Mégnin 1892)…

a quite extraordinary moult: under the cuticle of the nymphs (young octapod individuals that are not yet sexual) - and nymphs only, - is formed, not as is the case under normal circumstances an individual similar to them, just more advanced in development, but a completely different phoretic mite: its mouth parts are atrophied and it is provided with a group of suction cups under its abdomen and at its end which will enable it to stick to all lively creatures which will pass within its range and to flee a place of desolation where it would die of hunger like the ones that arrived inevitably as its adult parents did and young hexapod larvae, neither of which have the capacity to undergo this metamorphosis. When luck during the voyage brought the new phoretic mite to arrive at a place of abundance, it disembarks from its improvised carrier, shed its travel suit, regains its ancestral shape, and is put in a state to start a new colony. Following the discovery of this curious fact, we named this particular moult, a hypopal moult, because these phoretic mites, masked by their travel suits, were mistaken by acarologists for new species, which they had named *Hypopus*, ... and so on; these names have disappeared to make place for the adjective hypopal.

Mégnin even described two types of hypopi for the species *Falculifer rostratus* (Falculiferidae, Astigmata), the *nymph adventive or hypopiale première form* for large, male hypopal nymphs, which have been depicted in many textbooks, and the *nymph adventive* or *hypopiale deuxième form* for small, female hypopal nymphs giving the term heteromorphic a total different meaning (Robin and Mégnin 1877). However, Mégnin might himself have fallen victim of the difficulties of acarology. Graf Vitzthum contests that the large form represents both sexes of the hypopi and the small form is actually the hypopus of another species of another family, *Megninia columbae* (Analidae, Astigmata) (Fain and Laurence 1974; Graf Vitzthum 1933). At present there is still some controversy about the use of the term 'hypopus' for the morphologically modified phoretic deutonymphs of the Astigmata; some specialists prefer to call these nymphs heteromorphic deutonymphs (Houck and OConnor 1991; Walter and Proctor 1999).

Heteromorphic phoretic deutonymphs of Uropodidae (Mesostigmata) and the hypopi of Histiostomatidae (Astigmata) have evolved into aerodynamic flattened and sclerotised shapes with reduced legs normally folded inside lateral cavities during flight. They can even fold up their rudimentary, unfunctional gnatosoma or they might just have a reduced gnatosoma or have lost the chelicera. The anal region of the phoretic deutonymphs of these distantly related families is characteristically modified. The deutonymphs of uropodid species, such as *Uroseius* spp, secret a cement to build a pedicel that adheres to the exotegument of the host. The phoretic hypopi of Histiostomatidae on the other hand present strong anal suckers that fix the mites to the carrier.

Perhaps the phoretomorphic females are the most recently studied. The type was described in the Prostigmata family Tarsonemidea by Moser and Cross as 'a female specialised for riding insects' (Moser and Cross 1975). Again, these females can arise facultatively, which allows dimorphism in some species, with phoretic and non-phoretic females produced at the same time depending on the need for transportation to a suitable habitat.

5.5 Specificity of Phoretic and Carrier

Carrion as a microenvironment or microhabitat undergoes constant changes. Necrophilous insects and the many occasional visitors of carcasses may attract and associate with phoretic mites less specialised to their carriers. Necrophagous insects, specialists for fabric or hair and specific predators and parasitoids will often exhibit a much higher degree of carrier specificity. For many of the dipterans that arrive after death, the carcass is just one of the many different habitats that they are able to colonise, often having the choice of another suitable rotten food source. On the other hand, there is often a high degree of specificity for carrion beetles and the mites carried.

Euryxene mites will use different arthropod taxa while stenoxene ones will travel on different species of a same genus. Mites that are transported by only one species of host are oioxene mites (Athias-Binche 1994). And, according to Lindquist it is possible to discriminate up to four types of life cycle synchronization between phoront and carrier (Lindquist 1975).

Macrocheles muscaedomesticae is a wide spread phoretic mite of forensic interest. Although it was early classified as a eurexene phoretic (Athias-Binche 1984) it seems to be specialised to muscoid flies. The behaviour of macrochelid mites is particular unusual when it combines food with transportation. They feed on the progeny of their own carriers, Muscidae or Fanniidae flies (Perotti and Brasesco 1997). Its name refers to its close association as phoretic with *Musca domestica* (Diptera). It has evolved into a complex interaction of phoresy and predation at the same time. The life of this mite is one of the shortest known, properly adapted to fast-changing microhabitats and shortage of food, it only feeds on the eggs and first instar larva of the muscoid fly, never on older stages. In optimal environmental conditions it takes the species to develop from egg to egg only 3 days, pasing through larva, protonymph, deutonymph and adult (Rodriguez and Wade 1961; Wade and Rodriguez 1961). *M. muscaedomesticae* arrives at a decomposition site as a virgin female attached to the body of only female *Musca* spp. (Muscidae, Diptera) or *Fannia* spp. (Fanniidae, Diptera) (Perotti 1998). Coincidentally, *Musca* and *Fannia* commonly overlap in their breeding habitats. It has been proved that the mite is not host-density dependent, what already determines its selectivity, becoming a stenoxene or semi-specific phoront, attracted by a chemical gradient produced by the female carriers (Farish and Axtell 1971; Niogret et al. 2006; Perotti and Brasesco 1996).

For some species, the phoretic mites aggregate before the carrier develops to adulthood. *Myanoethus muscarum* (Histiostigmatidae, Astigmata) hypopi cluster on the anterior end of the pupa of the host fly, which is the location from which the imago will emerge (Greenberg and Carpenter 1960). Less than 1 h after pupation of the fly, *Muscina stabulans* (Diptera), the hypopi heavily aggregate on the anterior third of the pupa. Different experiments conducted by Greenberg have demonstrated that the mite nymphs are attracted to a volatile substance secreted or emitted from the anterior end of the pupa of several fly species including *Musca domestica*, *Muscina stabulans* and *Stomoxys calcitrans*. The grouped hypopi remain on the pupal case until eclosion, at which time they transfer to the emerging fly. After a brief excursion on the carrier the mites occupy characteristic locations for attachment. They rather prefer areas bearing setae and avoid smooth cuticle. Horizontal rows of mites build outward and toward the rear. They uniformly orient themselves toward the rear (Greenberg 1961). After dismounting, a minimum of 2 days at 24°C is required for the moulting into the next stage, the tritonymph, which may moult to adult in 1 day.

The relation between carrion beetles and their phoretic mites have been studied in detail (Schwarz and Koulianos 1998). *Nicrophorus* beetles are used by mites of the *Poecilochirus carabi* (Poecilochiridae, Acari incertae sedis) species complex, which are specialists that depend on the food provided by the carrier and its nests. The mites are not able to search for new carrion on their own. They have synchronised their life cycle with that of their beetle hosts. They arrive at a new carrion on the parental hosts, reproduce to such a strict time table that their offspring is ready to leave as soon as the parental male decides it is time to abandon the brood chamber in the search for a new breeding site. The female beetle follows the male with the remaining mites attached to her, and in cases where some passengers didn't get a seat, they have to wait until the larvae of the next generation has fully developed. *Poecilochirus* mites also switch between burying beetle species, they discriminate between adult beetles olfactorily and they choose those reproductively active. Almost every adult male carrion beetle arrives at a body transporting phoretic mobile deutonymphs of *Poecilochirus* and this phenomenon is expected worldwide. Within hours after arrival, the deutonymphs disembark, moult into adults and reproduce immediately. The non-phoronts live off microfauna associated with the carcass in the brood chamber. The favourite association of the deutonymphs are always the parent beetles. 100–250 deutonymphs can be found on adult males compared to less than five commonly observed on the larvae. In central Europe, one might expect three to four different species of carrion beetles at a big carcass at a time, which would contribute up to seven mesostigmatic and three astigmatic mites species to the carcass' microfauna, probably originated from diverse allopatric populations.

5.6 Hyperphoresy

Hitchhiking on a hitchhiker has been developed by some phoretic nematodes, fungi and mite species. Some nematode and fungi species take a ride on a phoretic mite. Mite species hitchhike on other mite species. The trait of hyperphoresy has evolved

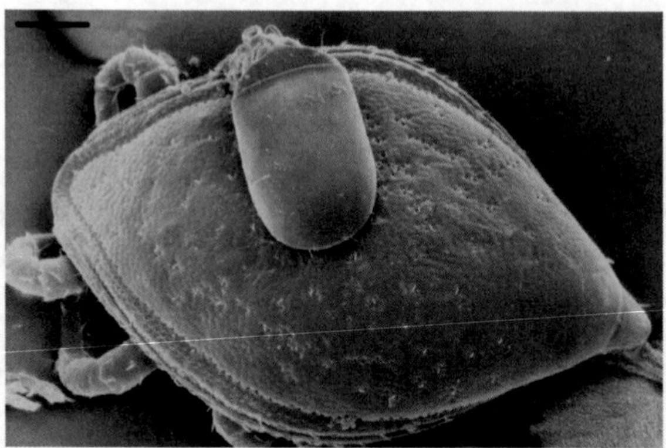

Fig. 5.2 Hyperphoresy. Scanning electronmicrograph of a Myianoetus sp. heteromorphic deutonymph or hypopus (Astigmata) attached to the dorsal shield of a modified deutonymph of Uroseius sp. (Mesostigmata). The hypopus is using its anal suckers for attaching the dorsum of Uroseius, which is 'glued' to the tegument of the carrier insect by a short secreted pedicel (*bottom right corner*). Scale bar = 100 μm

independently in several taxa. The hyperphoretic mites are specialised heteromorphic deutonymphs of either astigmatans or mesostigmatic uropodids. *Myanoethus* hypopi have been recorded attached to the dorsal shield of *Uroseius* (Fig. 5.2) and *Macrocheles muscaedomesticae* (Perotti 1998).

Hypopi of Acaridae (Astigmata) are hyperphoretic on *Hypoaspis* sp. (Laelapidae, Mesostigmata) carried by a queen of the bumblebee *Bombus dahlbohmi* (Garrido and Casanueva 1995). *Uropoda orbicularis* modified phoretic deutonymphs are carried by other deutonymphs of the same species and by a female of *Macrocheles glaber* on coprophagous beetles (Błoszyk and Bajerlein 2003).

Regarding carcasses there is an example of hyperphoresy, which escaped attention. *Poecilochirus subterraneus* associated with *Nicrophorus* also practice hyperphoresy when switching between congregated burying beetles. These are not heteromorphic deutonymphs and attach to the large mobile deutonymphs of *P. carabi* prompted to jump on a new beetle host (Korn 1983).

5.7 Phoretic Mites Separated by Carrier and Time

In the historic case detailed by Mégnin and the more recent case reported by Goff from Hawai'i, mites have been used to estimate the post mortem interval (Goff 1991; Mégnin 1894). Present cases often reveal large infestations with mites, e.g., case 7 in (Arnaldos et al. 2005), however the tools and expertise to identify these mites have still to be recruited and basic knowledge about the biology, especially the succession of these mites is still lacking. Population dynamics of mites in decomposing human or animal remains will prove to be of value in estimation of post-mortem intervals in cases of homicide, suicide, accidental death or unattended death due to natural causes where the post-mortem interval is greater than 17 days (Goff 1989). Perhaps the greatest value of phoretic mites might lie in increasing the resolution of entomological data during later stages of decomposition.

Already a good body of information about phoretic mites of necrophagous and necrophilous insects is scattered in the acarological literature.

One of the best studied systems is the association between phoretic mites with carrion beetles, *Poecilochirus* spp. and *Nicrophorus* spp. (Schwarz and Koulianos 1998; Wilson 1983). Due to their close interaction and high specificity, these mites might become as valuable as the beetles for forensic investigations.

The small scuttle or hump-backed flies (Phoridae) may be useful indicators of time of death when blow flies have disappeared (Disney 2005; Leclercq 1999). However, in some forensic cases these flies have presented problems of misidentification and variable pre-oviposition times, both of which can lead to wrong estimates of the time of death (Disney 2005). Phorids might visit carcasses to feed but they might not lay eggs (Centeno et al. 2002). Alex Fain alone has described a new genus and seven new species of phoretic mites (Prostigmata and Mesostigmata) on Phoridae and Ephydridae from around the world (Fain 1998a, b). Phorid flies are now known to carry six species of Mesostigmata, five species of Prostigmata and one species of Astigmata (Fain and Greenwood 1991). The flies could easily be linked to the occurrence of their specific phoretics and it would help to precise the identification. Some phoronts will land when the fly oviposits, some others when the fly is only feeding on the corpse.

Dermestes or skin beetles are usually present at later stages of decomposition or dry decay. These beetles transport Lardoglyphus mites (Lardoglyphidae, Astigmata). The mites are much specialised to their carriers and compete for food, cleaning the left soft tissue attached to the bones. In laboratory cultures the mites can grow out of control and the prevalence of the phoretic nymphs on the carrier insect can be deleterious (Iverson et al. 1996). These mites have also been isolated form the gut of human mummies excavated in Chile and the United States (Baker 1990).

Although associations between muscoid flies (Muscidae, Diptera) and phoretic mites have been widely studied and data of life history traits of several species are well known (e.g., Ciccolani 1979; Ciccolani et al. 1977; Rodrigueiro and do Prado 2004; Singh et al. 1967), the phoronts of blow flies have been poorly characterised. In a preliminary survey in Argentina, phoretic mites were examined in relation with the stages of decomposition and arrival of insects to a pig carcass (Centeno and Perotti 1999). *Parasitus fimetorum* phoretic deutonymphs were found at the very fresh stage of decomposition together with the first blow flies. *Macrocheles mamifer* females arrived 2 days after *P. fimetorum. M. mamifer* is a very widespread macrochelid in the Argentinean pampas that has not (yet) established a specific association with carrion beetles although uses them for transportation (Perotti 1998). *M. mamifer* is likely to be found traveling on filth, carrion flies and mammals (Krantz and Whitaker 1988). *M. muscaedomesticae* collected under carcasses of cats were correlated with the presence of *Musca domestica*. Both, the occurrence of mites and flies were synchronised at the beginning of decomposition, on day 3 and following day 21 (Goff 1989). *P. fimetorum* will also arrive during a second wave. Carrion beetles (Silphidae) will carry *P. fimetorum* to a carcass (Hyatt 1980). The influence of phoretic mites of burying beetles on the competition between *Nicrophorous quadripunctus* and the blow fly *Chrysomya pinguis* (Diptera) for mice carcasses revealed that the phoretic mites contributed significantly to the success of this beetle species in colonising carcasses by killing blow fly eggs and larvae (Satou et al. 2000).

The best life history stage to found a new colony is a non-gravid meaning non-mated female (Binns 1982; Southwood 1962). A haplodiploid genetic system also called arrhenotokous reproduction perfectly matches with such requirements and phoresy. The phoront for Macrochelidae is the virgin female. Females are diploids and males are the outcome of asexual reproduction, they develop from unfertilised eggs resulting in haploids. When the virgin female arrives at a new and suitable breeding site carried by its fly or beetle, she detaches and shortly after feeding commences laying asexually a very few haploid eggs. The handful sons will be enough to fertilise their own mother and later on their sisters, which will bring then daughters to the site. In the first wave only males will be produced, then only females, which will overlap with the few males of the first wave. In a third wave the second generation daughters will then produce a less sex ratio biased and stable but likely spanandrous (scarceness of males) population depending to the environmental conditions. The short life cycle and the succession of sex ratio biases of the macrochelid and other haplodiploid mites might provide a highly increased resolution of the time line.

Similarly, representatives of the families Histiostomatidae and Glycyphagidae (Astigmata) are known to disperse by heteromorphic deutonymphs and hypopus types. Mites of the genus *Pelzneria* (Acaridae, Astigmata) are found in large numbers under the elytra of *Nicrophorus* beetles and synchronise with the life cycle of their carrier. Due to the haplodiploid genetic system common to Histiostomidae, the specialised deutonymphs must be females and must disembark from the parent beetles. The newly developed females will give birth asexually to sons who will fertilise their mothers. The next generation of female phoronts need to be ready only at the moment of the emergence of the next generation of adult beetles.

A different scenario is observed for movile phoretic deutonymphs of *Poecilochirus* mites during the landing of carrion beetles. Most Parasitidae mites are diplodiploid species. Because they need to have sex to reproduce, the phoretic deutonymphs hurry to moult into the adult stage to mate just after disembarking. The beetle has to carry phoretic deutonymphs of both genders. They disperse always in mixed groups having always the chance to find a mate at the destination point. Again, looking at the succession in situ, in a matter of hours the populations of landing deutonymphs will be no longer be detectable and adults with an equal, unbiased sex ratio will have replaced them building subsequent generations with a stable and even sex ratio.

5.8 Phoretic Mites Separated Spatially

A study on carrion beetles in central New York state, USA, has shown that the size of the forest and its fragmentation had a large impact on the load of phoretic mites (Gibbs and Stanton 2001). The diversity of phoretic mites for a single carrier species is often so high that the investigation of its phoretic mites might contribute a lot of high resolution site-specific information for a carcass. More detailed studies on the phoretic mite species *Poecilochirus carabi* associated with carrion beetles of the genus *Nicrophorous* showed that what was perceived as a single species were actually two different and therefore undescribed mites species (Brown and Wilson 1992). The difference, morphologically suggestive, became evident in crossing experiments between the mites phoretic on two sympatric carrion beetles, *Nicrophorous orbicollis* and *N. tomentosus*. Two sites in lower Michigan, USA, Kellogg Biological Station in Hickory Corners and University of Michigan Biological Station in Pellston are separated by 315 km from each other. At Hickory Corners, two different mite species exhibited strict host specificity for *N. orbicollis* and *N. tomentosus*. At Pellston, in contrast, *P. carabi* was carried by *N. tomentosus* and *N. defodiens*, a species not found at Hickory Corners. A specialist mite species for *N. defodiens* could not be found at Pellston. *Nicrophorus* beetles reproduce on small carrion such as birds and mice. However, they are often found feeding on large carcasses such as humans, where they do not oviposit. During these feeding stops, *Poecilochirus* mites disembark to feed on the large carcass as well. Fascinatingly, most mites manage to return to their carrier before take off (Korn 1983). The phoretic mites of carrion beetles ovipositing on larger carcasses, e.g. *Oiceoptoma* species, have barely been described (Mašán 1999).

5.9 Ongoing Research

With support from the Leverhulme Trust a project on forensic acarology has recently been started at the Bangor University in Wales in collaboration with Bryan D. Turner of King's College London and Anne S. Baker of the Natural History Museum London. It will be a first approach on identifying mites associated with corpses and it includes not only the study of the occurrence of acarological fauna on the body along the stages of decomposition but also a survey on the phoretic mites arriving on arthropods.

Acknowledgement MAP and HRB wish to thank the Leverhulme Trust for support of this work.

References

Aguiar NO, Buhrnheim PF (1998) Phoretic pseudoscorpions associated with flying insects in Brazilian Amazonia. J Arachnol 26:452–459

Anderson GS, Vanlaerhoven SL (1996) Initial studies on insect succession on carrion in southwestern British Columbia. J Forensic Sci 41:617–625

Arnaldos MI, Garcia MD, Romera E, Presa JJ, Luna A (2005) Estimation of postmortem interval in real cases based on experimentally obtained entomological evidence. Forensic Sci Int 149:57–65

Athias-Binche F (1984) La phorésie chez les acariens uropodides (Anactinotriches), une stratégie écologique originale [Phoresy in uropodid mites (Anactinotrichida), an interesting ecological strategy]. Acta Oecol, Oecol Generalis 5:119–133

Athias-Binche F (1990) Sur le concept de de symbiose – l'example de la phorésie chez les acariens et son évolution vers le parasitisme ou le mtualisme [On the concept of symbiosis – the example of phoresy in mites and its evolution towards parasitism or mutualism]. Bull Soc Zool France 115:77–98

Athias-Binche F (1994) La Phorésie chez les Acariens – Aspects Adaptatifs et Evolutifs [Phoresy in acarina – adaptive and evolutionary aspects]. Editions du Castillet, Perpignan

Athias-Binche F, Morand S (1993) From phoresy to parasitism: the example of mites and nematodes. Res Rev Parasitol 53:73–79

Bajerlein D, Błoszyk J (2003) Two cases of hyperphoresy in mesostigmatic mites (Acari: Gamasida: Uropodidae, Macrochelidae). Biol Lett (Warsaw) 40:135–136

Baker AS (1990) Two new species of *Lardoglyphus* Oudemans (Acari, Lardoglyphidae) found in the gut contents of human mummies. J Stored Prod Res 26:139–148

Benecke M (2001) A brief history of forensic entomology. Forensic Sci Int 120:2–14

Bennett SG, Webb JP Jr (1985) A possible human infestation by *Eutrombicula belkini* (Gould) (Acari: Trombiculidae) in Laguna Beach, California. Bull Soc Vector Ecol 10:118–121

Binns ES (1982) Phoresy as migration – some functional aspects of phoresy in mites. Biol Rev Cambridge Philos Soc 57:571–620

Bloszyk J, Klimczak J, Lesniewska M (2006) Phoretic relationships between Uropodina (Acari: Mesostigmata) and centipedes (Chilopoda) as an example of evolutionary adaptation of mites to temporary microhabitats. Eur J Entomol 103:699–707

Błoszyk J, Bajerlein D (2003) Two cases of hyperphoresy in mesostigmatic mites (Acari: Gamasida: Uropodidae, Macrochelidae). Biol Lett 40:135–136

Bourel B, Martin-Bouyer L, Hedouin V, Cailliez J-C, Derout D, Gosset D (1999) Necrophilous insect succession on rabbit carrion in sand dune habitats in northern France. J Med Entomol 36:420–425

Braack LEO (1986) Arthropods associated with carcasses in the northern Kruger National Park. S Af J Wildlife Res 16:91–98

Brouardel P (1879). De la détermination de l'époque de la naissance et de la mort d'un nouveau-née, faite à l'aide de la présence des acares et des chenilles d'aglosses dans cadavre momifié [Determination of the time of birth and of death of a new-born child, made using the presence of mites and Aglossa caterpillars on the mummified corpse]. Annales d'Hygiène Publique et de Médecine Légale, (série 3) 2:153–158

Brown JM, Wilson DS (1992) Local specialization of phoretic mites on sympatric carrion beetle hosts. Ecology 73:463–478

Byrd JH, Castner JL (2001) Forensic Entomology: the Utility of Arthropods in Legal Investigations. CRC, Boca Raton

Centeno N, Maldonado M, Oliva A (2002) Seasonal patterns of arthropods occurring on sheltered and unsheltered pig carcasses in Buenos Aires Province (Argentina). Forensic Sci Int 126:63–70

Centeno N, Perotti MA (1999) Acaros vinculados a procesos de descomposición de cadáveres y sus posibles asociaciones foréticas [Acari linked to processes of decomposition of cadavers and their possible phoretic associations]. Paper presented at the XIX Reunión Argentina de Ecología, Tucumán, Argentina

Ciccolani B (1979) The intrinsic rate of natural increase in dung macrochelid mites, predators of *Musca domestica* eggs. Bolletino Zool 46:171–178

Ciccolani B, Passariello S, Petrelli G (1977) Influenza della temperatura sull'incremento di popolazione in *Macrocheles subbadius* (Acarina: Mesostigmata) [The infuence of temperature on the increment of populations of *Macrocheles subbadius* (Acarina: Mesostigmata)]. Acarologia 19:563–578

Colloff MJ, Hopkin SP (1986) The ecology, morphology and behavior of *Bakerdania elliptica* (Acari, Prostigmata, Pygmephoridae), a mite associated with terrestrial isopods. J Zool A 208:109–124

Colwell RK (2000) Rensch's rule crosses the line: Convergent allometry of sexual size dimorphism in hummingbirds and flower mites. Am Nat 156:495–510

de Moor FC (1999) Phoretic association of blackflies (Diptera: Simuliidae) with heptageniid mayflies (Ephemeroptera: Heptageniidae) in South Africa. Af Entomol 7:154–156
Disney RHL (1997) Fantastic flies and flights of fancy. J Biol Educ 31:9–48
Disney RHL (2005) Duration of development of two species of carrion-breeding scutte flies and forensic implications. Med Vet Entomol 19:229–235
Domrow R (1981) A small lizard stifled by phoretic deutonymphal mites (Uropodina). Acarologia 22:247–252
Eickwort GC (1990) Associations of mites with social insects. Ann Rev Entomol 35:469–488
Elzinga RJ, Rittenmeyer CW, Berghoff SM (2006) Army ant mites: the most specialized mites found on any social insect. Paper presented at the The IUSSI 2006 Congress, Washington, DC
Fain A (1998a) Description of mites (Acari) phoretic on Phoridae (Insecta: Diptera) with description of four new species of the genus *Uroseius* Berlese (Parasitiformes, Uropodina, Polyaspididae). Int J Acarol 24:213–220
Fain A (1998b) New mites (Acari) phoretic on Phoridae and Ephydridae (Diptera) from Thailand. Bull Inst R Sci Nat Belg Entomol 68:53–61
Fain A, Beaucournu J-C (1993) Hypopi of astigmatic mites (Acari) phoretic on fleas (Siphonaptera) of mammals and birds. Bull Entomol 63:77–93
Fain A, Greenwood MT (1991) Notes on a small collection of mites Acari phoretic on Diptera mainly Phoridae from the British Isles. Bull Inst R Sci Nat Belg Entomol 61:193–197
Fain A, Laurence BR (1974) Guide to heteromorphic deutonymphs or hypopi (Acarina, Hypoderidae) living under skin of birds with description of *Ibisidectes debilis* Gen and sp nov from scarlet ibis. J Nat Hist 8:223–230
Farish DJ, Axtell RC (1971) Phoresy redefined and examined in *Macrocheles muscaedomesticae* (Acarina: Macrochelidae). Acarologia 13:16–29
Garrido C, Casanueva ME (1995). Acaros foréticos e hiperforéticos sobre *Bombus dahlbohmi* Guerin, 1835 (Hym., Apidae) [Phoretic and hyperphoretic mites on *Bombus dahlbohmi* Guerin, 1835 (Hym., Apidae)]. Bol Soc Biol 66:53–55
Gaume L, Matile-Ferrero D, McKey D (2000) Colony formation and acquisition of coccoid trophobionts by *Aphomomyrmex afer* (Formicinae): co-dispersal of queens and phoretic mealybugs in an ant-plant-homopteran mutualism? Insect Soc 47:84–91
Gibbs JP, Stanton EJ (2001) Habitat fragmentation and arthropod community change: Carrion beetles, phoretic mites, and flies. Ecol Appl 11:79–85
Goff ML (1989) Gamasid mites as potential indicators of postmortem interval. In: Channabasavanna GP, Viraktamath CA (eds) Progress in Acarology, vol 1. Oxford and IBH Publishing, New Delhi, pp 443–450
Goff ML (1991) Use of acari in establishing a postmortem interval in a homicide case on the island of Oahu, Hawaii. In: Dusbábek E, Bukva V (eds) Modern Acarology, vol 1. SPB Academic Publishing, The Hague, pp 439–442
Goff ML (2009). Early postmortem changes and stages of decomposition in exposed cadavers. Exp Appl Acarol 49:21–36
Goff ML, Early M, Odom CB, Tullis K (1986) A preliminary checklist of arthropods associated with exposed carrion in the Hawaiian islands, USA. Proc Hawaiian Entomol Soc 26:53–57
Goff ML, Odom CB (1987) Forensic entomology in the Hawaiian Islands: three case studies. Am J Forensic Med Pathol 8:45–50
Graf Vitzthum H (1933) Die endoparasitische Deutonymphe von *Pterolichus nisi* [The endoparasitic deutonymph of *Pterolichus nisi*]. Z Parasitenkunde 6:151–169
Grassberger M, Frank C (2004) Initial study of arthropod succession on pig carrion in a central European urban habitat. J Med Entomol 41:511–523
Greenberg B (1961) Mite orientation and survival on flies. Nature 190:107–108
Greenberg B, Carpenter PD (1960) Factors in phoretic association of a mite and fly. Science 132:738–739
Hewadikaram KA, Goff ML (1991) Effect of carcass size on rate of decomposition and arthropod succession patterns. Am J Forensic Med Pathol 12:235–240

Houck MA, OConnor BM (1991) Ecological and evolutionary significance of phoresy in the Astigmata. Ann Rev Entomol 36:611–636

Hunter PE, Rosario RMT (1988) Associations of Mesostigmata with other arthropods. Ann Rev Entomol 11:393–418

Hunziker H (1919). Über die Befunde bei Leichenausgrabungen auf den Kirchhöfen Basels. Unter besonderer Berücksichtung der Fauna und Flora der Gräber [About the findings during excavations of corpses on the cemeteries of Basel, especially of the fauna and flora of graves]. Frankf Z Pathol 22:147–207

Hyatt KH (1980). Mites of the subfamily Parasitinae (Mesostigmata: Parasitidae) in the British Isles. Bull Brit Mus (Nat Hist) Zool 38:237–378

Iverson K, Oconnor BM, Ochoa R, Heckmann R (1996) *Lardoglyphus zacheri* (Acari: Lardoglyphidae), a pest of museum dermestid colonies, with observations on its natural ecology and distribution. Ann Entomol Soc Am 89:544–549

Jalil M, Rodriguez JG (1970) Studies of behaviour of *Macrocheles muscaedomesticae* (Acarina: Macrochelidae) with emphasis on its attraction to the house fly. Ann Entomol Soc Am 63:738–744

Korn W (1983) Zur Vergesellschaftung der Gamasidenarten *Poecilochirus carabi* G. u. R. Canestrini 1882 (= *P. necrophori* Vitzthum 1930), *P. austroasiaticus* Vitzthum 1930 und *P. subterraneus* Müller 1859 mit Aaskäfern aus der Familie der Silphidae [On the association between gamasid mites, *Poecilochirus carabi* G. u. R. Canestrini 1882 (= *P. necrophori* Vitzthum 1930), *P. austroasiaticus* Vitzthum 1930 and *P. subterraneus* Müller 1859 with carrion beetles (Silphidae)]]. Spixiana 6:251–280

Krantz GW, Poinar GO Jr (2004) Mites, nematodes and the multimillion dollar weevil. J Nat Hist 38:135–141

Krantz GW, Whitaker JO Jr (1988) Mites of the genus *Macrocheles* (Acari: Macrochelidae) associated with small mammals in North America. Acarologia 29:225–259

Leclercq M (1999). Entomologie et médecine légale. Importance des phoridés (Diptères) sur cadavres humains [Entomology and forensic medicine. Importance of phorid flies on human corpses]. Annales de la Société Entomologique de France 35 (Suppl.):566–568

Leclercq M, Vaillant F (1992) Entomologie et médecine légale: une observation inédite [Forensic entomology – an original case]. Ann Soc Entomol France 28:3–8

Leclercq M, Verstraeten C (1988a) Entomologie et médecine légale. Datation de la mort. Acariens trouvés sur des cadavres humains [Entomology and forensic medicine. Determination of the time of death. Acari found on human cadavers]. Bull Ann Soc R Belge Entomol 124: 195–200

Leclercq M, Verstraeten C (1988b). Entomologie et médicine légale. Datation de la mort: insectes et autres arthropodes trouvés sur les cadavres humains [Entomology and forensic medicine, determination of the time of death: insects and other arthropods on human cadavers]. Bull Ann Soc R Belge Entomol 124:311–317

Lindquist EE (1975) Associations between mites and other arthropods in forest floor habitats. Can Entomol 107:425–437

Lombardero MJ, Ayres MP, Hofstetter RW, Moser JC, Lepzig KD (2003) Strong indirect interactions of *Tarsonemus* mites (Acarina: Tarsonemidae) and *Dendroctonus frontalis* (Coleoptera: Scolytidae). Oikos 102:243–252

Lopez LCS, Filizola B, Deiss I, Rios RI (2005) Phoretic behaviour of bromeliad annelids (Dero) and ostracods (Elpidium) using frogs and lizards as dispersal vectors. Hydrobiologia 549:15–22

Luzzatto D, Pastorino G (2006) *Adelomelon brasiliana* and *Antholoba achates*: A phoretic association between a volutid gastropod and a sea anemone in Argentine waters. Bull Mar Sci 78:281–286

Mašán P (1999) Acarina associated with burying and carrion beetles (Coleoptera, Silphidae) and description of *Poecilochirus mrciaki* sp.n. (Mesostigmata, Gamasina). Biologia 54:515–524

McGarry JW, Baker AS (1997) Observations on the mite fauna associated with adult *Stomoxys calcitrans* in the UK. Med Vet Entomol 11:159–164

Mégnin P (1887) La faune des tombeaux [The fauna of graves]. C R Acad Sci Hebd 105: 948–951

Mégnin P (1889). Entomologie appliquée à la médecine légale à propos de la thèse de M. Georges Yovanovitch [Entomology applied to forensic medicine in connection with the thesis of Mr. George Yovanovitch]. Bull Soc Med France 21:249–251

Mégnin P (1892). Les Acariens Parasites [Parasitic acarina]. Edited by M. Léauté, Encyclopédie scientifique des Aide-Mémoire. Paris: G. Masson.

Mégnin P (1894) La Faune des Cadavres. Application de l'Entomologie à la Médecine Légale [The fauna of corpses. Application of entomology to forensic medicine]. G. Masson and Gauthier-Villars et Fils, Paris

Mertins JW, Hartdegen RW (2003) The ground skink, *Scincella lateralis*, an unusual host for phoretic deutonymphs of a uropodine mite, *Fuscuropoda marginata*, with a review of analogous mite-host interactions. Texas J Sci 55:33–42

Moser JC, Cross EA (1975) Phoretomorph: A new phoretic phase unique to the Pyemotidae (Acarina: Tarsonemoidea). Ann Entomol Soc Am 68:820–822

Moya Borja GE (1981) Effects of *Macrocheles muscadomesticae* (Scopoli) on the sexual behavior and longevity of *Dermatobia hominis*. Rev Bras Biol 41:237–241

Niogret J, Lumaret J-P, Bertrand M (2006) Semiochemicals mediating host-finding behaviour in the phoretic association between *Macrocheles saceri* (Acari: Mesostigmata) and *Scarabaeus* species (Coleoptera: Scarabaeidae). Chemoecology 16:129–134

OConnor BM (1982) Evolutionary ecology of astigmatid mites. Ann Rev Entomol 27:385–409

Ozdemir MH, Aksoy U, Akisu C, Sonmez E, Cakmak MA (2003) Investigating demodex in forensic autopsy cases. Forensic Sci Int 135:226–231

Pérez SP, Duque P, Wolff M (2005) Successional behavior and occurrence matrix of carrion-associated arthropods in the urban area of Medellín, Colombia. J Forensic Sci 50: 448–454

Perotti MA (1998) Interacciones entre ácaros (depredadores y foréticos) y dípteros muscoideos (presas y forontes) en hábitats rurales y suburbanos de la pendiente atlántica bonaerense [Predatory and phoretic interactions between mites and flies in the Argentinean pampas (ecology and physiology)]. Universidad Nacional de Mar del Plata, Mar del Plata

Perotti MA, Braig HR (2009) Phoretic mites associated with animal and human decomposition. Exp Appl Acarol 49:85–124

Perotti MA, Brasesco MJA (1996) Especificidad forética de *Macrocheles muscaedomesticae* (Acari: Macrochelidae) [Phoretic specificity of *Macrocheles muscaedomesticae* (Acari: Macrochelidae)]. Ecol Austral 6:3–8

Perotti MA, Brasesco MJA (1997) Orientación química de *Macrocheles muscaedomesticae* Scopoli (Acari: Macrochelidae) y percepción a distancia de posturas de *Musca domestica* (Diptera: Muscidae) y *Calliphora vicina* (Diptera: Calliphoridae) [Chemo-orientation of *Macrocheles muscaedomesticae* Scopoli (Acari: Macrochelidae) and detection of distant eggs of *Musca domestica* (Diptera: Muscidae) and *Calliphora vicina* (Diptera: Calliphoridae)]. Revista Soc Entomol Arg 56:67–70

Petersson E, Sivinski J (2003) Mating status and choice of group size in the phoretic fly *Norrbomia frigipennis* (Spuler) (Diptera: Sphaeroceridae). J Insect Behav 16:411–423

Polak M (2003) Heritability of resistance against ectoparasitism in the *Drosophila-Macrocheles* system. J Evol Biol 16:74–82

Prat N, Anon-Suarez D, Rieradevall M (2004) First record of Podonominae larvae living phoretically on the shells of the water snail *Chilina dombeyana* (Diptera: Chironomidae/Gastropoda: Lymnaeidae). Aquat Insects 26:147–152

Prichard JG, Kossoris PD, Leibovitch RA, Robertson LD, Lovell FW (1986) Implications of trombiculid mite bites: reports of a case and submission of evidence in a murder trial. J Forensic Sci 31:301–306

Pugh PJA, Llewellyn PJ, Robinson K, Shackley SE (1997) The associations of phoretic mites (Acarina: Chelicerata) with sand-hoppers (Amphipoda: Crustacea) on the South Wales coast. J Zool 243:305–318

Richards EN, Goff ML (1997) Arthropod succession on exposed carrion in three contrasting tropical habitats on Hawaii Island, Hawaii. J Med Entomol 34:328–339

Robin C, Mégnin P (1877) Mémoire sur les Sarcoptides plumicoles [Notes about feather mites]. J Anat Physiol 13:402–406

Rodrigueiro TSC, do Prado AP (2004) *Macrocheles muscaedomesticae* (Acari, Macrochelidae) and a species of *Uroseius* (Acari, Polyaspididae) phoretic on *Musca domestica* (Diptera, Muscidae): effects on dispersal and colonization of poultry manure. Iheringia Sér Zool 94:181–185

Rodriguez JG, Wade CF (1961) The nutrition of *Macrocheles muscaedomesticae* (Acarina: Macrochelidae) in relation to its predatory action on the house fly egg. Ann Entomol Soc Am 54:782–788

Satou A, Nisimura T, Numata H (2000) Reproductive competition between the burying beetle *Nicrophorus quadripunctatus* without phoretic mites and the blow fly *Chrysomya pinguis*. Entomol Sci 3:265–268

Saul-Gershenz LS, Millar JG (2006) Phoretic nest parasites use sexual deception to obtain transport to their host's nest. Proc Natl Acad Sci USA 103:14039–14044

Sazima I, Grossman A (2006) Turtle riders: remoras on marine turtles in Southwest Atlantic. Neotrop Ichthyol 4:123–126

Schwan TG (1993) Sex ratio and phoretic mites of fleas (Siphonaptera, Pulicidae and Hystrichopsyllidae) on the Nile grass rat (*Arvicanthus niloticus*) in Kenya. J Med Entomol 30:122–135

Schwarz HH, Koulianos S (1998) When to leave the brood chamber? Routes of dispersal in mites associated with burying beetles. Exp Appl Acarol 22:621–631

Shalaby OA, deCarvalho LML, Goff ML (2000) Comparison of patterns of decomposition in a hanging carcass and a carcass in contact with soil in a xerophytic habitat on the island of Oahu, Hawaii. J Forensic Sci 45:1267–1273

Siepel H (1994) Life-history tactics of soil microarthropods. Biol Fert Soils 18:263–278

Siepel H (1995) Applications of microarthropods life-history tactics in nature management and ecotoxicology. Biol Fert Soils 19:75–83

Singh P, McEllistrem MT, Rodriguez JG (1967) The response of some macrochelids (*Macrocheles muscaedomesticae* (Scopoli), *M. merdarius* (Berl.) and *M. subbadius* (Berl.)) to temperature and humidity (Acarina: Macrochelidae). Acarologia 9:1–20

Sivinski J, Marshall S, Petersson E (1999) Kleptoparasitism and phoresy in the Diptera. Fla Entomol 82:179–197

Southwood TRE (1962) Migration of terrestrial arthropods in relation to habitat. Biol Rev Cambridge Philos Soc 37:171–214

Sudhaus H (1981) Über die Sukzession von Nematoden in Kuhfladen [Succession of nematodes in cow droppings]. Pedobiologia 21:271–297

Sudhaus H, Rehfeld K, Schulter D, Schweiger J (1988) Beziehungen zwischen Nematoden, Coleoptera und Dipteren in der Sukzession beim Abbau von Kuhfladen [Interrelationships of nematodes, beetles and flies in the succession of cow pats during decomposition]. Pedobiologia 31:305–322

Tschapka M, Cunningham SA (2004). Flower mites of *Calyptrogyne ghiesbreghtiana* (Arecaceae): Evidence for dispersal using pollinating bats. Biotropica 36:377–381

Vance GM, VanDyk JK, Rowley WA (1995) A device for sampling aquatic insects associated with carrion in water. J Forensic Sci 40:479–482

Vaz-De-Mello FZ, Louzada JNC (1997) Considerações sobre forrageio arbóreo por Scarabaeidae (Coleoptera, Scarabaeoidea) e dados sobre sua ocorrência em floresta tropical do Brasil [Considerations on arboreal foraging by Scarabaeidae (Coleoptera, Scarabaeoidea), and data on their occurrence in tropical forests of Brazil]. Acta Zool Mex (n s) 72:55–61

Wade CF, Rodriguez JG (1961) Life history of *Macrocheles muscaedomesticae* (Acarina, Mesostigmata), a predator of the house fly. Ann Entomol Soc Am 54:776–781

Walter DE, Proctor HC (1999) Mites – ecology, evolution and behaviour. CABI, Wallingford

Webb JP Jr, Loomis RB, Madon MB, Bennett SG, Greene GE (1983) The chigger species *Eutrombicula belkini* Gould (Acari: Trombiculidae) as a forensic tool in a homicide investigation in Ventura County, California. Bull Soc Vector Ecol 8:141–146

Wilson DS (1983) The effect of population structure on the evolution of mutualism: a field test involving burying beetles and their phoretic mites. Am Nat 121:851–870

Yovanovitch GP (1888) Entomologie Appliquée à la Médicine Légale [Entomology applied to forensic medicine]. Ollier-Henry, Paris

Zeh DW, Zeh JA (1992) On the function of harlequin beetle riding in the pseudoscorpion *Cordylochernes scorpoides* (Pseudoscorpionida: Chernetidae). J Arachnol 20:47–51

Chapter 6
Indoor Arthropods of Forensic Importance: Insects Associated with Indoor Decomposition and Mites as Indoor Markers

Crystal L. Frost, Henk R. Braig, Jens Amendt, and M. Alejandra Perotti

6.1 Introduction

It is not surprising with the great diversity of arthropods that some members have evolved to take advantage of the sheltered habitat that we provide or to take advantage of us and our products. Anthropophilic arthropods like cockroaches (Blattodea), silverfish (Thysanura), house flies (Diptera) and house and dust mites (Acari) have moved their habitat inside human dwellings to become part of the human biocenose. These arthropods, however, do not directly depend on humans. Synanthropic insects like filth flies, biting midges, no-see-ums, punkies, mosquitoes (Diptera) feed off humans directly through blood sucking or off excrements and garbage produced by humans. Some of these insects have adopted an endophilic lifestyle entering our homes to feed and rest. Like some stored product pests, some of the hematophagous insect species have lost their natural or peridomestic habitat and have become entirely dependant on domestic harbourage and humans. Forensic implications can be found in any area of entomology or acarology associated with human habitation. Forensic entomology is receiving much attention (Byrd and Castner 2010; Erzinçlioğlu 2002; Gennard 2007; Goff 2001; Greenberg and Kunich 2005; Gunn 2009; Hall and Huntington (2010); Hall and Haskell 1995). It is often dominated by the medicolegal or medicocriminal aspect investigating human decomposition. It also covers situations such as child neglect, child abuse and neglect of the elderly, events that normally occur indoors but that are covered elsewhere in the book and by Benecke et al. 2004. Forensic entomology encompasses as well situations that involve urban, structural and stored products entomology. Unusual cases as that of a container of materials shipped to the Antartic catch the attention. The presence of

---

C.L. Frost (✉) and H.R. Braig
Bangor University, School of Biological Sciences, Bangor, Wales, UK

J. Amendt
Institute of Forensic Medicine, Frankfurt am Main, Germany

M.A. Perotti
University of Reading, School of Biological Sciences, England, UK

J. Amendt et al. (eds.), *Current Concepts in Forensic Entomology*,
DOI 10.1007/978-1-4020-9684-6_6,

only second instar larvae and subsequent stages of the scuttle fly *Megaselia scalaris* revealed that the contamination of chicken eggs in the container occurred in Australia and not during transit in South Africa (Nickolls and Disney 2001). Hall and Huntington have beautifully illustrated that forensic entomology can even reach out to the world of musea, ancient Mexican ceramics and figurines (Hall and Huntington 2010; Pickering et al. 1998), and the use of necrophagous insects in archaeology is accepted as a helpful tool (Panagiotakopulu 2004; Nystrom et al. 2005).

In this chapter insects associated with indoor decomposition of human remains are reviewed. The closer association of indoor arthropods with living humans highlights a greater potential of indoor arthropods as forensic evidence in itself. This is underpinned by the huge diversity of mite species associated with human habitation.

6.2 Indoor Decomposition

According to Mann et al. access for arthropods to the body is the second most important variable after temperature affecting the decomposition of a body (Mann et al. 1990). This should be most pronounced in any kind of concealment, wrapping or relatively well sealed spaces such as containers, cars, car trunks, trucks, or closed kitchen appliances (Anderson 2010). The limiting factor here should be the diameter of the opening such as the 1 cm hole caused by the missing front door handle of the kitchen appliance. Shelter in itself provided by a roof should have little influence on the fauna of the corpse. In a comparison in Argentina, no differences between sheltered and exposed to open sky were found in summer; only in winter, the sheltered pig carcass attracted the secondary screwworm, *Cochliomyia macellaria* (Calliphoridae), and the rare *Phaenicia cluvia* (Calliphoridae) in addition to the common bluebottle *Calliphora vicina* (Calliphoridae) (Centeno et al. 2002). The hypothesis put forward is that the level of endophyly or exophyly of a species should determine the likelihood of its appearance at a decomposing corpse and that the level of concealment that human habitation provides might be of minor importance. It has been found that decomposition generally occurs faster in outdoor environments in comparison to indoor environments. This delay could broaden the range of PMI values if not fully understood.

A recent retrospective of the past 10 years (1998–2008) at the Institute of Forensic Medicine in Frankfurt/Germany shows that 81.9% of the 364 corpses, which were infested by insects, were found indoors. This highlights the need to analyse the insect fauna of indoor situations.

Goff undertook the most explicit comparison of decomposition between indoor and outdoor situations (Goff 1991). Covering a range of 2–21 days post mortem, 14 cases of accidental death, suicide, homicide, and unattended deaths over 8 years were compared to 21 cases retrieved from outdoor locations that were of a corresponding post mortem age. It should be mentioned that all these results suffer from the fact that they were not designed or controlled succession studies but case reports. Therefore a period of e.g. 5–6 days post-mortem might be correct but will depend on the reliability and quality of the general forensic investigation in these cases. At the same time, the example "5–6 days post-mortem" does not rule out the possibility that a certain species might have been there on days 1–4 and still could be there on days 7–10 as well: case data often just represent a small cutout of the whole case history.

One might expect that the numbers of insects are greater in outdoor situations than in indoor locations, but this has not yet been systematically investigated. Under the tropical conditions of Hawai'i, the peak of insect diversity is reached faster indoors, between days 6 and 7, compared to between days 8 and 10 for outside corpses. At later stages, few indoor species contrast to a wide diversity of outdoor species. The indoor carcass fauna was richer in Diptera species while the outdoor fauna was characterised by a higher abundance of Coleoptera species. Table 6.1 serves as an illustration of how often particular species have been reported per number of indoor cases (reports/cases indoors), and how often particular species have been reported from indoor cases and from outdoor cases (indoor/outdoor reports) in the study in Hawai'i.

In Goff's study, insects were not generally associated with decomposing remains discovered indoors above the sixth floor. However, in several cases from temperate regions, insect infestation of human remains have been reported from far taller buildings, e.g., an 11-story apartment building in Gdansk, Poland, and the 18th floor of an apartment complex in Canada (Anderson 2010; Piatkowski 1991).

In most outdoor situations blow flies (Calliphoridae) will dominate the first weeks of decomposition. The large numbers of eggs laid by blow flies will obscure any larva of flesh flies (Sarcophagidae) present. Nuorteva describes blow flies in four indoor cases in Helsinki (Nuorteva et al. 1967). However, in indoor cases it is

Table 6.1 Examples of species associated with human remains found indoors in Hawai'i

Dayspostmortem	Species	Family	Reports/cases indoors	Indoor/ outdoor reports
2,5,6,7	*Chrysomya rufifacies* (hairy maggot blow fly)	Calliphoridae	4/6	4/20
3,7,8	*C. megacephala* (oriental latrine fly)		3/3	3/14
4,6	*Sarcophaga (Boettcherisca) peregrina* (flesh fly)	Sarcophagidae	3/4	3/3
5,6,14–21	*S. (Bercaea) haemorrhoidalis* (red-tailed flesh fly)		4/9	4/0
6	*Fannia pusio* (chicken dung fly)	Fanniidae	1/3	1/0
6	*Hydrotaea (Ophyra) chalcogaster* (grave fly)	Muscidae	1/3	1/0
6,7	*Stomoxys calcitrans* (stable fly)	Muscidae	3/4	3/0
6,7	*Dermestes maculates* (hide beetle)	Dermestidae	2/4	2/11
7,8	*Musca domestica* (house fly)	Muscidae	2/2	2/1
14–21	*Megaselia scalaris* (scuttle fly)	Phoridae	5/5	5/0

more likely that flesh flies might prevail. The diversity of flies (Diptera) that can be collected from a flat even in a temperate zone is overwhelming. Schumann collected 2,148 flies belonging to 150 species and 46 families in a flat in the outskirts of Berlin between April and October. *Fannia canicularis* was the most frequent species with 726 specimens, followed by *Drosophila melanogaster*, *Culex pipiens*, *Lucilia (Phaenicia) sericata*, *Sarcophaga carnaria*, *Calliphora vicina*, *Muscina stabulans* and *F. manicata*, accounting for 55% of the caught flies (Schumann 1990). This study gives again an indication for a possible bias: while most of our knowledge about species diversity in outdoor situations is based on many succession studies with pig cadavers and other carcasses or baits, our knowledge about indoor diversity relies mainly on real cases, which gives just a limited insight in the whole process and time scale of decomposition and insect colonisation. Moreover, one should differentiate strictly between adult and immature stages. In a recent study comparing the colonisation of two pig cadavers placed indoors and outdoors at the same period of the year, we found indeed a higher diversity regarding the adult stages on the pig placed outdoors, but no differences occurred when analysing the numbers of species which colonised the cadavers (J Amendt et al. unpublished). This again illustrates the need for more indoor studies.

Table 6.2 gives a geographically more widespread representation of insect species found on indoor remains. The table limits itself to examples of human cases. Controlled studies of the indoor decomposition of pigs are rare; a recent example is the comparison of indoor and outdoor decomposition in Parma, Italy (Leccese 2004), but this study just deals with small baits (56–90 g pieces of pork meat), definitely not comparable with a human cadaver.

Comparison between summer and winter in the insect fauna on human corpses from 117 domestic cases around Hamburg, Germany, found *Calliphora vicina* and scuttle flies as all-year species, *C. vomitoria*, *Muscina stabulans*, and *Dermestes* species as spring and autumn species, whereas *Lucilia sericata*, *Phormia regina*, and *Sarcophaga* species as typical summer species (Schroeder et al. 2003).

None of the listed insect species can be considered as exclusively indoors. The various levels of endophily exhibited by insect species will always remain just a bias towards one or the other environment. This bias is expected to change between geographic regions.

Insects may be attracted not by a cadaver, but will infest it secondarily. Some flies such as the false stable fly, *Muscina stabulans*, and the lesser house fly, *Fannia canicularis*, are drawn by a wide range of decaying organic matter and are commonly found in human quarters. These flies, for example, are much more attracted to human feces than to the corpse itself. This has led to the use of these flies as an indication for possible neglect (Benecke and Lessig 2001). The oriental latrine fly, *Chrysomya megacephala*, is equally attracted to feces and can lead to errors in estimates of the postmortem interval in similar cases of neglect (Goff et al. 1991). Similarly, Anderson suggests that the presence of normal household garbage might have been the determining factor for the attraction of several species in her study (Anderson 2010). Additionally, keeping pets indoors could be attractive for these insects, especially in summer.

Table 6.2 Examples of insect species reported from indoor remains

Species	Family	Indoor prevalence	Place
	Diptera		
Calliphora vicina	Calliphoridae	Occasional	Auckland, Belgium, Br. Columbia1,2, Cieza, Hamburg, Helsinki, USA1,2
C. vomitoria			Hamburg, Leipzig
Protophormia terraenovae		Occasional	Br. Columbia1
Chrysomya rufifacies		Occasional	Oahu
C. megacephala		Occasional	Oahu
Lucilla(=Phaenicia) sericata		Common	Auckland, Br. Columbia1, Gdansk, Hamburg, USA1
Phaenicia regina		Common	Br. Columbia1, Hamburg, USA1
Drosophila spp.	Drosophilidae		Hamburg
	Fanniidae		Canada
Fannia spp.			Canada, Hamburg
F. cannicularis			Gdansk, Leipzig
F. pusio			Oahu
Musca domestica	Muscidae	Common	Oahu
Muscina stabulans			Gdansk, Hamburg, Leipzig
Hydrotaea spp.		In- and outdoors	Br. Columbia1
H. (=Ophyra) capensis			Eraclea
H. (=Ophyra) chalcogaster			Oahu
Stomoxys calcitrans			Oahu
Synthesiomyia nudiseta			USA2
Megaselia abdita	Phoridae		USA1
M. scalaris		Common	Oahu
Piophila spp.	Piophilidae	In- and outdoors	Br. Columbia1
Piophila casei			France, Victoria
Sarcophaga spp.	Sarcophagidae		Hamburg
S. (Bercaea) haemorrhoidalis		Common	Oahu
S. (Boettcherisca) peregrina		Equal	Oahu
Thanatophilus lapponicus	Silphidae	In- and outdoors	Br. Columbia1
	Sphaeroceridae		Canada
Leptocera caenosa			England1
	Hymenoptera		
Tachinaephagus zealandicus	Encyrtidae		Eraclea
Nasonia vitripennis	Pteromalidae		Eraclea, Hamburg
	Coleoptera		
Necrophilus hydrophiloides	Agyrtidae	In- and outdoors	Br. Columbia1

(continued)

Table 6.2 (continued)

Species	Family	Indoor prevalence	Place
Necrobia rufipes	Cleridae		Eraclea, Hamburg, Victoria
N. violacea			Eraclea
	Dermestidae		Canada
Dermestes lardarius			Denmark, USA2
D. haemorrhoidalis			Denmark
D. maculates		Occasional	Eraclea, Germany, Oahu
Hister cadaverinus	Histeridae		Hamburg
	Lepidoptera		
Hofmannophila (=Borkhausenia) pseudospretella	Oecophoridae		England2

Auckland, New Zealand (Smeeton et al. 1984)
Belgium (Leclerq 1969)
British Columbia, Canada 1 (Anderson 1995), 2 (Anderson 2010)
Canada (Anderson and Cervenka 2002)
Cieza, Murcia, Spain (Arnaldos et al. 2005)
Denmark (Voigt 1965)
England 1 (Erzinçlioğlu 1985), 2 (Forbes 1942)
Eraclea, Venice, Italy (Turchetto and Vanin 2004)
France (Mégnin 1894)
Gdansk, Poland (Piatkowski 1991)
Germany (Schroeder et al. 2002)
Hamburg, Germany (Schröder et al. 2001)
Helsinki, Finland (Nuorteva et al. 1967)
near Leipzig, Germany (Benecke and Lessig 2001)
Oahu, Hawai'i (Goff 1991)
USA 1 (Greenberg and Kunich 2005), 2 (Lord 1990)
Victoria, Australia (Archer et al. 2005)

6.3 House and Dust Mites

Mites are the most prevalent invertebrate inhabitants of our homes. Houses contain a diversity of mites that has not fully been explored. Representatives of all orders of the Acari are expected given that human habitation offers so many small niches and habitats that practically all lifestyle requirements can be accommodated for as in strong contrast to insects. The huge difference in size between mites and insects makes this possible.

Humans have been storing food and have been building homes for a much shorter period than animals such as birds and other mammals. The large number of acarine taxa that inhabit our homes suggests that mites are taking advantage of a habitat similar to that in which they have evolved. House dust mites of the genus *Dermatophagoides* (Pyroglyphidae, Astigmata) have been associated with bird's nests; their natural habitats are likely the nest and lair of birds and mammals

(Kniest 1994; Walter and Proctor 1999). Stored product mites of the genus *Tyrophagus* (Acaridae, Astigmata) are also inhabitants of dead leaf litter (Binotti et al. 2001; Walter and Proctor 1999). The short generation time of many mite species allows a fast evolution towards a synanthropic lifestyle.

Historically, the house dust mites or allergen mites have received notorious attention because of their medical importance as allergen producers and their involvement in chronic respiratory diseases or disorders. The population dynamics of some cosmopolitan and synanthropic species have been studied to such an extant that their behaviour on our cloth and in our mattresses can be precisely predicted in most climatic regions.

Based on the large amount of information available on the house dust mites (the allergen mites), the most abundant species indoors seem to belong to the family of Pyroglyphidae (Astigmata), which feed on the skin flakes and other dander found in our homes. In a global survey, *Dermatophagoides pteronyssinus*, *D. farinae* and *Euroglyphus maynei* can account for up to 90% of the house dust fauna (Blythe et al. 1974; Crowther et al. 2000). However, we feel that this represents the result of a repeated sampling bias toward skin feeding allergen mites. The majority of the investigators have used extraction and sampling methods for house mites associated with beds, mattresses, carpets of bedrooms bathrooms and living rooms where skin flakes are normally accumulated. On the other hand, the cabinets of the kitchen and/or the carpet of the dinning room could reflect a completely different scenario. Early on Hughes has catalogued 25 species of Astigmata, 9 of Prostigmata and 20 of Mesostigmata that are found in storage premises containing human food, such as kitchens or pantries (Hughes 1976). For example, *Carpoglyphus lactis* (Carpoglyphidae, Astigmata), *Melichares agilis* (Ascidae, Mesostrigmata) and *Blattisocius mali* (Ascidae, Mesostigmata) might be common inhabitants of the Christmas pudding, while several species of *Tyrophagus* and *Acarus* (Acaridae, Astigmata) are well known as the major cheese mites. *Lardoglyphus* spp. (Lardoglyphidae, Astigmata) will engage in serious competition for dried meat, bones, and hides with the beetles of the genus *Dermestes*. And every detritivorous mite will attract its own mite predator species, which in part explains the high diversity of the indoor mite fauna.

Colonisation of new homes could be by movement of infested furniture or soft furnishings, through the use of another animal for transport (phoresy), airborne, or simply brought in by humans and their pets (Bischoff and Kniest 1998; Warner et al. 1999). Clothing can carry a large number of mites and could be a potential source (Bischoff et al. 1998). House dust mites do not like to establish populations in new houses until humans are present (Warner et al. 1999).

Larger households are associated with higher allergen levels than smaller households (van der Hoeven et al. 1992). The average level of mite infestation is proportional to the number of members in the household (Arlian 1989). Bigger buildings are likely to produce more food for house and dust mites and offer more constant conditions of temperature and humidity.

Astigmatid mites, which include the house dust mites and stored product mites, have no respiratory system, so gas exchange and water intake occurs through the cuticle. This means that in a dry habitat they are subject to desiccation. The relative humidity

(RH) is usually required to be above 55–60% for mites to uptake water, though stored product mites require higher humidity levels than house dust mites. Humidity sufficient for water uptake is only required for about 1–3 h a day, during which the mites can take up enough water to survive for the rest of the day (Schei et al. 2002).

The concentrations of dust mites in beds and on floors peak in the summer and decreases in winter, with a corresponding peak of allergens in the autumn and a decrease in spring (Chew et al., 1999a, b). The lag between peaks and troughs of allergens and mite concentrations is explained with the persistence of allergens after the mites' death.

Because of their small size, mites can take advantage of small clines in environmental factors. These clines are common in houses due to their fragmented nature offering varied microhabitats that mites can colonise. Each room may have a slightly different microcosm, causing variation of mite occurrences and densities between rooms. The richness of the mite fauna in dust collected from different rooms secures a high level of specificity.

6.3.1 Kitchen

The environmental conditions of the kitchen usually favour the establishment of a large community of arthropods. In kitchens, dust may not gather as in other rooms due to the lack of soft furnishings. Dust may accumulate in cupboards and behind heavy electrical equipment such as fridges, freezers and washing machines. Extractor fans reduce the concentration of mite allergen in the kitchen, bedroom and in the living room. This effect is linked to a reduction in the relative humidity of the home environment (Luczynska et al. 1998).

Comparing the rooms of 134 houses in Chile, the highest prevalence of dust mites was found in the kitchen (Franjola and Rosinelli 1999). The most abundant and prevalent species was *Glycyphagus domesticus*, followed by *Tyrophagus putrescentiae* and *G. destructor*. The highest density was recorded in bedrooms caused by Pyroglyphidae mites. But the number of mite positive samples from houses might be significantly higher than the number of mite positive samples from kitchens of the same houses (Ezequiel et al. 2001). The kitchen might be considered as a habitat with the potential of being a source for re-colonization of the entire house. A study in Brazil showed that of the 190 mites collected in pantries, 141 (74%) belonged to the family Acaridae, with the stored product mite species *Tyrophagus putrescentiae* predominant (Binotti et al. 2001). The other mites observed belonged to the families Tarsonemidae (7%), Pyroglyphidae (3%), Glycyphagidae (3%), Cheyletidae (2%), Eriophyidae (2%), and also to the order Mesostigmata (8%) and the order Oribatida (1%).

In the United Kingdom, mites were counted in 14% of 727 kitchens (Turner and Bishop 1998). Booklice are one of the most expected inhabitants of flour or grain products stored in kitchen cabinets. With the sole aim to collect psocids from domestic cupboards, two surveys were conducted using bait traps (Turner and

Maude-Roxby 1989). Nevertheless, the mites were the predominant fauna. Of the whole mite collection, 55.5% constituted *Tyrophagus putrescentiae*, followed by *Glycyphagus domesticus* accounting for 25%. The remaining 19.5% included 10 species of which the skin feeders belonging to the family Pyroglyphidae made only 5.7%. The species collected in British kitchen cupboards included *Carpoglyphus lactis*, *Dermatophagoides pteronyssinus*, *Euroglyphus maynei*, *Acarus siro*, *Lepidoglyphus destructor*, *Euroglyphus longior*, *Glycyphagus privatus*, *Cheyletus eruditus*, *Dermatophagoides farinae* and *Dermanyssus* sp.

Cereal based foods stored in a kitchen for 6 weeks after purchase contained a predominance of *Acarus siro* (49.6%). Just one species of pyroglyphid mites was recovered, *Dermatophagoides pteronyssinus*, increasing the numbers by only 1.2% (Thind and Clarke 2001).

6.3.2 Pantries

Pests such as mice, rats and invertebrates can inhabit pantries bringing mites with them. Warner et al. found mites of the order Mesostigmata, which are commonly associated with rats and insect pests (Warner et al. 1999). Mites that eat plants can also be found in pantries and kitchens. The families Tarsonemidae and Eriophyidae (Prostigmata) are obligate plant parasites (Walter and Proctor 1999). Binotti et al. found 9% of plant mites in Brazilian pantries, possibly brought in along with fresh herbs and vegetables (Binotti et al. 2001).

Mites, especially stored product mites, are also found in groceries in high concentrations (Harju et al. 2006; Koistinen et al. 2006). The same species of mites were found in different stores; over 60% out of 949 specimens belonged to the Acaridae (*Acarus* sp. and *Tyrophagus* sp.), which are common stored product mites; and just one house dust mite (Pyropgyphidae) was found (0.1%).

6.3.3 Bedroom

The areas of highest concentration of mites and mite antigens within the bedroom are often the sleeping area, this includes the mattress and bedding (Sidenius et al. 2002). This is due to the favourable conditions in the bed. Skin flakes and other dander found in the bed are replenished every night when the bed is used by its owner.

Washing bedding above 40°C reduces allergen concentrations of carpets as well as of beddings. Washing at 40°C does not affect the allergen concentrations in the mattress and does not seem to destroy the mites (Arlian et al. 2003; Luczynska et al. 1998). In Ohio, 13 houses were examined for the presence of mites on different parts of the bed (Yoshikawa and Bennett 1979). The majority of mites were recorded from the mattress, followed by the blanket and the pillows. The mattress edge harboured the highest concentrations when compared with the head sites of the bed.

Foam mattresses can be four times more likely to contain mites than sprung mattresses (Schei et al. 2002). Traditional sprung mattresses had a thicker covering which is more impermeable to mites than the thinner covering on foam mattresses.

Other areas in the bedroom in which mites thrive are textiles such as the curtains and the carpet. The concentration of mites in bedding affects concentrations of mites in these textiles, as the concentrations of allergens in the carpet are related to the age of the bedding (Luczynska et al. 1998). Curtains are an area of the house that often includes other interesting mite species. Binotti et al. found phytophagous Eriophyidae (Prostigmata) and ground mites of the order Oribatida in curtains in Brazil (Binotti et al. 2005). The authors suggest that wind may have a large part to play in the mite contamination of curtains. However, the association of Oribatida with curtains is not surprising. *Phauloppia lucorum*, the window sill mite, aggregates in large numbers around windows inside houses. Although this species is associated with different habitats outdoors, such as lichens, indoors it has its specific distribution (Hughes 1976).

Stored product mites are also common inhabitants of our bedrooms, they are found in association with fungi that grow here. Mould from residual damp on the walls or fungi that are found in the bed can be a source of food for stored product mites. The fungi or molds are often not obvious to the naked eye. In the tropics, stored product mites can become the most abundant mite in the bedroom. A study in Singapore found that *Blomia tropicalis* (Glycyphagidae, Astigmata) was the most abundant mite in the bedroom, whereas in temperate climates, it is usually thought to be *Dermatophagoides pteronyssinus* (Chew et al., 1999a, b). This may be due to the warm and humid nature of the tropics encouraging fungal growth, changing the constituents of the dander in the bedroom in comparison to temperate climates. This would favour stored product mites over house dust mites.

Predatory mites such as species within the Cheyletidae (Prostigmata) are found in homes. They are found in association with the mites on which they feed. If we assume that the bedroom contains the highest numbers of mites, it may be safe to propose that they also contain the highest concentration of predatory mites. Warner et al. found Cheyletidae mites in 22% of homes in a study in Sweden (Warner et al. 1999). Chew et al. (1999a, b) found that homes with predatory mites contained relatively few Astigmata mites, which are stored product and house dust mites, suggesting they play a part in balancing the ecosystem in our homes.

Concrete floors in the bedroom are thought to increase humidity, corresponding to higher allergen level in mattresses. The level in a house, on which bedrooms are found also affect concentrations of allergens. Luczynska et al. (1998) showed that having a bedroom on the ground floor greatly increased concentrations of allergens. This may be due to increased humidity or reduced ventilation.

Mites can be found in great numbers on clothing. Bischoff et al. (1998) propose that between 20,000 and 30,000 house dust mites can be found in various items of clothing. They also showed that washing at low temperatures only removes around 70% of the population, leaving some to carry on the next generation. All this implies that mite numbers in a large wardrobe could make millions.

6.3.4 Living Room

The living room can contain comparatively large numbers of mites. The situation is similar to that in the bedroom, where soft textiles contain the highest numbers of mites. The highest concentrations are often found in the most commonly used sofa or upholstered chair. Curtains, carpets and rugs can also be refuges for mites (Sidenius et al. 2002). In Ohio, a living room carpet was able to hold seven times more house dust mites than a bedroom mattress (Arlian 1989). Stored product mites are found in living rooms where fungi can grow. 38% of living rooms contained stored product mites in the survey of Warner (Warner et al. 1999). Having an open fireplace in the living room reduces the concentration of *D. pteronyssinus*. Houses containing smokers tend to have decreased concentrations of mites in the living room (Luczynska et al. 1998).

6.3.5 Bathroom

The lack of soft furnishings discourages dust mites as there is less accumulation of dust on hard flooring. It has been shown that uncarpeted floors usually contain less mites than carpeted floors (Sidenius et al. 2002). In Chile, a comparison between the rooms in 134 houses revealed not a single mite in the bathrooms, whereas in the study of Warner almost one third of the dust samples had mites (Franjola and Rosinelli 1999; Warner et al. 1999).

6.3.6 Indoor Pools

The water that we bathe in may contain mites. The swimming pool mite, *Hydronothus crispus* occurs in large numbers in indoor pools in Japan. It completes its entire life cycle in indoor pools (Robinson 2005). In a study on indoor pools, 53% tested positive for mites with a highest density around 50 individuals/m^2. Two other species seem to compete with *H. crispus*, *Trimalaconothrus maniculatus* and *Histiostoma ocellatum*. Individual pools were mainly colonised by a single species (Kazumi et al. 1992).

6.3.7 Store Room, Attic, Basement

Areas where belongings are stored are not often cleaned and are known for large accumulations of dust. The absence of humans from these areas may mean that the constituents of the dust differ from house dust found in other rooms, therefore changing the habitat and possibly selecting for different species of dominant mites. The humidity of the storeroom may determine the presence of fungi and therefore stored product mites. The fact that basements are at the lowest level and attics at the highest level of a house or building will greatly influence the humidity, and therefore the diversity of mite species in these rooms.

6.3.8 Study, Office

The soft furnishings of an office are usually similar to that of a living room with the obvious additions of papers, books and bookshelves, with a desk and often a computer. Bookshelves contain relatively small amounts of mites in comparison to carpets and other soft furnishings (Sidenius et al. 2002).

6.3.9 Pet's Room

The affinity that many people have developed with animals has led to the inclusion of pets within our homes. Over 250 species of mites are known to cause problems for humans and domesticated animals. Much more important, however, are the countless species of mites that do not cause any obvious pathology to animals. The number of microarthropods brought into the home by our pets may be greatly underestimated.

Birds, for example, have been found to have a considerable number of mites that reside within their feathers. Feather mites are a widespread group of mites that have contributions from 33 families. More than 440 species of feather-related mites are known. All areas of the feather have been exploited. Mites can be found within the shaft, in the base and on the feather surface. Walter and Proctor indicated that only three families in two orders of birds do not have associated feather mites, the Dromaiidae (emu's) and the Causuariidae (cassowaries) from the order Struthioniformes and the Spheniscidae from the order Sphenisciformes (penguins) (Walter and Proctor 1999). However, a relatively recent study reports feather mites on a cassowary as well (Proctor 2001).

Mite species are found associated with fish and amphibians, and mite families such as Pterygosomatidae (Prostigmata) and Omentolaelapidae (Mesostigmata) have been found with lizards and snakes. The snake mite *Ophionyssus natricis* (Macronyssidae, Mesostigmata) is the most common ectoparasite of captive snakes.

Members of the family Demodicidae (Prostigmata) are found in the hair follicles of most dogs, some cats, gerbils, hamsters and humans, and cause no problems in the large majority of mammals. The mites that live on our pets may often be found in our homes and could be described as house mites.

6.3.10 Conservatory, Plant Room

Conservatories are warm humid habitats. Regular watering of plants also increases relative humidity. This would make the conservatory an ideal place for mites. Mites of the family Tetranychidae (Prostigmata), known as spider mites, are plant parasites and have been found in 24% of homes in Sweden (Warner et al. 1999). Soika and Labanowski report spider mites found on ornamental trees and shrubs, which may be used in a conservatory. Plants may also affect the environment for house dust mites (Soika and Labanowski 2003). Anyone who keeps house plants attracts also dust. Therefore, houseplants can also be a habitat for house dust and stored product mites. With ornamental plants comes soil and compost, which will bring their own mite fauna with them. Warner et al. 1999) found Cryptostigmata (Oribatida) otherwise known as soil or beetle mites in homes of asthmatic children.

6.3.11 Mites as High Resolution Markers

We hope that this gross sketch of the high diversity of mite species in indoor environments might inspire forensic investigators to have a second look at dust and especially at mites associated with cloth and home furniture. Fibres of any kind have gained great importance as trace evidence, however, they lack DNA. As living organisms, the mites living upon our clothes and our clothed furniture can be molecularly characterized. We would like to propose that mites might add a useful contribution to forensic evidence, where fibres are not available or are not sufficient to link a suspect with a crime scene or a victim. In the domestic environment dust mites are ubiquitous; more importantly, their populations vary in composition. This diversity is in direct relation with human living habits and housing characteristics. The development of mite molecular markers able to differentiate between two human mite populations will particularly contribute to forensic investigations with data of high resolution.

Acknowledgement MAP and HRB wish to thank the Leverhulme Trust for support of this work.

References

Anderson GS (1995) The use of insects in death investigations: an analysis of forensic entomology cases in British Columbia over a five year period. J Can Soc Forensic Sci 28:277–292

Anderson GS (2010) Factors that influence insect succession on carion. In: Byrd JH, Castner JL (eds) Forensic entomology: the utility of arthropods in legal investigations, 2nd edn. CRC, Boca Raton, pp 201–250

Anderson GS, Cervenka VJ (2002) Insects associated with the body: their use and analyses. In: Haglund WD, Sorg MH (eds) Advances in forensic taphonomy. CRC, Boca Raton, pp 173–200

Archer MS, Bassed RB, Briggs CA, Lynch MJ (2005) Social isolation and delayed discovery of bodies in houses: The value of forensic pathology anthropology, odontology and entomology in the medico-legal investigation. Forensic Sci Int 151:259–265

Arlian LG (1989) Biology and ecology of house dust mites *Dermatophagoides* spp. and *Euroglyphus* spp. Immunol Allergy Clin North Am 9:339–356

Arlian LG, Vyszenski-Moher DL, Morgan MS (2003) Mite and mite allergen removal during machine washing of laundry. J Allergy Clin Immunol 111:1269–1273

Arnaldos MI, Garcia MD, Romera E, Presa JJ, Luna A (2005) Estimation of postmortem interval in real cases based on experimentally obtained entomological evidence. Forensic Sci Int 149:57–65

Benecke M, Lessig R (2001) Child neglect and forensic entomology. Forensic Sci Int 120:155–159

Benecke M, Josephi E, Zweihoff R (2004) Neglect of the elderly: forensic entomology cases and considerations. Forensic Sci Int 146:S195–S199

Binotti RS, Oliveira CH, Muniz JRO, Prado AP (2001) The acarine fauna in dust samples from domestic pantries in southern Brazil. Ann Trop Med Parasitol 95:539–541

Binotti RS, Oliveira CH, Santos JC, Binotti CS, Muniz JRO, Prado AP (2005) Survey of acarini fauna in dust samplings of curtains in the city of Campinas, Brazil. Braz J Biol 65:25–28

Bischoff ERC, Kniest FM (1998) Differences in the migration behaviour of two most widespread species of house dust mites (HDM). J Allergy Clin Immunol 101:S28

Bischoff ERC, Fischer A, Liebenberg B, Kniest FM (1998) Mite control with low temperature washing II. Elimination of living mites on clothing. Clin Exp Allergy 28:60–65

Blythe ME, Williams JD, Smith JM (1974) Distribution of pyroglyphid mites in Birmingham with particular refrence to *Euroglyphus maynei*. Clin Allergy 4:25–33

Byrd JH, Castner JL (2010) Forensic entomology: the utility of arthropods in legal investigations, 2nd edn. CRC, Boca Raton

Centeno N, Maldonado M, Oliva A (2002) Seasonal patterns of arthropods occurring on sheltered and unsheltered pig carcasses in Buenos Aires Province (Argentina). Forensic Sci Int 126:63–70

Chew FT, Zhang L, Ho TM, Lee BW (1999a) House dust mite fauna of tropical Singapore. Clin Exp Allergy 29:201–206

Chew GL, Higgins KM, Gold DR, Muilenberg ML, Burge HA (1999b) Monthly measurements of indoor allergens and the influence of housing type in a northeastern US city. Allergy 54:1058–1066

Crowther D, Horwood J, Baker N, Thomson D, Pretlove S, Ridley I et al (2000) House dust mites and the built environment: a literature review. University College London, London

Erzinçlioğlu YZ (1985) The entomological investigation of a concealed corpse. Med Sci Law 25:228–230

Erzinçlioğlu Z (2002) Maggots, murder, and men: memories and reflections of a forensic entomologist Thomas Dunne Books, New York

Ezequiel OdS, Gazeta GS, Amorim M, Serra-Freire NM (2001). Evaluation of the acarofauna of the domiciliary ecosystem in Juiz de Fora, State of Minas Gerais, Brazil. Memorias do Instituto Oswaldo Cruz 97:911–916

Forbes G (1942) The brown house moth as an agent in the destruction of mummified human remains. Police J London 15:141–148

Franjola TR, Rosinelli MD (1999) Acaros del polvo de habitaciones enla ciudad de Punta Arenas, Chile [Mites in the house ducts of the city of Punta Arenas, Chile]. Boletín Chileno de Parasitología 54:82–88

Gennard D (2007) Forensic entomology: an introduction. Wiley, Chichester

Goff ML (1991) Comparison of insect species associated with decomposing remains recovered inside dwellings and outdoors on the island of Oahu, Hawaii. J Forensic Sci 36:748–753

Goff ML (2001) A fly for the prosecution: how insect evidence helps solve crimes, New edited editionth edn. Harvard University Press, Cambridge

Goff ML, Charbonneau S, Sullivan W (1991) Presence of fecal material in diapers as a potential source of error in estimates of postmortem interval using arthropod development rates. J Forensic Sci 36:1603–1606

Greenberg B, Kunich JC (2005) Entomology and the law: flies as forensic indicators, New edited edn. Cambridge University Press, Cambridge

Gunn A (2009) Essential forensic biology: animals plants and microorganisms in legal investigations. Wiley, Chichester

Hall RD, Huntington TE (2010) Perception and status of forensic entomology. In: Byrd JH, Castner JL (eds) Forensic entomology: the utility of arthropods in legal investigations. CRC, Boca Raton, pp 1–16

Hall RD, Haskell NH (1995) Forensic entomology: applications in medicolegal investigations. In: Wecht C (ed) Forensic sciences. Matthew Bender, New York

Harju A, Husman T, Merikoski R, Pennanen S (2006) Exposure of workers to mites in Finnish groceries. Ann Agric Environ Med 13:341–344

Hughes AM (1976) The mites of stored food and houses. Her Majesty's Stationary Office, London

Kazumi T, Takaya I, Jyun-ichi H, Masamichi I, Kenji F (1992) Occurrence of aquatic oribatid and astigmatid mites in swimming pools. Water Res 26:1549–1554

Kniest FM (1994) Are storage mites different from house-dust mites? Atemswegs- und Lungenkrankheiten 20:40–45

Koistinen T, Ruoppi P, Putus T, Pennanen S, Harju A, Nuutinen J (2006) Occupational sensitization to storage mites in the personnel of a water-damaged grocery store. Int Arch Occup Environ Health 79:602–606

Leccese A (2004) Insects as forensic indicators: methodological aspects. Aggrawal's Internet J Forensic Med Toxicol 5:26–32

Leclerq M (1969) Entomological parasitology: the relations between entomology and the medical sciences. Pergamon, Oxford

Lord WD (1990) Case histories of the use of insects in investigations. In: Catts EP, Haskell NH (eds) Entomology and death: a procedural guide. Joyce's Print Shop, Clemson, pp 9–37

Luczynska C, Sterne J, Bond J, Azima H, Burney P (1998) Indoor factors associated with concentrations of house dust mite allergen, Der p 1, in a random sample of houses in Norwich, UK. Clin Exp Allergy 28:1201–1209

Mann RW, Bass WM, Meadows L (1990) Time since death and decomposition of the human-body – Variables and observations in case and experimental field studies. J Forensic Sci 35:103–111

Mégnin P (1894) La Faune des Cadavres. Application de l'Entomologie à la Médecine Légale [The fauna of corpses. Application of entomology to forensic medicine]. Gauthier-Villars et Fils and G. Masson, Paris

Nickolls P, Disney RHL (2001) Flies discovered at Casey station. Aus Antarctic Mag 1:54

Nuorteva P, Isokoski M, Laiho K (1967) Studies on the possibilities of using blowflies (Dipt.) as medicolegal indicators in Finland. I. Report of four indoor cases from the city of Helsinki. Annales Entomologici Fennici 33:217–225

Nystrom KC, Goff A, Goff ML (2005) Mortuary behaviour reconstruction through paleoentomology: a case study from Chachapoya, Peru. Int J Osteoarch 15:175–185

Panagiotakopulu E (2004) Dipterous remains and archaeological interpretation. J Arch Sci 31:1675–1684

Piatkowski S (1991) Synanthropic flies in an 11-story apartment house in Gdansk. Wiadomosci Parazytologiczne 37:115–117

Pickering RB, Ramos J, Haskell NH, Hall RD (1998). El significado de las cubiertas de crisalidas de insectos que aparecen en las figurillas del occidente de Mexico [The meaning of insect puparia appearing on figurines from the West of Mexico]. Paper presented at the El Occidente de Mexico: Arquelogia, Historia y Medio Ambiente. Perspectivas Regionales. Acta del IV Coloquio de Occidentalistas, Universidad de Guadalajara, Mexico

Proctor HC (2001). *Megninia casuaricola* sp. n. (Acari: Analgidae), the first feather mite from a cassowary (Aves: Struthioniformes: Casuariidae). Aus J Entomol 40:335–341

Robinson WH (2005) Urban insects and arachnids, a handbook of urban entomology. Cambridge University Press, Cambridge

Schei MA, Hessen JO, Lund E (2002) House-dust mites and mattresses. Allergy 57:538–542

Schröder H, Klotzbach H, Oesterhelweg L, Gehl A, Püschel K (2001) Artenspektrum und zeitliches Auftreten von Insekten an Wohnungsleichen im Großraum Hamburg [Species diversity and temporal occurrence of insects on indoor corpses in Greater Hamburg]. Rechtsmedizin 11:59–63

Schroeder H, Klotzbach H, Oesterhelweg L, Puschel K (2002) Larder beetles (Coleoptera, Dermestidae) as an accelerating factor for decomposition of a human corpse. Forensic Sci Int 127:231–236

Schroeder H, Klotzbach H, Püschel K (2003) Insects' colonization of human corpses in warm and cold season. Legal Med 5:S372–S374

Schumann H (1990) Über das Vorkommen von Dipteren in Wohnräumen [The occurence of Diptera in living quarters]. Angewandte Parasitologie 31:131–141

Sidenius KE, Hallas TE, Brygge T, Poulsen LK, Mosbech H (2002) House dust mites and their allergens at selected locations in the homes of house dust mite-allergic patients. Clin Exp Allergy 32:1299–1304

Smeeton WMI, Koelmeyer TD, Holloway BA, Singh P (1984) Insects associated with exposed human corpses in Auckland, New Zealand. Med Sci Law 24:167–174

Soika G, Labanowski G (2003) Spider mites (Tetranychidae) recorded on ornamental trees and shrubs in nurseries. J Plant Prot Res 43:105–112

Thind BB, Clarke PG (2001) The occurrence of mites in cereal-based foods destined for human consumption and possible consequences of infestation. Exp Appl Acarol 25:203–215

Turchetto M, Vanin S (2004) Forensic evaluations on a crime case with monospecific necrophagous fly population infected by two parasitoid species. Aggrawal's Internet J Forensic Med Toxicol 5:12–18

Turner BD, Bishop J (1998) An analysis of the incidence of psocids in domestic kitchens: the PPFA 1997 household survey (What's bugging your kitchen). Environ Health J 106:310–314

Turner BD, Maude-Roxby H (1989) The prevalence of the booklouse *Liposcelus bostrychophilus* Badonnel (Liposcelidae, Psocoptera) in British domestic kitchens. Int Pest Control 31:93–97

van der Hoeven W, de Boer R, Bruin J (1992) The colonisation of new houses by house dust mites (Acari: Pyroglyphidae). Exp Appl Acarol 16:75–84

Voigt J (1965) Specific post-mortem changes produced by larder beetles. J Forensic Med 12:76–80

Walter DE, Proctor HC (1999) Mites – ecology, evolution and behaviour. CABI, Wallingford

Warner A, Bostrom S, Moller C, Kjellman NIM (1999) Mite fauna in the home and sensitivity to house-dust and storage mites. Allergy 54:681–690

Yoshikawa M, Bennett PH (1979) House dust mites in Columbus, Ohio, USA. Ohio J Sci 79:280–282

Chapter 7
Contemporary Precision, Bias and Accuracy of Minimum Post-Mortem Intervals Estimated Using Development of Carrion-Feeding Insects

Martin H. Villet, Cameron S. Richards, and John M. Midgley

7.1 Introduction

Medicocriminal forensic entomology focuses primarily on providing evidence of the amount of time that a corpse or carcass has been exposed to colonization by insects, which helps to estimate the post mortem interval (PMI). Specifically, the estimate is of a *minimum* post mortem interval (PMI_{min}), because death may occur a variable amount of time before colonization (Fig. 7.1); the maximum post mortem interval (PMI_{max}) is estimated using the time that the person was last seen alive. Forensic entomology derives the bulk of its evidence from two sources: the ecological succession of carrion insect communities and the development of immature insects (Byrd and Castner 2001; Catts and Haskel 1990; Smith 1986). This chapter is concerned with assessing the confidence that can be placed in the accuracy of estimates derived from insect development. (Schoenly et al. 1996) dealt with this theme in succession-based estimates of PMI_{min}.

A PMI_{min} based on development is estimated by calculating the age of the oldest immature insect on a corpse using various mathematical models (Grassberger and Reiter 2001; Higley and Haskell 2001; Reiter 1984). The most popular of these is the thermal accumulation model (Higley and Haskell 2001), which takes into account linear effects of temperature on species-specific growth rates to enhance its accuracy. Other models of even greater sophistication have been designed for even greater accuracy (Byrd and Allen 2001; Ieno et al. 2010). Models are commonly implemented on computers and their equations can generate a spurious level of precision – eight or more significant figures – that far exceeds the realities of the biology underlying them. For at least this reason, estimates of a PMI_{min} need to be

M.H. Villet (✉), C.S. Richards (✉) and J.M. Midgley (✉)
Southern African Forensic Entomology Research Laboratory, Department of Zoology and Entomology, Rhodes University, Grahamstown, 6140, South Africa
e-mail: m.villet@ru.ac.za; cam.richards@gmail.com; johnmidge@gmail.com

J. Amendt et al. (eds.), *Current Concepts in Forensic Entomology*,
DOI 10.1007/978-1-4020-9684-6_7,

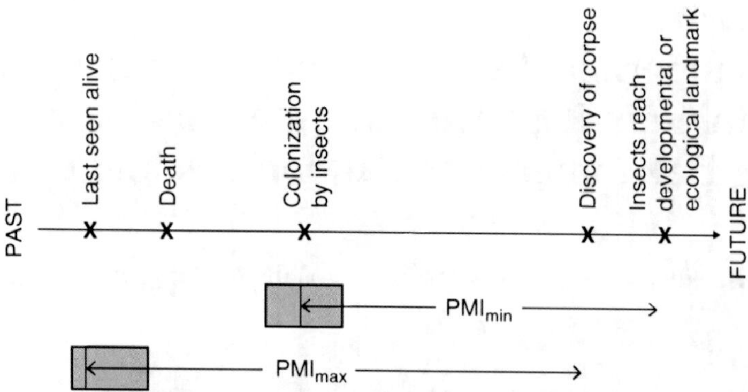

Fig. 7.1 Time line summarizing events in a generalized death investigation, and indicating maximum (PMI_{max}) and minimum (PMI_{min}) estimates of the post mortem interval. The grey boxes associated with each interval are their 'windows' of prediction, which may be asymmetrical. The accuracy of the estimates is reflected in how close the windows are to the actual events they estimate, while the precision of the estimates is reflected in the width of the window. The spacing of the events is arbitrary; some events could be practically simultaneous

framed by a 'window' of prediction (Fig. 7.1) that gives a measure of the precision of the estimate (Catts and Haskel 1990). This window can be estimated from the statistical confidence interval of the model, but it will probably need further qualification based on information about the biology of the relevant insects and the weather conditions around the putative date of oviposition.

Several authors have provided useful introductions to the growing field of factors confounding estimates of PMI_{min} (Campobasso et al. 2001; Catts 1992; Greenberg and Kunich 2002; Higley and Haskell 2001). The following discussion first examines the concepts of precision, bias and accuracy. It then reviews variables that affect the use of insect development to estimate a PMI_{min}, particularly in terms of their likelihood of occurrence and the magnitude of their effects on precision, bias and accuracy, and suggests ways to take them into account. The discussion concludes with some general comments about making estimates based on insect development.

7.2 Precision, Bias and Accuracy

Precision is the converse of variability and relates to the spread in the data (Fig. 7.2). It may arise from three sources: the size of the sample (*sample precision*), the resolution of the measuring equipment (*measurement precision*) or the inherent variability of the quantity being measured (*estimate precision*). Sample precision can be improved by enlarging a sample to the point where estimate precision predominates

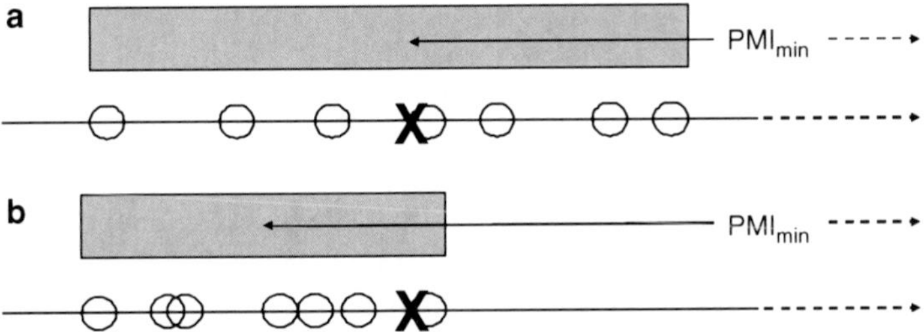

Fig. 7.2 Two time lines bearing circles representing estimates (based on individual observations) of a true time of death (**X**) and associated estimates of PMI_{min} with windows of prediction. Estimate A is less biased but also less precise than estimate B

(Sokal and Rohlf 2005). Unfortunately, the improvement is generally one of reciprocally diminishing returns because each additional observation contributes proportionately less to the overall estimate. Estimate precision is determined by the inherent variation in the variable, e.g. inherent variation in the lengths of maggots, and cannot be ameliorated without biasing the sample. Measurement precision depends on how coarsely or finely the measuring equipment is calibrated, and is therefore not influenced by the size of the sample. Ideally, the resolution of measurements should exceed the precision required of the estimate by at least one significant figure, i.e. one order of magnitude. This can be though of as setting the signal-to-noise ratio to at least 10:1. Measures of precision (or more strictly, imprecision) include the (sample) standard deviation (s) and the coefficient of variation (C.V.). The latter is a particularly useful measure because it is scaled as a percentage to be independent of the measuring units (Sokal and Rohlf 2005) and can therefore be used to compare the precision of variables measured in different units e.g. the masses and lengths of maggots. (Schoenly et al. 1996) used the inclusive range of estimates of PMI_{min}, which they termed PMI_{width}, as a simple measure of precision.

Bias is the difference between the mean of the measurements and the true value (Fig. 7.2), and may be negative (underestimation) or positive (overestimation). It too can arise from sampling, measurement or statistical estimation. *Sample bias* occurs when unrepresentative observations are collected e.g. when only the largest larvae on a corpse are sampled. Enlarging a sample will not reduce sampling bias unless more representative observations are included. *Measurement bias* arises from ill-calibrated equipment, and, like measurement precision, it remains even when sample sizes are increased. *Estimate bias* or systematic error is due to the method of estimation. For example, calculating the variance of a sample using the population variance (σ^2) gives a statistically biased estimate (and hence the 'n – 1' correction in the denominator of the formula for unbiased sample variance, s^2: (Sokal and Rohlf 2005)). Estimate bias usually decreases with increasing sampling intensity, e.g. the population and sample variances converge at large sample sizes. If one knows the true value being estimated (as can happen in an experiment), bias can be quantified by the mean difference between the estimates and the true value, sometimes termed the mean error (ME) or bias; unbiased estimates will have measures of bias that are very near zero. This measure of bias can be re-scaled to a percentage like the coefficient of variation by dividing it by the true value being estimated, with the same benefits.

Accuracy is the result of bias and precision combined (Fig. 7.2), and therefore cannot be directly decomposed into sources traceable purely to sampling, measurement or statistical estimation. If one knows the true value, the accuracy of data can be quantified, for instance by the mean squared error (which is algebraically equivalent to the sum of the variance and the square of the bias Casella and Berger 1990: 303), or the scaled mean absolute error.

There are other measures of precision, bias and accuracy. They should all be interpreted as suggestive, rather than prescriptive, assessments of the reliability or performance of an estimate because they do not always agree because of their differing underlying approaches and assumptions. Fortunately, sampling, measurement and statistical estimation of the PMI_{min} generally produce uncertainty that needs little more than qualitative assessment because of a suite of confounding variables that are discussed in the next section. The assessment of precision, bias and accuracy in measuring these variables, and their integration into predictive models, are major challenges facing contemporary forensics entomology.

7.3 Sources of Inaccuracy

7.3.1 Promptness of Colonisation

Corpses do not usually emit oviposition cues for insects immediately after death (Leblanc 2010), so there may be a delay before a corpse is colonised (Fig. 7.1), although this may not be true in cases of neglect or wounding. Even once suitable cues evolve, the activity of flies may be delayed by the inaccessibility of a corpse or inimical environmental conditions.

Flies may be very specific in the odour cues to which they will respond (Bänziger and Pape 2004). Wounds may be promptly infested if gravid female flies are in the area because suitable odour cues are available immediately. Some calliphorids and muscids can arrive within 1–3 h after death, and other calliphorids and sarcophagids a day or two later (Anderson and VanLaerhoven 1996; Hall and Doisy 1993; Rodriguez and Bass 1983). Eggs may be laid within 1 h of death, depending on species (Anderson and VanLaerhoven 1996; Hall and Doisy 1993; Shean et al. 1993). Silphid beetles may also arrive on the first day, and are suspected of breeding promptly (Midgley and Villet 2009b; Midgley et al. 2010). Odour cues may be modified by burning of a body, but the reported experimental effect varied from a delay of 3 days to arrival a day earlier than on control carcasses (Avila and Goff 1998; Campobasso et al. 2001; Catts and Goff 1992). The development of cues following drowning (Payne and King 1972), hanging and freezing may also be modified in ways that require simulation of particular cases.

In cases of neglect, various sores may become infested with fly larvae before death, a condition termed myiasis (Zumpt 1965). Myiasis occurs primarily in indoor cases because sufferers are generally helpless infant or elderly people or animals, and may set in hours to weeks before death (Goff et al. 1991b; Anderson and Huitson 2004; Benecke 2004; Benecke et al. 2004). Cases may be recognised by the nature of the sores, the presence of dressings, and the presence of species that are uncharacteristic of post-mortem infestations (Benecke 2004; Benecke and Lessig 2001; Benecke and Lessig 2002; Benecke et al. 2004). In some cases it is possible to estimate both the minimum period of neglect and the PMI_{min} because of the different species involved (Benecke and Lessig 2001; Benecke and Lessig 2002).

Access may be delayed by physical means such as burial (Gaudry 2010; Payne and King 1968; Turner and Wiltshire 1999; Weitzel 2005; Wyss et al. 2003), wrapping (Goff 1992; Lord 1990), or confinement in a building (Frost et al. 2010) or car (Voss et al. 2008). The delay caused by these factors is highly contingent on the details of the case. For instance, the accessibility of buried corpses is affected by soil characteristics (VanLaerhoven and Anderson 1996; VanLaerhoven and Anderson 1999), and attraction to a corpse indoors depends on open windows, chimneys, or ventilation blocks, and the size of the room. Simulations (Faucherre et al. 1999; Turner and Wiltshire 1999) are recommended where possible.

Apart from physical inaccessibility, the arrival of flies may be delayed by their physiological incapacity due to, for example, bad weather (Archer 2004; Bass 1997; Leclercq and Watrin 1973), snow (Wyss et al. 2003), or nightfall. In these situations, environmental conditions may be too cold or too hot for ectothermic insects to be active or to lay eggs. For instance, *Calliphora croceipalpis* Jaenicke has a minimum temperature threshold for muscular activity as much as 8°C lower than five *Chrysomya* species (Richards et al. 2009a). The degree of delay depends not only on the prevailing conditions, but also on the physiological tolerances of particular species, and for this reason they may be modelled using physiological and weather data, or a contentious case may be simulated (Faucherre et al. 1999). The modelling approach can be complicated by the ability of insects to bask, so that ambient temperatures are not necessarily representative of insects' body temperatures, leading to activity that would not be anticipated on the basis of ambient temperature alone. This complication is absent at night.

Nocturnal oviposition has proved contentious, some authors denying that it ever happens, thus facing the problem of proving a negative. While nocturnal oviposition is very unlikely under the outdoor nocturnal weather conditions characteristic of temperate regions, it definitely occurs under other conditions (Brown 2006; Faucherre et al. 1999; Kirkpatrick 2004; Singh and Bharti 2001; Singh and Bharti 2008; Williams et al. 2008), providing Popperian falsification of its non-occurrence. To circumvent the falsification problem in a particular investigation, one needs to show that the case's contingencies militate against nocturnal oviposition. Although significantly depressed by circadian rhythms, there is generally a low level of adult flight activity in the dark (Smith 1983; Smith 1987). At worst, the occurrence of nocturnal oviposition may affect an estimate of PMI_{min} by 10–14 h at most latitudes, with the inaccuracy being less when nocturnal oviposition is more likely (in warmer conditions, when nights are generally shorter). Again, physiological thresholds are relevant, and nocturnal oviposition appears to be more likely in cold-adapted species such as *Calliphora vicina* Robineau-Desvoidy (Brown 2006; Faucherre et al. 1999; Singh and Bharti 2001; Singh and Bharti 2008), and under indoor conditions. In addition, light intensity, ambient temperature, wind speed, oviposition site humidity and circadian rhythms are relevant to assessing the probability of an insect moving to and actually laying on a corpse (Pyza and Cymborowski 2001; Vogt and Woodburn 1985; Williams et al. 2008; Wooldridge et al. 2007). During a rainy period in July (winter) in Grahamstown, South Africa, *Chrysomya chloropyga* (Wiedemann) was found to fly to a carcass during breaks in the rain during the day and to shelter there overnight (Villet pers. obs.), so that the important consideration in some cases may not be whether insects can reach a corpse, but rather whether conditions are suitable for oviposition.

7.3.2 Precocious Development

Precocious egg development refers to the starting of embryonic development in a single egg in the common oviduct of a gravid female fly, before it is laid. This phenomenon, first mentioned by (Smith 1986), occurs when a fertilized female has sufficient time to convert a protein meal into mature eggs but, for several days, lacks opportunity to lay them. During this period, one egg will be pushed into the common oviduct, become fertilised as it passes the opening of the spermathecal duct, and begin developing to the point where it may even hatch almost as soon as it is laid (Wells and King 2001). Because it is commonly advised to use the largest larvae available (Fig. 7.3) to estimate the PMI_{min} (Amendt et al. 2007), precocious development will lead to overestimates.

This phenomenon is very common in sarcophagids (where many eggs may be involved, which is often misinterpreted as viviparity), but has also been reported in the calliphorids *Aldrichina grahami* (Aldrich), *Calliphora terraenovae* Macquart, *Calliphora nigribarbis* Vollenhoven, *C. vicina*, and perhaps also *Lucilia sericata* (Meigen) (Erzinçlioglu 1990; Wells and King 2001). It is not known if precocious eggs occur in beetles. (Wells and King 2001) reported that 34 (62%) of 55 wild, gravid females of *C. terraenovae* had a virtually mature embryo in their common oviduct, but this rate certainly varies with the availability of protein meals, the speed at which meals are metabolised to eggs, and oviposition opportunity, which in turn probably all vary seasonally.

Precociously developing larvae would lead to an overestimate of PMI_{min} that may be as long as the entire length of embryonic development in calliphorids, and perhaps longer in sarcophagids. This can be 7–323 h (0.5 – 13.5 days), depending on species and temperature (Davies and Ratcliffe 1994; Higley and Haskell 2001; Melvin 1934). (Wells and King 2001) suspected that calliphorid larvae that hatch within their mother are expelled, but sarcophagids appear to retain such larvae for some time.

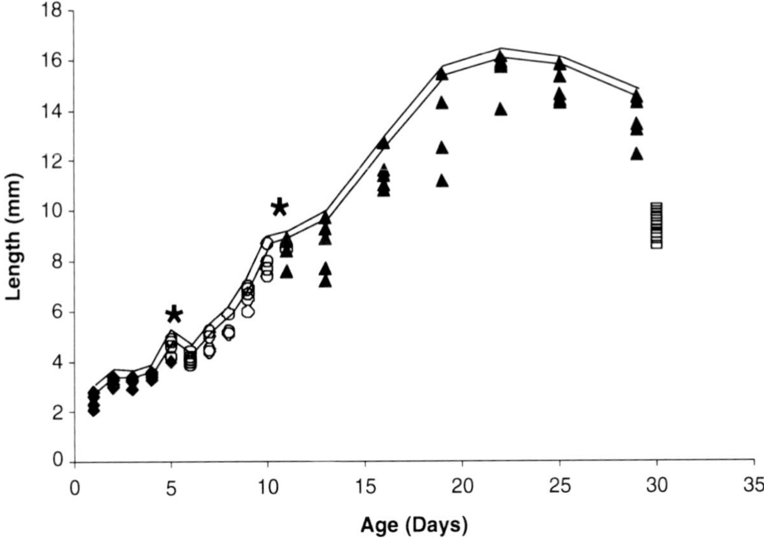

Fig. 7.3 Growth curve of Chrysomya chloropyga at 20°C, showing the typical decelerations in growth rate at ecdysis (*) and the decrease in size during the wandering phase. The two lines enclose the area where the largest larvae occur. Using larvae selected by such a narrow criterion to estimate the PMI_{min} allows more temporal precision in the estimate. Their measurements will have smaller relative errors, and they are less likely to be influenced by competition, thermoregulation, parasitoids or other disturbances (cf. Tarone and Foran 2006), although they may be precocious (Erzinçlioglu 1990; Wells and King 2001). Examination of the crop size will distinguish wandering larvae (to the right of the peak) from feeding larvae of the same size (to the left of the peak) (Greenberg and Kunich 2002; Reiter and Hajek 1984). ♦ – first instar larva; ○ – second instar larva; ▲ – third instar larva; □ – puparium

Managing this source of inaccuracy must take contingencies into account. (Erzinçlioglu 1990) pointed out that if several hundred female flies laid on a carcass, there may be substantial numbers of precociously developed larvae. He suggested that, '[i]nstead of using the age of the largest larva as the basis for time of death estimation, it would be more accurate to base the estimate on the age of the larval stage that is present in the largest numbers'. This approach assumes that female flies oviposit within a few hours of each other, and may be applicable when only one or two flies are involved. When larger numbers contribute to the maggot population, they may lay over a few days, and the modal age of the resulting larvae will underestimate the PMI_{min}. In these situations, an interim solution is to err conservatively and 'to subtract the time required for embryonic development when calculating the minimum possible age of a bluebottle larva' (Wells and King 2001). The assumption here is that the very largest larvae are likely to be precocious. The reference to a 'minimum possible age' refers to the lower bound of the window of prediction (Fig. 7.1); the upper bound would assume that the larvae were not precocious. The strategy that is appropriate to a particular case can be refined by estimating how many females contributed to the pool of larvae and empirically examining the occurrence (and maturity) of precocious eggs in a sample of gravid, conspecific, wild females trapped under comparable seasonal conditions.

7.3.3 Measuring Developmental Maturity

The age of immature insects can be estimated from their developmental maturity, represented by their physical growth or by the developmental milestones they have reached. Embryogenesis and metamorphosis involved little growth but extensive and complex changes in morphology that provide a rich suite of developmental events or 'milestones' for estimating the age of eggs (since fertilization, not necessarily since oviposition) (Bourel et al. 2003) and puparia (Denlinger and Zdárek 1994; Greenberg and Kunich 2002).

On the other hand, the larval stage is dominated by growth and larvae pass through fewer physical developmental events – essentially just two ecdyses. With the exception of preserved specimens in which the new spiracles and mouthparts can be seen through the cuticle (indicating imminent ecdysis), the timing of ecdysis is observed only in live larvae because it is a brief event. The age of dead larvae can be estimated from their mass, length or width (Day and Wallman 2006a; Donovan et al. 2006; Grassberger et al. 2003; Queiroz 1996). Mass is difficult to measure, especially if specimens have been preserved (because lipids may leave the body, and water may flux), and few data are available for this variable. Length and width can be measured quickly, accurately and precisely on live or preserved larvae using the geometrical micrometer devised by (Villet 2007) or a calibrated microscope. The measurement precision possible is about 0.1 mm, but the accuracy is compromised by changes in size due to the method of killing and preservation (Adams and Hall 2003; Midgley and Villet 2009a; Tantawi and Greenberg 1993). It has been suggested that width is more reproducible than length in fly larvae (Day and Wallman 2006a), but because the relative error will be greater using the same measuring equipment, the measurement precision is lower. Midgley and Villet (2009a) found that the variation produced by preservation in a beetle species is about 25–30% of live length, depending on which statistical method is used to summarise the variation, and that the distribution is often asymmetrical (i.e. biased) relative to live length. Sample precision therefore outweighs measurement precision by about an order of magnitude, especially for older larvae.

However, the estimate precision associated with size may be even more limiting. As in humans, conspecific larvae of the same size may be very different ages (Fig. 7.3), and larvae of the same age may differ considerably in size (Fig. 7.3), so that size is not precisely correlated to physiological or chronological age (Dadour et al. 2001; Richards et al. 2008). Larvae of the same age show coefficients of variation of 10–35% for length and mass, even when large samples are measured carefully. This can be ameliorated to some extent by selecting a biased sample of only the largest larvae to measure (Fig. 7.3; (Amendt et al. 2007)), but requires larger samples and the absence of precocious larvae. Size is therefore not favoured by some investigators, who use developmental milestones instead (Dadour et al. 2001; Gaudry et al. 2001). Given the extensive suite of developmental events in embryogenesis and metamorphosis, it is better to estimate the age of eggs and larvae from material preserved as soon after discovery as possible, because the passage of more time between collection and preservation will only multiply the uncertainties of estimation. On the other hand, if one wishes to use developmental events to estimate the age of larvae, they will have to be kept alive until they reach a recognisable event. However, since larvae are commonly presented as preserved material, it is likely that their age will have to be estimated from their length or width, taking into account the instar and the methods of killing and preservation. Ideally, if specimens are to be preserved in investigations and experiments, fly larvae should be killed in hot water (Amendt et al. 2007) and beetle larvae in ethanol (Midgley and Villet 2009a) and their length measured to maximise sample and measurement precision.

A recent development is the investigation of physiochemical, as opposed to physical, developmental processes, which promise greater temporal resolution in the larval stage. These include the ontogeny of cuticular hydrocarbons in eggs (Roux et al. 2008), larvae (Roux et al. 2008; Zhu et al. 2006), puparia (Roux et al. 2008; Ye et al. 2007; Zhu et al. 2007) and adults (Roux et al. 2008; Trabalon et al. 1992), and changes in pupal steroids (Gaudry et al. 2006). The age of adults can also be estimated by the accumulation of the red pigment pteridine in their eyes, the development of the ovaries (in females!) and even to some extent by the degree of wear on the wing margins (Hayes et al. 1998; Wall et al. 1991). The use of hydrocarbons and pteridine requires calibration for geographical (and other) variation, but the factors affecting the precision and bias of these estimates are largely unquantified; (Gaudry et al. 2006) discuss the measurement and interpretation of ecdysteroid levels.

7.3.4 Maggot-Generated Heat

Maggots are gregarious and tens of thousands may inhabit a carcass simultaneously. Their collective metabolism generates enough heat to raise their temperatures and that of parts of the carcass to 45–52°C in sheep, dogs, pigs and humans, even in cold weather (Anderson and VanLaerhoven 1996; Deonier 1940; Marchenko 1988; O'Flynn 1983; Payne 1965; Reed 1958; Richards and Goff 1997; Waterhouse DF 1947). This may be 25°C above the type of ambient air temperatures measured at weather stations, and even above the temperature of a standard black body in full sun (Fig. 7.4), and represents a significant measurement bias. Because the rate of development of ectothermic larvae is correlated with the microenvironmental temperatures they experience, the influence of maggot-generate heat may need careful consideration when estimating a PMI_{min}.

Several important points about the bias caused by maggot-generated heat are illustrated in Fig. 7.4. On the temporal scale of the life cycle, it becomes increasingly

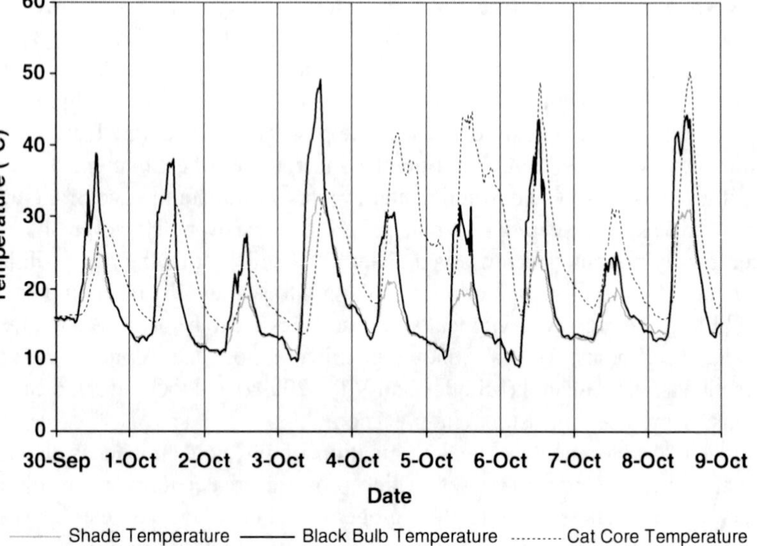

Fig. 7.4 Graph illustrating heat generated by maggots of Chrysomya chloropyga in the carcass of a ginger cat in Grahamstown, South Africa, that died on 29 September. The dotted line is the temperature in the core of the abdominal cavity; the upper continuous black line is the temperature of air inside a standard black body (a hollow copper sphere, 150 mm in diameter, spray-painted matt black) placed in full sun beside the carcass, 0.3 m above ground; and the lower continuous grey line is the air temperature in shade 1 m from the carcass, 1 m above ground. Temperatures were recorded every 30 min using a BAT 12 digital data logger and three calibrated copper/constantan thermocouples that were fixed in place. The heat generated by maggots in the abdomen warmed it 12°C above the temperature of the black body by Day 5, by 20°C on the night of Day 6; and 24°C warmer than shade temperature on Days 6 and 7. On Day 9 the carcass temperature reached 50°C. Cooling of the carcass lagged behind air temperatures by about 5 h, even on Days 1–4 when the maggots were small or not active around the thermocouple

important as larvae grow (Hanski 1976), especially once the volume of aggregations of maggots exceeds 20 cm^3 (Slone and Gruner 2007). It is also not consistent over the temporal scale of a day because solar radiation makes a contribution to elevating metabolism beyond the effect of maggot-generated heat alone. Solar heating may dissipate only slowly at night (Fig. 7.4) due to the thermal inertia of corpses. Nor is maggot-generated heat spatially uniformly distributed in a corpse; it is focussed where the maggots are currently feeding. To compound the problem, maggots thermoregulate to some degree (see days 7 and 9 in Fig. 7.4) by moving around (and probably also by evaporative cooling using saliva and urine), and almost certainly do not spend their entire larval life in the hottest part of the carcass (Slone and Gruner 2007). In fact, maggots fare best at temperatures nearer 25 – 30°C (Hanski 1976; Richards et al. 2008), and may be physiologically compromised by living at higher temperatures for long periods (e.g. (Richards et al. 2008)), so they can be expected to be 'sensible' about thermoregulating. *Chrysomya albiceps* (Wiedemann)

Fig. 7.5 Segregation of larvae of *Chrysomya albiceps* (dark maggots on left; temperature 33.2°C) and *C. marginalis* (whitish maggots on right; temperature 40.8°C) on a warthog carcass. Similar differentiation has been reported between *C. rufifacies* and *C. megacephala* (Greenberg and Kunich 2002) (photograph by C.S. Richards)

characteristically inhabits the periphery of carrion, while *Chrysomya marginalis* (Wiedemann) inhabits the core (Fig. 7.5), apparently because the latter has a higher tolerance for maggot-generated heat (Richards et al. 2009a). In early instars and small clumps of more mature maggots, maggot-generated heat may be a trivial issue, and it seems unlikely to affect beetle larvae significantly, but its true prevalence needs refined quantification (Slone and Gruner 2007). Wandering larvae are not affected by this problem, and buried pupae may be at cooler and more stable temperatures than ambient conditions.

The magnitude of the biasing effect of maggot-generated heat is potentially large. In Australia, an excess temperature of about 27°C has been recorded in dead sheep (Deonier 1940; Waterhouse DF 1947), and this is not unusual (Fig. 7.4). If one was making a hypothetical estimate of PMI_{min} using a routine thermal accumulation model (Higley and Haskell 2001), and assuming a (conceptually fictional) lower developmental threshold (D_0, also called the base temperature, T_b) of about 10°C (a rough compromise between the values characteristic of *Calliphora* and *Chrysomya*: (Higley and Haskell 2001)), average ambient temperatures of 20°C (which is generously high for many areas) and an excess temperature of 20°C (i.e. a carcass temperature of only 40°C), the calculations would fall short by 720/240 h °C or 300% a day. This worst-case scenario is restricted to the later part of the larval phase, and no consideration has been given to thermoregulation; (Marchenko 1988) has suggested that in practice the estimate might fall short of the true development time by 100%.

Ideally, the temperatures of maggot aggregations should be measured before the corpse is removed from the site of discovery and compared to ambient and black-body temperatures to account for maggot-generated heat and solar radiation in an investigation. The temperature of a maggot mass is a biased estimate of maggot-generated heat because is includes solar radiation (Fig. 7.4). One may then design a correction to retrospective weather station data (see Fig. 7.4), or assume that the maximum temperature routinely reached 45°C (cf. Anderson and VanLaerhoven 1996; Deonier 1940; Marchenko 1988; O'Flynn 1983; Payne 1965; Reed 1958; Richards and Goff 1997; Waterhouse DF 1947). However, while the maggots found near the warmest parts of the corpse may have grown fastest, they may also be stunted by competition and thermal stress, which makes their size harder to interpret. However, thermoregulation (by migration or evaporative cooling) may be practically impossible to account for. Obvious ways around this are to sample larvae from smaller aggregations that generate less heat, and to focus on species like *C. albiceps* (Fig. 7.4) that inhabit thermal niches on the corpse that are more similar to ambient conditions. One might also use the development of species that inhabit different thermal niches to cross-validate estimates. This emphasises the value of the recommended standard procedure of recording the location of samples on a corpse (Amendt et al. 2007).

Even in thermostat-regulated incubators, maggots may not be at the set temperature (Nabity et al. 2007) because of maggot-generated heat. It is recommended that maggots' temperatures should be measured directly even in such controlled environments to improve the measurement precision of models.

7.3.5 Diet

Blowfly larvae have an assimilation efficiency of about 80% (Hanski 1976), and mature larvae weigh 75–100 mg, so each one needs at least 90–120 mg of food to complete development. There is circumstantial evidence that this figure is in fact about 200 mg of pig's liver, 500 mg of pig's muscle or 1,000 mg of pig's brain in *Calliphora vomitoria* (Linnaeus) (Ireland and Turner 2005). However, *C. vicina* and *L. sericata* developed best on pig's lung (and brain or heart), and worst on pig's liver (Clark et al. 2006; Kaneshrajah and Turner 2004). *Calliphora augur* (Fabricius) and *Lucilia cuprina* (Wiedemann) grew significantly better on sheep's meat and brain than on sheep's liver (Day and Wallman 2006b). *Chrysomya albiceps* grew larger on horse meat than on sardines (Cardoso Ribeiro and Veira Milward-de-Azevedo 1997), while *L. sericata* grew faster and larger on pig than on cow tissues (Clark et al. 2006). These studies imply that the nutritive value of different tissues and different animals varies, that different insects respond differently to that variation. This is probably also true of human tissues, but the magnitude of the effect is unknown.

There are reports of variations in mean developmental time of up to 2 day (out of 6–8 day) at 20°C (Ireland and Turner 2005; Kaneshrajah and Turner 2004) and 30 h (out of 70–120 h) at 25°C (Clark et al. 2006), and in mature larval length of 2 mm (out of 12–16 mm) due to dietary differences (Clark et al. 2006; Day and Wallman 2006b; Ireland and Turner 2005; Kaneshrajah and Turner 2004). This means that variation in diet can limit overall precision to 12–43% of the estimated PMI_{min}.

Two ways of managing these disparities are obvious: develop substrate-specific developmental models, or be selective about which tissues are sampled for insects. In either case, there is a need to formulate developmental models for human tissues, and to test which animals' tissues are the best substitutes. The effects of diet reinforce the need to record the location of samples on a corpse as a standard procedure (Amendt et al. 2007).

7.3.6 *Drugs*

Drugs in an insect's food may cause a variety of disruptions to its development ((Introna et al. 2001); Lopes de Carvalho 2010), accelerating or extending the duration of the larval and/or puparial stages.

The effects of drugs may be species-specific. For instance, heroin (and/or its metabolite, morphine) was associated with accelerated larval growth in *Sarcophaga peregrina* (Robineau-Desvoidy) (Goff et al. 1991a), but morphine retarded larval growth in *Lucilia sericata* Meigen and *C. vicina* (Bourel et al. 1999). Diazepam (and/or its metabolites, nordiazepam and oxazepam) accelerated larval growth and puparial metamorphosis in *C. albiceps* and *Chrysomya putoria* (Wiedemann) (Lopes de Carvalho et al. 2001), while nordiazepam (and oxazepam) administered at lower concentrations had no effect on *Calliphora vicina* (Pien et al. 2004). The dose that larvae experience is complicated by drug tropism, and experimental protocols need to take into account metabolism, sequestration and excretion of the administered drug over time (Introna et al. 2001), and the effects of the metabolites. Perhaps for these reasons, some drugs' effects are restricted to only parts of development (Kharbouche et al. 2008; O'Brien and Turner 2004). It is clear that the effects of one drug do not predict those of another (Sadler et al. 1997a; Sadler et al. 1997b; Sadler et al. 1997c), but whether a drug's effects are widespread or consistent across species, whether larval responses have a phylogenetic component, and whether they are affected by temperature all need to be clarified.

The effects of drugs may bias the mean duration of development to eclosion by as much as 109 h (an underestimate of 40%) at 27°C (Lopes de Carvalho et al. 2001, 2010), but are generally much less extreme (Introna et al. 2001). The initial effects of some drugs pass off later in development (Kharbouche et al. 2008; O'Brien and Turner 2004) and development returns to schedule. Few studies have reported measures of precision, but those that do (e.g. (Lopes de Carvalho et al. 2001; Musvasva et al. 2001)) indicate a slight increase in the coefficients of variation compared to control treatments, which is expected from the addition of an extra experimental variable (the drug effect). Most entomotoxicological experiments are done under temperatures in the region of 22–27°C, which is lower than the temperatures of many natural maggot aggregations, so that the absolute error may be smaller under natural conditions and precision may improve, but it is probable that the relative error changes little. Some studies were done at slightly variable temperatures in the 22–27°C range, where blowfly development is unfortunately most sensitive to changes in temperature, and it is not clear how this affected their findings.

From a practical perspective, a direct correlation between the drug content of larvae or puparia and that of their food is generally poor or absent (Campobasso et al. 2004; Pien et al. 2004; Sadler et al. 1997a; Sadler et al. 1997b; Sadler et al. 1997c; Tracqui et al. 2004), although not always (Kharbouche et al. 2008). For this reason, when an effect is dose-dependent and the drug is not sequestered, it is currently difficult to work backwards from knowing the concentrations of drugs in larvae to applying a correction factor to the PMI_{min}. Even if species-specific pharmacological models for each drug become available that take into account the metabolism, sequestration or excretion of drugs by larvae, and the effects of their metabolites, temperature and their interactions, they may be difficult to apply because the duration of exposure (i.e. the age of the insect) is a key unknown variable that would have to be estimated from the size of the insect. Fortunately, sophisticated statistical models have been developed (Ieno et al. 2010) that can help to improve the precision of such models and take into account the non-linearity of pharmacodynamics. The drug concentrations in the food in such situations may be irrelevant if there is significant tropism and larval migration within the corpse. However, not all effects are dose-dependent (Kharbouche et al. 2008; O'Brien and Turner 2004), and in these cases the corrections are simpler to make. In casework and experiments, sample sizes of 30 or more of the largest larvae should be measured to improve sample precision (Fig. 7.3). It is also recommended that research is directed to distinguishing the effects of drug tropism from the effects of the tissue type itself (Cardoso Ribeiro and Veira Milward-de-Azevedo 1997; Clark et al. 2006; Ireland and Turner 2005; Kaneshrajah and Turner 2004), and that particular attention is given to controlling for the confounding effects of temperature. The validation of toxicological methods is discussed in much greater detail by (Peters et al. 2007).

7.3.7 Interspecific Competition

Maggots predigest their food with secretions (Mackerras and Freney 1933), and this process is more effective when more maggots are involved (dos Reis et al. 1999). Maggots grow faster in groups, probably due to maggot-generated heat and improved pre-digestion, which explains their gregariousness. However, escalating numbers lead to intraspecific competition, increased mortality, earlier emigration and pupariation and smaller-bodied adults (dos Reis et al. 1999; Ireland and Turner 2005; Saunders and Bee 1995; Shiao and Yeh 2008; Smith and Wall 1997). Interspecific competition has similar effects (dos Reis et al. 1999; Shiao and Yeh 2008; Smith and Wall 1997). These conditions lead to the anomaly that the PMI_{min} will be underestimated if based on the size of the dwarfed larvae, but overestimated if based on the onset of emigration.

Competition, especially amongst conspecifics, is pervasive. It sets in at low densities and increases fairly linearly with increasing maggot densities (dos Reis et al. 1999; Ireland and Turner 2005; Shiao and Yeh 2008). Because the assimilation efficiency of maggot may vary between tissues (Ireland and Turner 2005), competition will be exacerbated on certain parts of a corpse or carcass.

Maggots in their third instar pass through a brief phase of obligate feeding and then a much longer phase of facultative feeding (Denlinger and Zdárek 1994), which allows feeding to stop well before the larvae are fully grown. Larval development can end about 25 h early in *Chrysomya megacephala* (Fabricius) and 34 h early in *Chrysomya rufifacies* (Macquart) under interspecific competition at high densities; when these species compete with one another, development can end about 54 h early (Shiao and Yeh 2008). Similar results were found by (Kheirallah et al. 2007) using *L. sericata* and *C. albiceps*. Generally, the effects of interspecific competition in carrion are reported to be more extreme than those of intraspecific competition. The effect of competition on size has generally been quantified as the mass of the dried adult (e.g. (Ireland and Turner 2005; Shiao and Yeh 2008)), so that the magnitude of the effect on larval size, and therefore on the PMI_{min}, has not been established, but is expected to not exceed a few millimetres of length, which translates into a day or two's growth. It is not known how competition affects the duration of the pupal stage.

This effect is one of the elephants in the room for estimating PMI_{min}: until recently, it was recognised as a problem but not discussed in practical terms. One might manage it by comparing estimates of PMI_{min} based on size and on onset of wandering, and averaging the results, which will have opposing biases, although it is not known whether they have comparable magnitudes. Unfortunately, this solution is not possible if the specimens have already started to pupariate or are preserved, because it is not possible to rear them to a developmental milestone.

7.3.8 Chilling and Diapause

Diapause is a phase of arrested growth or metamorphosis that allows individuals to synchronise their development with environmental conditions. It is primarily triggered by chilling (Cragg & Cole 1952; Denlinger 1978; Bell 1994; Anderson and VanLaerhoven 1996) and shorter day lengths experienced by the larvae after hatching or by their mother before oviposition (Saunders 1987). When diapause occurs, it is usually late in the final instar of calliphorids and in the pupal phase of sarcophagids (Denlinger 1978; Greenberg and Kunich 2002); data for carcass beetles is lacking. It is genetically regulated and more prevalent in autumn in populations from higher latitudes, affecting all or only part of a cohort (Saunders and Cymborowski 2003), so it is not a pervasive source of bias. Malnourished maggots can evade diapause (Saunders 1987). Diapause may last months and lead to serious underestimates of a PMI_{min}.

Cold spells near the developmental threshold temperature (D_0, or base temperature) that do not trigger diapause may still disrupt development in different ways in different species (Johl and Anderson 1996; Myskowiak and Doums 2002; Ames and Turner 2003), which is a concern if live larvae are chilled, for instance in winter weather, through mortuary refrigeration (Huntington et al. 2007), or in an attempt to stop growth during transit to a forensic entomologist. Disruption of development by up to 56 h (907 h°C) have been measured experimentally in *Protophormia terraenovae* (Robineau-Desvoidy) (Myskowiak and Doums 2002), and leading to both under- or overestimates of PMI_{min}, depending on which life stage is chilled. In one investigation, an inappropriate assumption of chilling led to an estimate of PMI_{min} that predated the last time the person was seen alive by a month (Ames and Turner 2003).

(Greenberg and Kunich 2002) advise that, 'luckily for the entomologist, diapausing larvae and pupae are rarely the key forensic indicators.' When it is suspected to have occurred (e.g. when PMI_{min} exceeds PMI_{max}), alternative sources of evidence should be sought. It is also recommended that, at least until more is known about the response of life stage of each species, live insects should not be chilled after collection to stop development while they are in transit to the laboratory (Myskowiak and Doums 2002). The effects of chilling appear to be highly repeatable ((Myskowiak and Doums 2002); Ames and Turner 2003), so that once suitable data are available, reliable corrections to estimates are a possibility. It is likely that precision will be much lower than usual.

7.3.9 The Wandering Phase

The onset of migration from the food resource follows the cessation of feeding and the consequent emptying of the crop, and fly larvae may spend several days crawling metres before they find a satisfactory place to pupariate (Gomes et al. 2006; Greenberg 1990). While larvae are still actively migrating, the larvae or puparia furthest from the food source should be amongst the oldest present, but paradoxically, smaller (and therefore apparently younger) larvae tend to be found furthest from the food (Gomes et al. 2006). This is because larvae lose mass through exertion (Gomes et al. 2006), dehydration, and the gradual depletion of their crops' contents (Denlinger and Zdárek 1994). Additionally, at least in *Calliphora vicina*, extended wandering adds disproportionately to duration of the puparial stage (Arnott and Turner 2008).

Larvae in laboratory settings do not generally have to go far to find suitable pupariation sites as they do in the field, and so that laboratory measurements of the duration of this stage are biased relative to natural situations (Arnott and Turner 2008). Migration in the field may be extended if larvae have difficulty finding refuges, and particularly if larvae of predatory species such as *C. rufifacies* or *C. albiceps* are also present (Gomes et al. 2006; Greenberg and Kunich 2002). Larvae that have settled down and contracted may resume wandering if they are disturbed, for instance by other larvae, so that laboratory results can be both biased and unrepresentatively precise.

Wandering lasts about 150–400 h °C in *C. albiceps* (Richards et al. 2008) and 600–900 h °C in *C. chloropyga* (Richards and Villet 2008; Richards et al. 2009b) under laboratory conditions, a level of imprecision that is a substantial proportion of the mean. Except in a few species like *C. vicina*, migration is seldom protracted by more than 3 or 4 days (Arnott and Turner 2008; Gomes et al. 2006; Greenberg 1990), after which larvae tend to die. In *C. vicina*, if the wandering phase exceeded a threshold of 5 h (at 21°C), the physiological duration of the puparial stage increased (i.e. was biased) by roughly 700–1,500 h °C (Arnott and Turner 2008).

Because larvae and puparia of different species and different sizes are found in spatial patterns around a food source, (Gomes et al. 2006) suggested a protocol for sampling them in the area around corpses. Although the shrinkage of the crop as it empties after feeding ends is not a smooth process (Greenberg and Kunich 2002; Reiter and Hajek 1984), the associated decline in body length and mass is (Fig. 7.3), and both are useful sources of estimates of PMI_{min}. The slope of the relationship relating size to age is rather shallow (Fig. 7.3), so that the estimates always have poor estimate precision due to the natural variation amongst larvae, but parts of the curve for crop emptying are steeper and thus provide more precision (Greenberg and Kunich 2002). Nonetheless, these processes do not continue indefinitely, so other signs of additional bias due to extended wandering need to be sought in the natural history of specific cases and taken into account.

7.3.10 Circadian Rhythms

Circadian rhythms are clock-like cycles in physiological processes, and therefore in behaviour, that arise from an endogenous, temperature-compensated genetic oscillator that is naturally entrained by light and temperature cycles each day (Johnsson and Engelmann 2008; Saunders 2008; Saunders et al. 2002). They effectively allow an organism to predict cyclic environmental phenomena, giving them a means to prepare for and synchronise with predictable exogenous conditions, and often have periods that are not exactly 24 h long. They can be reset by exposure to light or heat (Johnsson and Engelmann 2008; Joplin and Moore 1999; Saunders et al. 2002), which leads to pauses or jumps in the tick of the developmental clock that needs to be considered when estimating PMI_{min}.

In calliphorids, circadian rhythms affect processes like oviposition (Tessmer et al. 1995), larval growth (Saunders 1972), onset of wandering (Kocárek 2001; Smith et al. 1981), eclosion (Joplin and Moore 1999; Smith 1987), adult activity (Smith 1983; Smith 1987) and even the termination of larval diapause (Tachibana and Numata 2004). Furthermore, eclosion can be entrained by light-dark cycles experienced prior to pupariation, even if the puparia are subsequently kept under constant lighting (Joplin and Moore 1999), for instance by burying. In short, the influence of circadian rhythms in natural settings is practically inevitable. In air-conditioned indoor situations where lights are left on, rhythms may become uncoordinated. When *Phormia regina* (Meigen) was reared under constant lighting, so that its circadian rhythm was allowed to run freely, variation (imprecision) in overall adult developmental time increased and there was a significant delay (bias) in development compared with cyclic light (Nabity et al. 2007), indicating that the cycle length was greater than 24 h.

The peaks in behaviour produced by circadian rhythms may be sharp or broad (Kocárek 2001; Smith et al. 1981), but offer no clue to the true duration of the preceding developmental stage. Specimens may wait the whole daylight period before wandering, leading to a maximum imprecision of about 12 h if peaks are broad, and more if they are sharp (Smith et al. 1981). This distortion of the developmental clock probably sets the limit to precision for most estimates of PMI_{min} and introduced a variable bias that is not easy to estimate.

Little can be done about adjusting a PMI_{min} if there is evidence that circadian rhythms have been in play. They are a prime reason for defining an asymmetrical window of prediction (Fig. 7.1) for an estimate. The window can be adjusted by other considerations such as weather conditions on the estimated day(s) of oviposition.

7.3.11 Modelling

There is a variety of graphical, mathematical and physical models for estimating the age of insects from their development, taking temperature into account. Isomegalen and isomorphen diagrams (Grassberger and Reiter 2001; Midgley and Villet 2009b; Reiter 1984; Richards et al. 2009b) provide a simple, empirical graphical model (Fig. 7.6). Thermal accumulation models use a linear regression model as the basis for a simple, mathematical way of accumulating development at varying temperatures across an insect's whole developmental period (Higley and Haskell 2001; Higley et al. 1986). More sophisticated statistical models are becoming available (Byrd and Allen 2001; (Ieno et al. 2010; Oliveira-Costa et al. 2010)). The most popular physical model is a dead pig, which has been used with success (Schoenly et al. 1991; Schoenly et al. 1996; Schoenly et al. 2005; Schoenly et al. 2007; Shahid et al. 2003), as has minced meat (Faucherre et al. 1999; Turner and Wiltshire 1999).

Potential estimate biases in models arise from their assumptions, which all models have (Higley et al. 1986), and the degree to which a model's assumptions are violated biases its accuracy. For example, thermal accumulation models that assume a straight line relationship between growth rate and temperature perform poorly when temperatures fluctuate predominantly to one side of the optimal developmental temperature because of a systematic bias known as the Kauffman effect (Higley and Haskell 2001; Worner 1992); most current models assume that air temperature in the shade is representative of the temperature experienced by larvae; and dead animals are assumed to be representative of corpses. Only in some cases have these assumptions been validated as unbiased (e.g. (Schoenly et al. 2005; Schoenly et al. 2007; Tarone and Foran 2006; VanLaerhoven 2008)). The omission of a variable from a model is formally equivalent to assuming that it carries no weight.

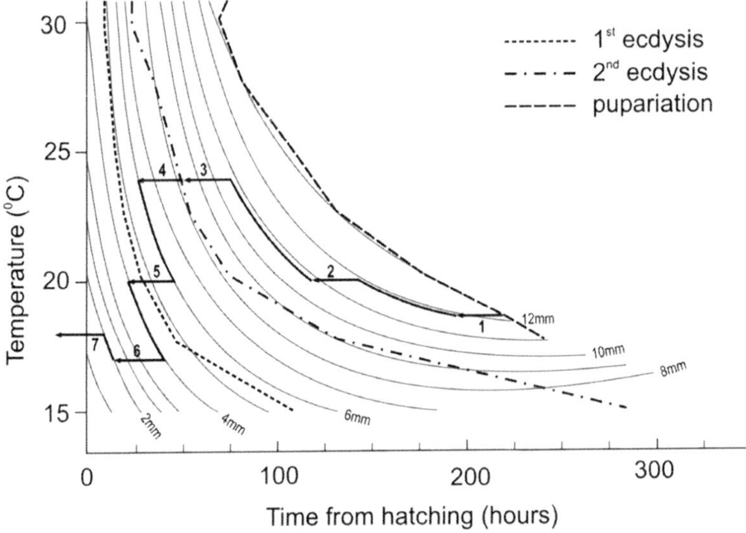

Fig. 7.6 A combined isomegalen and isomorphen diagram, illustrating how to estimate the age of a larvae assuming that it measured 12.2 mm in length just before pupariation and experienced sequential daily environmental mean temperatures (going backwards from the date of discovery) of 19°C, 20°C, 24°C, 24°C, 20°C, 17°C, and 18°C. The length of each horizontal arrow is 24 h, and the ends of the arrows are joined by starting on the right and moving along the length contour to each new daily temperature. In this example, the PMI_{min} is estimated to be 6.4 days. If retrospective temperature data are available at a better temporal resolution (e.g. hourly), the length of the arrows can be adjusted to this precision

Sometimes it is not even clear if variables are 'real': Ames and Turner (2003) suggested that 'The whole concept of D[evelopmental] T[hreshold] T[emperature] and its input to the [thermal summation model] calculation needs an urgent review', but Trudgill et al. (2005) gave it theoretical support. At present many experimentally determined errors are reported in chronological time (usually hours) rather than in physiological time (degree-hours). While this can be expected to change as more data are gathered, those same data may fuel the creation of statistical Mixed Models (Ieno et al. 2010) that estimate time directly, without the problematic incorporation of a developmental threshold temperature.

Because each model has its own inventory of assumptions, the wide variety of models of development precludes a more detailed discussion of their estimate biases here. Users should make the assumptions of their preferred model explicit and validate them as rigorously as possible, consulting a mathematician or biometrician if necessary.

Sampling bias occurs in models when, for example, parameter values for one species are substituted for unknown values of another species. This cannot be done reliably even when the species are very close relatives such as *Calliphora vicina* and *C. vomitoria* (Ireland and Turner 2005; Kaneshrajah and Turner 2004) or

Chrysomya chloropyga and *C. putoria* (Richards et al. 2009b). It has also proved unwise to substitute data derived from different geographical areas, although some of the geographical effects might be predictable (Richards and Villet 2009; Richards et al. 2008).

The estimate precision of models can also be manipulated to some degree. For instance, more precise thermal summation parameters (K and D_0) can be derived using the regression model developed by (Ikemoto and Takai 2000) than by more traditional methods (Higley and Haskell 2001; Trudgill et al. 2005). It is also important to avoid rounding off values prematurely in the computations. Again, it is best to consult a statistician about how to improve the estimate precision of a particular model.

The measurement precision of models of development often hinges on the ensuring that the process is sampled often enough in time (Richards and Villet 2008; Richards and Villet 2009). Measurements that exceed the precision of the required estimate by one significant figure (i.e. an order of magnitude) are desirable, and can be achieved by ensuring that the resolution of the measurements is about 10% of the likely value of the estimate (Richards and Villet 2008; Richards and Villet 2009). For example, if a developmental process takes 30 h on average, samples should be taken every 3 h.

Sampling precision can be improved by enlarging a sample, although the improvement is generally inversely proportional to the square root of the sample size. General Additive Models require more replication than regression models but making better use of it (Ieno et al. 2010). As a general guideline, samples of about 30 individuals representing each combination of factors in an experiment is highly desirable (due in part to the Central Limit Theorem), but not often practical.

Guidelines for managing sample and measurement precision in experimental calibrations of development for model building has been published by (Richards and Villet 2008). The same guidelines should be applied to sampling in case studies to keep levels of precision commensurate with experimental data.

7.4 Discussion

Many variables that influence the rate of development of insects are not usually incorporated into the commonly-applied models for estimating PMI_{min}, largely because of the paucity of relevant data. At present forensic entomologists generally estimate a PMI_{min} using a reliable, simplified model and then account for confounding variables using qualitative estimates of their bias and precision (Table 7.1) derived from peer-reviewed scientific literature, ad hoc experimentation and their experience.

Fortunately, not all of the confounding variables are equally prevalent or equally significant (Table 7.1). For example, inaccuracy arising from using incubator temperature setting instead of direct measurement of maggots' temperatures overshadows any affect that lighting has on development (Nabity et al. 2007).

Table 7.1 Reported qualitative worst-case estimates of the bias and precision in PMI_{min} associated with different confounding variables affecting insect development in corpses, ranked by the magnitude of the error. The effects are not necessarily additive, and can be ameliorated under some conditions

Source	Bias	Minimum precision
Modelling	Variable	Variable
Circadian Rhythms	Variable	Hours
Light-induced variability	Variable	Hours
Precocious development	Overestimated (temperature-dependent)	Hours, up to duration of embryonic development
Preservation and measurement	Variable	Hours to days
Promptness of arrival	Underestimated in general	Hours to days
	Overestimated with myiasis	Days to weeks
Diet	Variable (tissue- and species-dependant)	Days
Drugs	Variable (drug- and species-dependant)	Days
Competition	Underestimated by size	Days
	Overestimated by developmental events	Days
Wandering	Underestimated	Days
Maggot-generated heat	Overestimated	Days
Developmental plasticity	Variable	Days
Chilling	Underestimated	Days
	Overestimated	Weeks
Diapause	Underestimated	Months

Ruling out some variables is easier in earlier stages of development because they have not had time to manifest themselves, while variation has also had less time to accumulate, which simplifies estimation. The disadvantage of working with early stages is that an error of, say, 24 h may represent a *relative error* of 100% in the first day of development, but only 4.5% by the time of eclosion. Thus, precocious development is most problematic in young larvae (because of the large relative error), while maggot-generated heat becomes an increasingly serious problem as larvae age. This emphasises the collection of data at an appropriate resolution to minimise relative error (Richards and Villet 2008; Richards and Villet 2009) and processing entomological evidence as soon after its discovery as possible.

The sources of error are not necessarily additive; lesser errors may be subsumed within greater ones. In particular, the resetting of the developmental clock by circadian rhythms can effectively resynchronise development at several points in insects' life cycles, helping to eradicate the accumulating variation due to other factors. Furthermore, errors counteract one another in some cases because of their opposing biases (Table 7.1). Maggot-generated heat and competition may have opposing effects on growth rates as maggot densities increase, and errors in the estimation of the parameters of thermal accumulation models tend to cancel out because the two model parameters (K and D_0) are inversely correlated (Trudgill et al. 2005).

There is scope for research on the interaction of confounding variables that operate on comparable time scales.

The complexity of integrating all of the parameters discussed in this chapter into an accurate mathematical model is daunting; assessing the interactions amongst their varying precisions is even more of a Gordian knot. However, it is undue cause for gloom. First, akin to Zeno's paradoxes, there is a finite envelope of possibility for diapause-free insect development, outlined quite literally in isomorphen diagrams (Grassberger and Reiter 2001; Midgley and Villet 2009b; Richards and Villet 2009; Richards et al. 2008; Richards et al. 2009b). In many worst-case situations, one can still estimate bounds on a PMI_{min}. Second, the physical model provided by dead pigs achieves all of this integration without needing as much parameterization (Turner and Wiltshire 1999). Pigs provide simulations of ecological succession that appear to be both repeatable and representative of the processes on human corpses (Schoenly et al. 1991; Schoenly et al. 1996; Schoenly et al. 2005; Schoenly et al. 2007; Shahid et al. 2003), and can very probably serve the same role in providing a valid model for development. Even simulations using minced meat can provide realistic estimates (Faucherre et al. 1999). Third, statistical models are becoming increasingly sophisticated (Ieno et al. 2010), and await suitable data to test their capabilities (Richards and Villet 2008; Richards et al. 2008; VanLaerhoven 2008). Even given these reasons for optimism, it is likely that the inherent variation (or estimate precision) of insect development will limit the accuracy of estimates of PMI_{min} to about 5–15% of the true elapsed time. Whether this is satisfactory will depend on the contingencies of the particular case.

Finally, there is great scope for the development and validation of standard methods and best practices (Amendt et al. 2007; Peters et al. 2007). Some progress has been made in this direction in terms of rearing techniques (e.g. (Ireland and Turner 2005; Nabity et al. 2007; Tarone and Foran 2006)), entomotoxicological validation (Peters et al. 2007) and experimental design (Richards and Villet 2008). It is particularly desirable that the sampling, measuring and estimation methods used to handle case material are closely similar to those used to generate benchmark data, so that the validity, precision, bias and reliability of estimates of PMI_{min} are not disputed.

Acknowledgements We thank Lucy Glover and Kerri-Lynne Kriel for helping with the experiment in Fig. 7.4; Kendall Crous, Martin Grassberger, Mervyn Mansell, Iain Patterson and Ben Price for their input on various topics in this chapter; and Susan Abraham for help with graphics.

References

Adams ZJO, Hall MJR (2003) Methods used for the killing and preservation of blowfly larvae, and their effect on post-mortem larval length. Forensic Sci Int 138:50–61

Amendt J, Campobasso CP, Gaudry E, Reiter C, LeBlanc HN, Hall MJR (2007) Best practice in forensic entomology – standards and guidelines. Int J Legal Med 121:90–104

Ames C, Turner BD (2003) Low temperature episodes in development of blowflies: implications for postmortem interval estimation. Med Vet Entomol 17:178–186

Anderson GS, Huitson NR (2004) Myiasis in pet animals in British Columbia: the potential of forensic entomology for determining duration of possible neglect. Can Vet J 45:993–998

Anderson GS, VanLaerhoven SL (1996) Initial studies on insect succession on carrion in southwestern British Columbia. J Forensic Sci 41:617–625

Archer MS (2004) Annual variation in arrival and departure times of carrion insects at carcasses: implications for succession studies in forensic entomology. Aus J Zool 51:569–576

Arnott S, Turner B (2008) Post-feeding larval behaviour in the blowfly, *Calliphora vicina*: effects on post-mortem interval estimates. Forensic Sci Int 177:162–167

Avila FW, Goff ML (1998) Arthropod succession patterns onto burnt carrion in two contrasting habitats in the Hawaiian Islands. J Forensic Sci 43:581–586

Bänziger H, Pape T (2004) Flowers, faeces and cadavers: natural feeding and laying habits of flesh flies in Thailand (Diptera: Sarcophagidae, *Sarcophaga* spp.). J Nat Hist 38:1677–1694

Bass WM (1997) Outdoor decomposition rates in Tenessee. In: Haglund WD, Sorg MH (eds) Forensic taphonomy: the fate of human remains. CRC, Boston, pp 181–186

Bell CH (1994) A review of diapause in stored-product insects. J Stored Prod Res 30:99–120

Benecke M (2004) Forensic entomology: arthropods and corpses. In: Tsokos M (ed) Forensic pathology review. Humana, Totowa, New Jersey, pp 207–240

Benecke M, Lessig R (2001) Child neglect and forensic entomology. Forensic Sci Int 120: 155–159

Benecke M, Lessig R (2002) El maltrato cadaver t y la entomologia forense. Rev Jalisci Ciencias Forenses 2:32–37

Benecke M, Josephi E, Zweihoff R (2004) Neglect of the elderly: forensic entomology cases and considerations. Forensic Sci Int Suppl 146:195–199

Bourel B, Hedouin V, Martin-Bouyer L, Becart A, Tournel G, Deveaux M, Pharm D, Gosset D (1999) Effects of morphine in decomposing bodies on the development of *Lucilia sericata* (Diptera: Calliphoridae). J Forensic Sci 44:354–358

Bourel B, Callet B, Hédouin V, Gosset D (2003) Flies eggs: a new method for the estimation of short-term post-mortem interval? Forensic Sci Int 135:27–34

Brown APS (2006) Assessing the use of a camera to gain greater accuracy in the study of the relationship between temperature and humidity for first oviposition of blowflies in an urban environment and a forensic context. University of Bradford, Bradford 95

Byrd JH, Allen JC (2001) Computer modeling of insect growth and its application to forensic entomology. In: Byrd JH, Castner JL (eds) Forensic entomology: the utility of arthropods in legal investigations. CRC, Boca Raton, pp 303–329

Byrd JH, Castner JL (2001) Forensic entomology: the utility of arthropods in legal investigations. CRC, Boca Raton

Campobasso CP, Di Vella G, Introna FJ (2001) Factors affecting decomposition and Diptera colonization. Forensic Sci Int 120:18–27

Campobasso CP, Gherardi M, Caligara M, Sironi L, Introna FJ (2004) Drug analysis in blowfly larvae and in human tissues: a comparative study. Int J Legal Med 118:210–214

Cardoso Ribeiro R, Veira Milward-de-Azevedo EM (1997) Dietas naturais na criação de *Chrysomya albiceps* (Wiedemann, 1819; Diptera: Calliphoridae): estudo comparado. Ciência Rural 27:641–644

Casella G, Berger RL (1990) Statistical inference. Duxbury, Belmont

Catts EP (1992) Problems in estimating the post-mortem interval in death investigations. J Agric Entomol 9:245–255

Catts EP, Goff ML (1992) Forensic entomology in criminal investigations. Ann Rev Entomol 37:253–272

Catts EP, Haskel NH (1990) Entomology and death: a procedural guide. Joyce's Print Shop, Clemson, South Carolina

Clark K, Evans L, Wall R (2006) Growth rates of the blowfly, *Lucilia sericata*, on different body tissues. Forensic Sci Int 156:145–149

Cragg JB, Cole P (1952) Diapause in *Lucilia sericata* (Meigen) Diptera. J Exp Biol 29:600–604

Dadour IR, Cook DF, Fissioli JN, Bailey WJ (2001) Forensic entomology: application, education and research in Western Australia. Forensic Sci Int 120:48–52

Davies L, Ratcliffe GG (1994) Development rates of some pre-adult stages in blowflies with reference to low temperatures. Med Vet Entomol 8:245–254

Day DM, Wallman JF (2006a) Width as an alternative measurement to length for post-mortem interval estimations using *Calliphora augur* (Diptera: Calliphoridae) larvae. Forensic Sci Int 159:158–167

Day DM, Wallman JF (2006b) Influence of substrate tissue type on larval growth in *Calliphora augur* and *Lucilia cuprina* (Diptera: Calliphoridae). J Forensic Sci 51(657–663):1221

Denlinger DL (1978) The developmental response of flesh flies (Diptera: Sarcophagidae) to tropical seasons: variations in generation time and diapause in East Africa. Oecol 35:105–107

Denlinger DL, Zdárek J (1994) Metamorphosis behavior of flies. Ann Rev Entomol 39:243–266

Deonier CC (1940) Carcass temperatures and their relation to winter blowfly populations and activity in the Southwest. J Econ Entomol 33:166–170

Donovan SE, Hall MJR, Turner BD, Moncrieff CB (2006) Larval growth rates of the blowfly, *Calliphora vicina*, over a range of temperatures. Med Vet Entomol 20:106–114

dos Reis SF, von Zuben CJ, Godoy WAC (1999) Larval aggregation and competition for food in experimental populations of *Chrysomya putoria* (Wied.) and *Cochliomyia macellaria* (F.) (Dipt., Calliphoridae). J Appl Entomol 123:485–489

Erzinçlioglu YZ (1990) On the interpretation of maggot evidence in forensic cases. Med Sci Law 30:65–66

Faucherre J, Cherix D, Wyss C (1999) Behavior of *Calliphora vicina* (Diptera, Calliphoridae) under extreme conditions. J Insect Behav 12:687–690

Frost et al (2010) in J. Amendt et al (eds) Current Concepts in Forensic Entomology Springer Science + Business Media B.V. pp. 93–108

Gaudry E (2010) in J. Amendt et al (eds) Current Concepts in Forensic Entomology Springer Science + Business Media B.V. pp. 273–311

Gaudry E, Myskowiak JB, Chauvet B, Pasquerault T, Lefebvre F, Malgorn Y (2001) Activity of the forensic entomology department of the French Gendarmerie. Forensic Sci Int 120:68–71

Gaudry E, Blais C, Maria A, Dauphin-Villemant C (2006) Study of steroidogenesis in pupae of the forensically important blow fly *Protophormia terraenovae* (Robineau-Desvoidy) (Diptera: Calliphoridae). Forensic Sci Int 160:27–34

Goff ML (1992) Problems in estimation of post-mortem intervals resulting from wrapping of the corpse: a case study from Hawaii. J Agric Entomol 9:237–243

Goff ML, Brown WA, Hewadikram KA, Omori AI (1991a) Effect of heroin in decomposing tissues on the development rate of *Boettcherisca peregrina* (Diptera: Sarcophagidae) and implications of this effect on estimation of post-mortem intervals using arthropod development patters. J Forensic Sci 36:537–542

Goff ML, Charbonneau S, Sullivan W (1991b) Presence of fecal material in diapers as a potential source of error in estimations of postmortem interval using arthropod development rates. J Forensic Sci 36:1603–1606

Gomes L, Godoy WAC, Von Zuben CJ (2006) A review of postfeeding larval dispersal in blowflies: implications for forensic entomology. Naturwiss 93:207–215

Grassberger M, Reiter C (2001) Effect of temperature on *Lucilia sericata* (Diptera: Calliphoridae) development with special reference to the isomegalen- and isomorphen-diagram. Forensic Sci Int 120:32–36

Grassberger M, Friedrich E, Reiter C (2003) The blowfly *Chrysomya albiceps* (Wiedemann) (Diptera: Calliphoridae) as a new forensic indicator in central Europe. Int J Legal Med 117:75–81

Greenberg B (1990) Behaviour of postfeeding larvae of some Calliphoridae and a muscid (Diptera). Ann Entomol Soc Am 83:1210–1214

Greenberg B, Kunich JC (2002) Entomology and the law: flies as forensic indicators. Cambridge University Press, Cambridge

Hall RD, Doisy KE (1993) Length of time after death: effect on attraction and oviposition or larviposition of midsummer blow flies (Diptera: Calliphoridae) and flesh flies (Diptera: Sarcophagidae) of medicolegal importance in Missouri. Ann Entomol Soc Am 86:589–593

Hanski I (1976) Assimilation by *Lucilia illustris* (Diptera) larvae in constant and changing temperatures. Oikos 27:288–299

Hayes EJ, Wall R, Smith KE (1998) Measurement of age and population age structure in the blowfly, *Lucilia sericata* (Meigen) (Diptera: Calliphoridae). J Insect Phys 44:895–901

Higley LG, Haskell NH (2001) Insect development and forensic entomology. In: Byrd JH, Castner JL (eds) Forensic entomology: the utility of arthropods in legal investigations. CRC, Boca Raton

Higley LG, Pedigo LP, Ostlie KR (1986) DEGDAY: a program for calculating degree-days, and assumptions behind the degree-day approach. Envir Entomol 15:999–1016

Huntington TE, Higley LG, Baxendale FP (2007) Maggot development during morgue storage and its effect on estimating the post-mortem interval. J Forensic Sci 52:453–458

Ieno I (2010) in J. Amendt et al (eds) Current Concepts in Forensic Entomology Springer Science + Business Media B.V. pp. 139–162

Ikemoto T, Takai K (2000) A new linearised formula for the law of total effective temperature and the evaluation of line-fitting methods with both variables subject to error. Environ Entomol 29:671–682

Introna FJ, Campobasso CP, Goff ML (2001) Entomotoxicology. Forensic Sci Int 120:42–47

Ireland S, Turner B (2005) The effects of larval crowding and food type on the size and development of the blowfly, *Calliphora vomitoria*. Forensic Sci Int 159:175–181

Johl HK, Anderson GS (1996) Effects of refrigeration on development of the blow fly, *Calliphora vicina* (Diptera: Calliphoridae) and their relationship to time of death. J Entomol Soc BC 93:93–98

Johnsson A, Engelmann W (2008) The biological clock and its resetting by light. In: Björn L (ed) Photobiology: the science of life and light. Springer, New York, pp 321–388

Joplin KL, Moore D (1999) Effects of environmental factors on circadian activity in the flesh fly, *Sarcophaga crassipalpis*. Physiol Entomol 24:64–71

Kaneshrajah G, Turner BD (2004) *Calliphora vicina* larvae grow at different rates on different body tissues. Int J Legal Med 118:242–244

Kharbouche H, Augsburger M, Cherix D, Sporkert F, Giroud C, Wyss C, Champod C, Mangin P (2008) Codeine accumulation and elimination in larvae, pupae, and imago of the blowfly *Lucilia sericata* and effects on its development. Int J Legal Med 122:205–211

Kheirallah AM, Tantawi TI, Aly AH, El-Moaoty ZA (2007) Competitive interaction between larvae of *Lucilia sericata* (Meigen) and *Chrysomya albiceps* (Wiedemann) (Diptera: Calliphoridae). Pakistan J Biol Sci 10:1001–1010

Kirkpatrick RS (2004) Nocturnal light and temperature influences on necrophageous, carrion-associating blow fly species (Diptera: Calliphoridae) of forensic importance in central Texas. MSc Thesis. Texan AandM University

Kocárek P (2001) Diurnal patterns of postfeeding larval dispersal in carrion calliphorids (Diptera, Calliphoridae). Eur J Entomol 98:117–119

LeBlanc H (2010) in J. Amendt et al (eds) Current Concepts in Forensic Entomology Springer Science + Business Media B.V. pp

Leclercq J, Watrin P (1973) Entomologie et médecine legale. Acariens et insectes trouvé sur une cavadre humain en Decembre 1971. Bull Ann Soc Roy Belge Entomol 109:195–200

Lopes de Carvalho LM (2010) in J. Amendt et al (eds) Current Concepts in Forensic Entomology Springer Science + Business Media B.V. pp

Lopes de Carvalho LM, Linhares AX, Trigo JR (2001) Determination of drug levels and the effect of diazepam on the growth of necrophagous flies of forensic importance in southeastern Brazil. Forensic Sci Int 120:140–144

Lord WD (1990) Case histories of the use of insects in investigations. In: Catts EP, Haskell NH (eds) Entomology and death: a procedural guide. Joyce's Print Shop, Clemson, South Carolina

Mackerras M, Freney M (1933) Observations on the nutrition of maggots in Australian blow-flies. J Expt Biol 10:237–246

Marchenko MI (1988) Medico-legal relevance of cadaver entomofauna for the determination of the time since death. Acta Med Leg Soc 38:257–302

Melvin R (1934) Incubation period of eggs of certain muscoid flies at different constant temperatures. Ann Entomol Soc Am 27:406–410

Midgley JM, Villet MH (2009a) Development of *Thanatophilus micans* (Fabricius 1794) (Coleoptera: Silphidae) at constant temperatures. Int J Legal Med 123:285–292

Midgley JM, Villet MH (2009b) Effect of the killing method on post-mortem change in length of larvae of *Thanatophilus micans* (Fabricius 1794) (Coleoptera: Silphidae) stored in 70% ethanol. Int J Legal Med 123:103–108

Midgley JM, Richards CS, Villet MH (2010) The utility of Coleoptera in forensic investigations. in J. Amendt et al. (eds) Current Concepts in Forensic Entomology Springer Science + Business Media B.V. pp. 57–68

Musvasva E, Williams KA, Muller WJ, Villet MH (2001) Preliminary observations on the effects of hydrocortisone and sodium methohexital on development of *Sarcophaga* (*Curranea*) *tibialis* Macquart (Diptera: Sarcophagidae), and implications for estimating post mortem interval. Forensic Sci Int 120:37–41

Myskowiak JB, Doums C (2002) Effects of refrigeration on the biometry and development of *Protophormia terraenovae* (Robineau-Desvoidy) (Diptera: Calliphoridae) and its consequences in estimating post-mortem interval in forensic investigations. Forensic Sci Int 125:254–261

Nabity PD, Higley LG, Heng-Moss TM (2007) Light-induced variability in development of forensically important blow fly *Phormia regina* (Diptera: Calliphoridae). J Med Entomol 44:351–358

O'Brien CW, Turner BD (2004) Impact of paracetamol on *Calliphora vicina* larval development. Int J Legal Med 118:188–189

O'Flynn MA (1983) The succession and rate of development of blowflies in carrion in southern Queensland and the application of these data to forensic entomology. J Aus Entomol Soc 22:137–148

Oliveira-Costa et al (2010) in J. Amendt et al (eds) Current Concepts in Forensic Entomology Springer Science + Business Media B.V. pp

Payne JA (1965) A summer carrion study of the baby pig *Sus scrofa* Linnaeus. Ecology 46:592–602

Payne JA, King EW (1968) Arthropod succession and decomposition of buried pigs. Nature 219:1180–1181

Payne JA, King EW (1972) Insect succession and decomposition of pig carcasses in water. J Georgia Ent Soc 73:153–162

Peters FT, Drummer OH, Musshoff F (2007) Validation of new methods. Forensic Sci Int 165:216–224

Pien K, Laloup M, Pipeleers-Marichal M, Grootaert P, De Boeck G, Samym N, Boonen T, Vits K, Wood M (2004) Toxicological data and growth characteristics of single post-feeding larvae and puparia of *Calliphora vicina* (Diptera: Calliphoridae) obtained from a controlled nordiazepam study. Int J Legal Med 118:190–193

Pyza E, Cymborowski B (2001) Circadian rhythms in the behaviour and in the visual system of the blow fly, *Calliphora vicina*. J Insect Physiol 47:897–904

Queiroz MMC (1996) Temperature requirements of *Chrysomya albiceps* (Wiedemann, 1819) (Diptera: Calliphoridae) under laboratory conditions. Mem Inst Oswaldo Cruz 91:1–6

Reed HB (1958) A study of dog carcass communities in Tennessee, with special reference to the insects. Am Midl Nat 59:213–245

Reiter C (1984) Zum Wachstumsverhalten der Maden der blauen Schmeißfliege *Calliphora vicina*. Z Rechtsmed 91:295–308

Reiter C, Hajek P (1984) Zum altersabhängigen Wandel der Darmtraktfüllung bei Schmeißfliegenmaden – eine Untersuchungsmethode im Rahmen der forensischen Todeszeitbestimmung. Int J Legal Med 92:39–45

Richards EN, Goff ML (1997) Arthropod succession on exposed carrion in three contrasting tropical habitats on Hawaii Islands, Hawaii. J Med Entomol 34:328–329

Richards CS, Villet MH (2008) Factors affecting accuracy and precision of thermal summation models of insect development used to estimate post-mortem intervals. Int J Legal Med 122:401–408. http://dx.doi.org/10.1007/s00414-008-0243-5

Richards CS, Villet MH (2009) Data quality in thermal summation models of development of forensically important blowflies (Diptera: Calliphoridae): a case study. Med Vet Entomol 23:269–276

Richards CS, Paterson ID, Villet MH (2008) Estimating the age of immature *Chrysomya albiceps* (Diptera: Calliphoridae), correcting for temperature and geographical latitude. Int J Legal Med 122:271–279. http://dx.doi.org/10.1007/s00414-007-0201-7

Richards CS, Crous KL, Villet MH (2009b) Models of development for the blow fly sister species *Chrysomya chloropyga* and *C. putoria* (Diptera: Calliphoridae). Med Vet Entomol 23:56–61

Richards CS, Price BW, Villet MH (2009a) Thermal ecophysiology of seven carrion-feeding blowflies (Diptera: Calliphoridae) in southern Africa. Entomol Exp Appl 131:11–19

Rodriguez WC, Bass WM (1983) Insect activity and its relationship to decay rates of human cadavers in East Tennessee. J Forensic Sci 28:423–432

Roux O, Gers C, Legal L (2008) Ontogenetic study of three Calliphoridae of forensic importance through cuticular hydrocarbon analysis. Med Vet Entomol 22:309–317

Sadler DW, Robertson L, Brown G, Fuke C, Pounder DJ (1997a) Barbiturates and analgesics in *Calliphora vicina* larvae. J Forensic Sci 42:481–485

Sadler DW, Senevernatne C, Pounder DJ (1997b) Commentary on Goff ML, Miller ML, Paulson JD, Lord WD, Richards E, Omori AI. Effects of 3,4-methylenedioxymethamphetamine in decomposing tissues on the development of *Parasarcophaga ruficornis* (Diptera: Sarcophagidae) and detection of the drug in postmortem blood, liver tissue, larvae and pupae. J Forensic Sci 1997; 42(2):276–280; 42:1212–1213

Sadler DW, Chuter G, Senevernatne C, Pounder DJ (1997c) Commentary on 'D.W. Sadler, L. Robertson, G. Brown, C. Fuke, D.J. Pounder', Barbiturate and analgesics in *Calliphora vicina* larvae. J Forensic Sci 42:1214–1215

Saunders DS (1987) Maternal influence on the incidence and duration of larval diapause in Calliphora vicina. Physiol Ent 12:331–338

Saunders DS (2008) Photoperiodism in insects and other animals. In: Björn L (ed) Photobiology: the science of life and light. Springer, New York, pp 389–416

Saunders DS, Bee A (1995) Effects of larval crowding on size and fecundity of the blowfly *Calliphora vicina* (Diptera: Calliphoridae). Eur J Entomol 92:615–622

Saunders DS, Cymborowski B (2003) Selection for high diapause incidence in blow flies (*Calliphora vicina*) maintained under long days increases the maternal critical daylength: some consequences for the photoperiodic clock. J Insect Physiol 49:777–784

Saunders DS, Steel CGH, Vafopoulou X, Lewis RD (2002) Insect clocks, 3rd edn. Elsevier, Amsterdam

Schoenly KG, Griest K, Rhine S (1991) An experimental field protocol for investigating the postmortem interval using multidisciplinary indicators. J Forensic Sci 36:1395–1415

Schoenly KG, Goff ML, Wells JD, Lord WD (1996) Quantifying statistical uncertainty in succession-based entomological estimates of the postmortem interval in death scene investigations: a simulation study. Am Entomol 42:106–112

Schoenly KG, Shahid SA, Haskell NH, Hall RD (2005) Does carcass enrichment alter community structure of predaceous and parasitic arthropods? A second test of the arthropod saturation hypothesis at the Anthropology Research Facility in Knoxville, Tennessee. J Forensic Sci 50:134–141

Schoenly KG, Haskell NH, Hall RD, Gbur JR (2007) Comparative performance and complementarity of four sampling methods and arthropod preference tests from human and porcine remains at the Forensic Anthropology Center in Knoxville, Tennessee. J Med Entomol 44:881–894

Shahid SA, Schoenly KG, Haskell NH, Hall RD, Zhang W (2003) Carcass enrichment does not alter decay rates or arthropod community structure: a test of the arthropod saturation hypothesis at the Anthropology Research Facility in Knoxville, Tennessee. J Med Entomol 40:559–569

Shean BS, Messinger L, Papworth M (1993) Observations of differential decomposition on sun exposed versus shaded pig carrion in coastal Washington State. J Forensic Sci 38:938–949

Shiao A-F, Yeh T-C (2008) Larval competition of *Chrysomya megacephala* and *Chrysomya rufifacies* (Diptera: Calliphoridae): behaviour and ecological studies of two blow fly species of forensic significance. J Med Entomol 45:785–799

Singh D, Bharti M (2001) Further observations on the nocturnal oviposition behaviour of calliphorids (Diptera: Calliphoridae). Forensic Sci Int 120:124–126

Singh D, Bharti M (2008) Some notes on the nocturnal larviposition by two species of *Sarcophaga* (Diptera: Sarcophagidae). Forensic Sci Int 177:e19–e20

Slone DH, Gruner SV (2007) Thermoregulation in larval aggregations of carrion-feeding blow flies (Diptera: Calliphoridae). J Med Entomol 44:516–523

Smith PH (1983) Circadian control of spontaneous flight activity in the blowfly *Lucilia cuprina*. Physiol Entomol 8:73–82

Smith KGV (1986) A manual of forensic entomology. British Museum (Natural History), London

Smith PH (1987) Naturally occurring arrhythmicity in eclosion and activity in *Lucilia cuprina*: its genetic basis. Physiol Entomol 12:99–107

Smith KGV, Wall R (1997) Asymmetric competition between larvae of the blowflies *Calliphora vicina* and *Lucilia sericata* in carrion. Ecol Entomol 22:468–474

Smith PH, Dallwitz R, Wardaugh KG, Vogt WG, Woodburn TL (1981) Timing of larval exodus from sheep and carrion in the sheep blowfly *Lucilia cuprina*. Entomol Exp Appl 30:157–162

Sokal RR, Rohlf FJ (2005) Biometry, 4th edn. Freeman, New York

Tachibana S, Numata H (2004) Effects of temperature and photoperiod on the termination of larval diapause in *Lucilia sericata* (Diptera: Calliphoridae). Zool Sci 21:197–202

Tantawi TI, Greenberg B (1993) The effect of killing and preservative solutions on estimates of maggot age in forensic cases. J Forensic Sci 38:702–707

Tarone A, Foran D (2006) Components of developmental plasticity in a Michigan population of *Lucilia sericata* (Diptera Calliphoridae). J Med Entomol 43:1023–1033

Tessmer JW, Meek CL, Wright VL (1995) Circadian patterns of oviposition by necrophilous flies (Diptera: Calliphoridae) in southern Louisiana. Southwest Entomol 20:439–445

Trabalon M, Campan M, Clement JL, Lange C, Miquel MT (1992) Cuticular hydrocarbons of *Calliphora vomitoria* (Diptera): relation to age and sex. Gen Comp Endocrin 85:208–216

Tracqui A, Keyser-Tracqui C, Kintz P, Ludes B (2004) Entomotoxicology for the forensic toxicologist: much ado about nothing? Int J Legal Med 118:194–196

Trudgill DL, Honêk A, Van Straalen NM (2005) Thermal time – concepts and utility. Ann Appl Biol 146:1–14

Turner BD, Wiltshire P (1999) Experimental validation of forensic evidence: a study of the decomposition of buried pigs in a heavy clay soil. Forensic Sci Int 101:113–122

VanLaerhoven SL (2008) Blind validation of postmortem interval estimates using developmental rates of blow flies. Forensic Sci Int 180:76–80

VanLaerhoven SL, Anderson GS (1996) Determining time of death in buried homicide victims using insect succession. Report TR-02-96. Canadian Police Research Centre.

VanLaerhoven SL, Anderson GS (1999) Insect succession on buried carrion in two biogeoclimatic zones of British Columbia. J Forensic Sci 44:31–41

VanLaerhoven SL, Anderson GS (2001) Implications of using development rates of blow fly (Diptera: Calliphoridae) eggs to determine postmortem interval. J Entomol Soc Brit Columbia 98:189–194

Villet MH (2007) An inexpensive geometrical micrometer for measuring small, live insects quickly without harming them. Entomol Expt Appl 122:279–280. http://dx.dio.org//10.1111/j.1570-7458.2006.00520.x

Vogt WG, Woodburn TL (1985) The influence of weather and time of day on trap catches of males and females of *Lucilia cuprina* (Wiedemann) (Dipera: Calliphoridae). Bull Entomol Res 75:315–319

Voss SC, Forbes SL, Dadour IR (2008) Decomposition and insect succession on cadavers inside a vehicle environment. Forensic Sci Med Pathol 4:22–32

Wall R, Langley PA, Morgan KL (1991) Ovarian development and pteridine accumulation for age determination in the blowfly *Lucilia sericata*. J Insect Physiol 37:863–868

Waterhouse DF (1947) The relative importance of live sheep and of carrion as breeding grounds for the Australian sheep blowfly *Lucilia cuprina*. CSIR Bull 217:1–31

Weitzel M (2005) A report of decomposition rates of a special burial type in Edmonton, Alberta from an experimental field study. J Forensic Sci 50:641–647

Wells JD, King J (2001) Incidence of precocious egg development in flies of forensic importance (Calliphoridae). Pan-Pac Entomol 77:235–239

Williams KA, Villet MH, Mazungula DN (2008) Nocturnal oviposition in forensically important flies (Diptera: Calliphoridae, Sarcophagidae): laboratory and field studies. Abstract 203. 23rd International Congress of Entomology. Durban, 6–12 July 2008

Wooldridge J, Scrase L, Wall R (2007) Fight activity of the blowflies, *Calliphora vomitoria* and *Lucilia sericata*, in the dark. Forensic Sci Int 172:94–97

Worner SP (1992) Performance of phenological models under variable temperature regimes: consequences of the Kaufmann or rate summation effect. Env Entomol 21:689–699

Wyss C, Cherix D, Michaud K, Romain N (2003) Pontes de *Calliphora vicina*, Robineau-Desvoidy et de *Calliphora vomitoria*, (Linné) (Diptères, Calliphoridae) sur un cadaver humain enseveli dans la neige. Rev Int Crim Pol Tech Sci 56:112–116

Ye G, Li K, Zhu J, Zhu GH, Hu C (2007) Cuticular hydrocarbon composition in pupal exuviae for taxonomic differentiation of six necrophagous flies. J Med Entomol 44:450–456

Zhu GH, Ye GY, Hu C, Xu XH, Li K (2006) Development changes of cuticular hydrocarbons in *Chrysomya rufifacies* larvae: potential for determining larval age. Med Vet Entomol 20:438–444

Zhu GH, Xu XH, Yu XJ, Zhang Y, Wang JF (2007) Puparial case hydrocarbons of *Chrysomya megacephala* as an indicator of the postmortem interval. Forensic Sci Int 169:1–5

Zumpt F (1965) Myiasis in man and animals in the old world: a textbook for physicians, veterinarians and zoologists. Butterworths, London

Chapter 8
Analysing Forensic Entomology Data Using Additive Mixed Effects Modelling

Elena N. Ieno, Jens Amendt, Heike Fremdt, Anatoly A. Saveliev, and Alain F. Zuur

8.1 Introduction

Forensic pathologists and entomologists estimate the minimum post-mortem interval since a long time by describing the stage of succession and development of the necrophagous fauna (Amendt et al. 2004). From very simple calculations at the beginning, (Bergeret, see also Smith 1986) the discipline has evolved into a more mathematical one (e.g. Marchenko 2001; Grassberger and Reiter 2001, 2002) and tries to implement concepts like probabilities and confidence intervals (Lamotte and Wells 2000; Donovan et al. 2006; Tarone and Foran 2008, see also Villet et al. this book Chapter7). As pointed out by Tarone and Foran (2008) and Van Laerhoven (2008), the latter is one of the major tenets of the Daubert Standard (Daubert et al. v. Merrell Dow Pharmaceuticals (509 U.S. 579 (1993)).

Forensic Entomology deals with living systems and this means that we face problems related to influences depending e.g. on the time of the year, the ecosystem of the scene of crime or the geographic origin of the insects. Not surprisingly these varieties of possible impacts thwart the efforts to establish a statistically robust result in a forensic report, leading to strange situations in court, where the different methods and opinions of different experts may lead to different or inherent results and reports (Westerfield trial in San Diego, CA, USA (People v. Westerfield)). In a forensic context, this is simply a disaster for the reputation of the used method. The background of this problem is discussed in more detail by Villet et al. (this book Chapter7), and it can be stated that it is at least partly related to the application and misuse of statistical methods used in the past. In the present chapter we introduce methods which may help to better analyse forensic entomology data sets.

E.N. Ieno (✉) and A.F. Zuur
Highlands Statistics Ltd., 6 Laverock Road, AB41 6FN, Newburgh, Aberdeenshire, UK

J. Amendt and H. Fremdt
Institute of Forensic Medicine, Kennedyallee 104, 60596, Frankfurt am Main, Germany

A.A. Saveliev
Faculty of Geography and Ecology, Kazan State University, 18 Kremlevskaja Street, Kazan, 420008, Russia

J. Amendt et al. (eds.), *Current Concepts in Forensic Entomology*,
DOI 10.1007/978-1-4020-9684-6_8, © Springer Science+Business Media B.V. 2010

As mentioned above, several variables may influence the development of an insect, and certainly they sometimes don't just act simply by oneself, but in combination. This is valid as well for drugs, which are assimilated by maggots during feeding (see Lopes de Carvalho in this book Chapter 9). For the estimation of the post-mortem interval, it is always an important question to know if the presence of a particular drug causes an impairment of the development of the necrophagous insects. In a recent study, Fremdt (2008) showed the influence of an antemortem administration of a combination of two drugs on the insect successional patterns under natural conditions and the development on the blowfly *Calliphora vicina*. In this chapter, we use data from a similar study and show how to deal with interactions between three different types of drugs affecting the non-linear growth rates of the blowfly *Calliphora vicina*.

Our target audience is the reader who is familiar with linear regression models, tools which are still quite popular in forensic entomology. Traditional statistical methods like linear regression are based on a series of assumptions, some of which are violated for the data used here, as we will see later. We will discuss these assumptions; show how to verify them, and the implications of violating them. We also show how to solve the problems, and apply a series of methods that are all, in some way, extensions of linear regression. The underlying mathematics are not discussed here; the interested reader can consult Pinheiro and Bates (2000), Ruppert et al. (2003), or Wood (2006). Ecological applications can be found in Schabenberger and Pierce (2002), Crawley (2005), and Zuur et al. (2007, 2009), or for social science examples, see Keele (2008).

8.2 Data Introduction

8.2.1 Data Structure and Coding

Data coding may sound a trivial step, but it is a rather crucial part of the data analysis! The forensic entomological data used in this chapter consist of length values, recorded every 12 h during eight consecutives days. Hence, we have 16, regular spaced, measurements in time (also called: longitudinal), where a unit of 1 refers to 12 h. There are three drug treatment variables; Amino_FZ (a metabolit of Flunitrazepam), Propofol, and Ethanol. These are so-called nominal or categorical variables, and take the values "with" or "without"; hence there are eight combinations of them. From each combination, we have six replicates, which by the way, are from the same batch. Finally, different larval stages (with values L1, L2, L3, and the postfeeding stage prior pupation L3PF) were also measured. Table 8.1 shows how you have to structure and code the data into a spreadsheet.

8.2.2 Working Towards a Model

From the statistical point of view, length is the only response (dependent) variable. The first aim of this chapter is to develop a model that describes length as a function

Table 8.1 Illustration of the preparation of the spreadsheet. Eight columns were made that contained all the variables measured in the experiment. The first 48 rows are length measurements from series 1 (six replicates times eight treatments), and the next 48 rows from series 2. Notice that there should be 768 rows in total. However, due to missing values there are slightly less observations. We also created the nominal variable ID that indicates which observations are from the same batch. If there were no missing values, it would run from 1 to 128 (= 8 × 16), and each value would be repeated six times

No	Amino_FZ	Propofol	Ethanol	Series	Larvae	Length	ID
1	Without	Without	Without	1	L1	2.2	1
...							
6	Without	Without	Without	1	L1	3.0	1
7	With	Without	Without	1	L1	3.1	2
...							
12	With	Without	Without	1	L1	3.3	2
13	Without	With	Without	1	L1	2.9	3
...							
18	Without	With	Without	1	L1	2.9	3
19	Without	Without	With	1	L1	2.9	4
...							
24	Without	Without	With	1	L1	3.0	4
25	With	With	Without	1	L1	2.8	5
...							
30	With	With	Without	1	L1	3.3	5
31	With	Without	With	1	L1	2.8	6
...							
36	With	Without	With	1	L1	2.9	6
37	Without	With	With	1	L1	3.0	7
...							
42	Without	With	With	1	L1	1.4	7
43	With	With	With	1	L1	1.9	8
...							
48	With	With	With	1	L1	3.1	8
49	Without	Without	Without	2	L2	3.6	9
...							
718	With	With	With	16	L3P	9.3	128

Table 8.2 Available explanatory variables. The variables Amino_FZ, propofol, Ethanol and Larval stage are categorical. Series is a continuous explanatory variable that represents time

Variable	Nominal	Remark
Amino_FZ	Yes	Without and with
Propofol	Yes	Without and with
Ethanol	Yes	Without and with
Series	No	From 1 to 16 measured every 12 h
Larval stage	Yes	L1, L2, L3, L3PF

of the explanatory (independent) variables Amino_FZ, Propofol, and Ethanol, larval stage and time (called Series). Table 8.2 gives a description of all variables used in the study.

We now work towards a model of the form:

$$\text{Length} = \text{function}(\text{Amino\_FZ, Propofol, Ethanol, Series, Larval stage}) \quad (8.1)$$

This equation can be used to model the length *as a function* of the other variables.

The second question of interest is whether we can predict the age (time) of larvae removed from a body at the crime scene. This means that we get a sample from which we can measure the length, the larval stage, and drug combination, and we want to know the corresponding value of Series (post mortem interval). This is called inverse modelling (Draper and Smith 1998). Before addressing the inverse modelling problem, we need to find the best possible model in Eq. 8.1.

8.3 Data Exploration

The first step of any analysis is to explore the data. Zuur et al. (2007) divided the data exploration in three main steps using mainly graphical tools, namely:

a. Outliers in the response variable. We use Cleveland dotplots (Cleveland 1993) and boxplots. Cleveland dotplots are simple scatterplot of the observed values versus their index number (Zuur et al 2009b). An interesting publication dedicated to only the Cleveland dotplot can be found in Jacoby (2006). Outliers are points that have rather different values compared to the rest. Note that we do not advocate removing outliers; we just need to know whether such observations are present. If they are influential in the analysis, we can always remove them at that stage, if desired.
b. Spread of the data. If the spread is the same per strata (or along a gradient like time), we can safely assume homogeneity of variance. If the spread in the data is different, we talk about heterogeneity. Conditional boxplots and conditional Cleveland dotplots allow us to see whether we have similar or different variation per stratum in each explanatory variable. Linear regression assumes homogeneity of variance.
c. Collinearity is defined as explanatory variables being highly correlated with each other. Biologically, this means that we use explanatory variables that represent the same biological signal. Using collinear explanatory variables in a linear regression model (or any of its extensions) causes trouble with the model selection procedures and it also gives larger p-values (Montgomery and Peck 1992; Draper and Smith 1998; Zuur et al. 2007). Tools to detect collinearity are Pearson correlation coefficients between two continuous explanatory variables, scatterplots, or variance inflation factors (Montgomery and Peck 1992) if a large number of explanatory variables are used.

A boxplot and Cleveland dotplot of length are given in Fig. 8.1. The boxplot let you believe that there are outliers, but the dotplot shows that this is a whole group of observations that just have smaller length values. Hence, there are no observations with extreme large or small values. The advantage of a Cleveland dotplot is that it gives a more detailed overview of the spread of the data, compared to a boxplot.

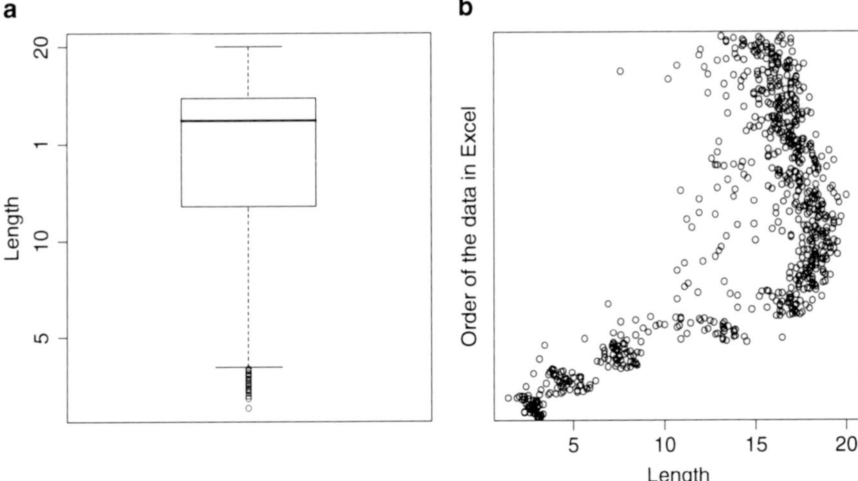

Fig. 8.1 (**a**) Boxplot of length. (**b**) Cleveland dotplot of length. The *x*-axis shows the value of an observation and the *y*-axis the order of the data as imported from the spreadsheet (which is in this case ordered by time). Note that there are no observations with extreme small or large values

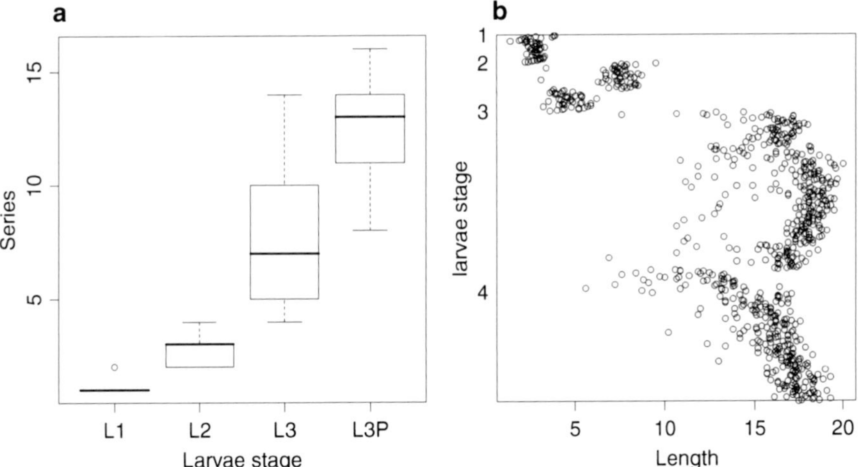

Fig. 8.2 (**a**): Boxpot of length conditional on larvae stage. (**b**) Cleveland dotplot conditional on larvae stage. The *x*-axis shows the value of an observation and the *y*-axis the order of the data grouped larvae stage

You should make the same graphs for all continuous explanatory variables in your data set. We also made boxplots and Cleveland dotplots for Series, but it did not show any problems.

Figure 8.2a shows a boxplot of length conditional on larval stage, and Fig. 8.2b shows the Cleveland dotplot of the length data again, but this time we grouped the observations by larval stage. Both graphs show the same problem, namely there is a difference in spread per larvae group, indicating potential heterogeneity problems (Larvae stage 3 shows more spread than the others). In Fig. 8.3, we again show a boxplot of length, but this time we used a boxplot for each value of series. We can see a difference in spread, and also a non-linear pattern.

The last point we discuss is collinearity. Due to the experimental design, there are no collinearity issues with the drug variables Amino_FZ, Propofol, and Ethanol. However, Fig. 8.4 shows that there is serious collinerarity between Series and larval stage; this is obvious, as larval stage increases with time.

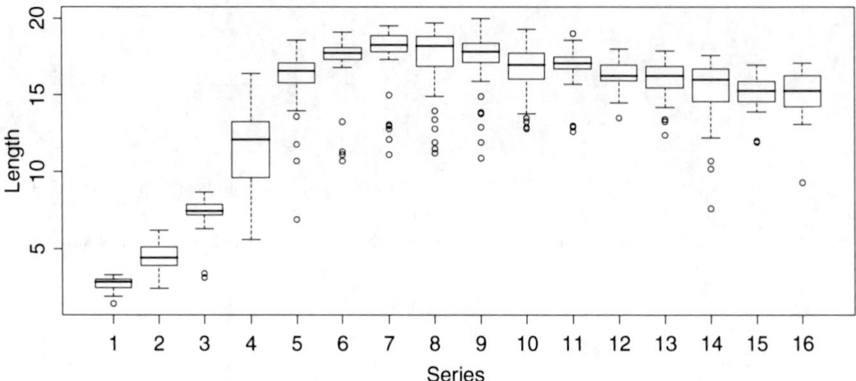

Fig. 8.3 Boxplot of length conditional on the time variable series. Note that there is a non-linear pattern over time

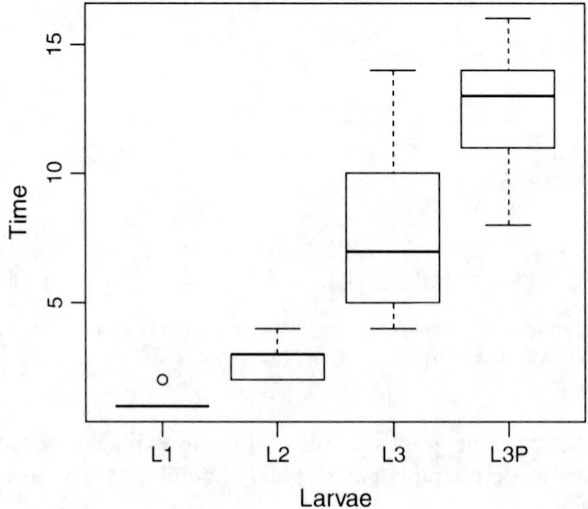

Fig. 8.4 Boxplot of Series conditional on larval stage. Note that the L3 and L3P stages correspond to higher values for time

8.4 Fine Tuning the Model Using Data Exploration Results

So, what does the data exploration tell us? There are no outliers, there is serious heterogeneity, and time and larval stage are collinear. Heterogeneity can be dealt with using a transformation of length. However, there are multiple arguments against using a transformation. First of all, it changes the relationship between the variables (Quinn and Keough 2002). Secondly, we end up with estimated values on the logarithmic or square root scale (or whichever transformation is used), and these are more difficult to interpret. Back transforming the results is not always easy neither! Another convincing argument against transformation was emphasised by Keele (2008); a transformation affects the entire Y–X relationship and as it can be seen in Fig. 8.3 the length – series relationship is partly linear and also partly non-linear. We therefore try to avoid applying a transformation.

Before discussing the collinearity issue, we shortly talk about the models that we will use in this chapter. They can broadly be written as

$$\text{Length} = \text{fixed part} + \text{random part} \tag{8.2}$$

The fixed part models the response variable Length in terms of the explanatory variables Amino_FZ, Propofol, Ethanol, Series and Larval stage. The random part is unexplained information, also called noise. If we apply a linear regression model, then (8.2) looks as follows:

$$\begin{aligned} Length &= \alpha + \beta_1 \times Amino\_FZ_i + \beta_2 \times Propofol_i \\ &+ \beta_3 \times Ethanol_i + \beta_3 \times Series_i + \beta_4 \times Larvae_i + \varepsilon_i \end{aligned} \tag{8.3}$$

The fixed bit contains the terms from α to $\beta_5 \times Larvae_i$, and this part can (and will) be made more complicated by adding interactions between the three treatment variables, Series or larval stage. The random bit is the ε_i, from which we assume it is independently, normally distributed with mean 0 and variance σ^2. It is common to write this as

$$\varepsilon_i \sim N(0, \sigma^2) \tag{8.4}$$

Due to the high collinearity between Series and larval stage, either larval stage or Series should be used in the fixed part of the model, but never both at the same time! High collinearity is causing problems with the model selection process and it also gives higher p-values (Montgomery and Peck 1992). The other problem we noticed from Fig. 8.3 is that the length – Series relationship is non-linear. This means that if we decide to use Series in the model (instead of Larval stage), modelling it as $\beta_4 \times \text{Series}_i$ is not a good option, and alternatives are (i) to use a quadratic function of Series and omit larval stage, (ii) drop series and use Larval stage as a categorical variable, or (iii) drop larval stage and use a smoothing function of Series. The third option gives a generalised additive model (GAM) of the form

$$Length = \alpha + \beta_1 \times Amino\_FZ_i + \beta_2 \times Propofol_i + \beta_3 \times Ethanol_i + F\,(Series_i) + \varepsilon_i \quad (8.5)$$

The notation $f(\text{Series}_i)$ stands for a smoothing function, and is further discussed in Section 8.6. Hence, we have lots of different options for the fixed part of the model. None of the options will give a perfect model; it just does not exist. But it is the task of the researcher to find the least bad model. There are multiple tools available for this task:

1. Choose a model that makes biological sense.
2. Choose a model that does not contain any residual information, and complies with all its underlying assumptions.
3. Use an information criterion like the Akaike Information Criterion (AIC, Akaike 1973) that measures goodness of fit and model complexity.

Obviously, any model needs to make biological sense, hence the first point. Now, for the models that do make biological sense, you also need to ensure that these comply at least with the most important assumptions. These are homogeneity and independence. If these assumptions are violated, then you cannot trust the statistical inference from the models. With this we mean that our biological conclusions are based on a model that gives us estimated parameters. These parameters are based on one sample, yet, we pretend that they hold for the larger population from which the samples were taken. As to the third point, you may have a series of models that make biological sense, and for which the important assumptions are valid. How do you choose among these models? An option is to put a number on each model, and choose the one with the best number. The golden question is then how to put a number on a model. The AIC can be used for this; it is a function that measures the goodness of fit (based on either the residual sum of squares or maximum likelihood criterion) and the number of parameters. There are different definitions of the AIC (e.g. a small sample adjustment) and alternative selection criteria exist, e.g. the Bayesian Information Criterion (BIC), which has a higher penalty for model complexity (Schwarz 1978). The use of the AIC is not without criticism; see for example Burnham and Anderson (2002).

If you think that for the random part in the model in Eq. 8.2, there are fewer options, then unfortunately you are wrong. We can clearly see heterogeneity in Fig. 8.2. In fact, the spread in length values differs per larval stage, which means that the assumption in Eq. 8.4 is incorrect: the residuals are normally distributed with mean 0 and variance σ^2. Incorrect can either refer to the Normal distribution or to the variance σ^2. Due to the design of linear regression in Eq. 8.3, the mean of the residuals is always 0. It is the variance σ^2 we are after; why would each residual have the same variance? It is mathematically easy to adopt something of the form

$$\varepsilon_i \sim N(0,\sigma_j^2)$$

where $j = 1,\ldots,4$. Such a model has four variances, one for each larval stage. Although this may look complicated, the underlying mathematics is reasonable simple, and modern statistical software allows one to extend linear regression and additive models with multiple variances (Pinheiro and Bates 2000).

So, the heterogeneity is one aspect of the random part in Eq. 8.2. Two other aspects are random effects (leading to linear mixed effects models) and violation of independence (which are actually related). Linear mixed effects models can be used if we have multiple observations from the same patient, beach, bird nest, school, or indeed, the same batch with meat. The simplest application on our data is of the form

$$Length_{is} = \alpha + \beta_1 \times Amino\_FZ_{is} + \beta_2 \times Propofol_{is} + \beta_3 \times Ethanol_{is} + \beta_3 \times Series_{is} + \beta_4 \times Larvae_{is} + a_i + \varepsilon_{is} \quad (8.6)$$

We swapped from a Length$_i$ to Length$_{is}$, where in the first notation i was the observation number (or row number in the spreadsheet), whereas in the second notation Length$_{is}$ is observation number s in batch i, where i is from 1 to 128. The random intercept a_i is assumed to be normally distributed with mean 0 and variance σ_a^2. The consequence of using the random intercept is that observations from the same batch are allowed to be correlated. We can even allow for more complicated correlation structures between sequential observations, if we repeatedly sample the same physical unit (Zuur et al 2009).

It is possible to test whether adding any of the random structures (heterogeneity, random effects, and correlation) improves the model.

8.4.1 Where to Start?

We have multiple options to proceed, namely:

1. Start with a GAM because the data exploration shows non-linear patterns.
2. Apply linear mixed effects models to allow for correlation between observations from the same batch.
3. Apply additive mixed modelling because we saw non-linear patterns and may have correlation over time, or between observations from the same batch.
4. Start with a basic linear regression model, see where we get stuck, and slowly improve the model step by step.

The first three approaches can be followed if you are familiar with GAMs and linear mixed effects models. But for pedagogical reasons, we will start with the last approach. After all, a linear regression model is so much easier to work with.

Therefore, we start with linear regression, followed by a section on GAM, and then generalised additive mixed modelling (GAMM). Each method is based on a series of assumptions, that all need to be verified, or else we cannot trust the inferences from the models. This verification process is also called the model validation. We will see how the failure of the model validation process of one model leads into another, more complicated model. Each approach contains a lot of numerical (anova tables, t-values and p-values) and graphical output. However, we only present these for the final model.

8.5 Linear Regression as a Starting Point

We will start very simple by fitting a linear regression model with interactions. Due to collinearity we use *Larvae stage* instead of *Series* in this first model. We also include the three-way interaction between the drugs. The model has the form:

$$\begin{aligned} Length_i = {} & \alpha + \beta_1 \times Amino\_FZ_i + \beta_2 \times Propofol_i + \beta_3 \times Ethanol_i \\ & + \beta_4 \times Amino\_FZ_i \times Propofol_i + \beta_5 \times Amino\_FZ_i \times Ethanol_i \\ & + \beta_6 \times Propofol_i \times Ethanol_i + \beta_7 \times Amino\_FZ_i \times Propofol_i \\ & \times Ethanol_i + \beta_8 \times LarvalStage_i + \varepsilon_i \end{aligned} \tag{8.7}$$

All the terms are fitted as nominal variables. Note that the notation for the first part implies that main terms, two-way interactions and the three-way interaction terms between the three drugs are included. For the residuals, we assume $\varepsilon_i \sim N(0, \sigma^2)$. This is just a straightforward linear regression, and it can also be called analysis of variance. Its main underlying assumptions are homogeneity of variance, independence, no residual patterns, and normality (although this is the least important assumption). To verify these assumptions, we need to fit the model, obtain the residuals, and inspect these for violation of the assumptions.

To verify homogeneity, you can plot the residuals versus fitted values, or the residuals versus each explanatory variable that was used in the model (and also those that were not used in the model); the spread (variation) of residuals should be the same along the gradients in these graphs. If not, violation occurs and action should be taken! Figure 8.5a shows a graph of the residuals versus fitted values. The fact that there is a gap along the horizontal axis is not important; it is the difference in spread we are after. The residuals from observations with *Larval stage* 3 have more variation (this can be seen by using different colours depending on larval stage in Fig. 8.5a).

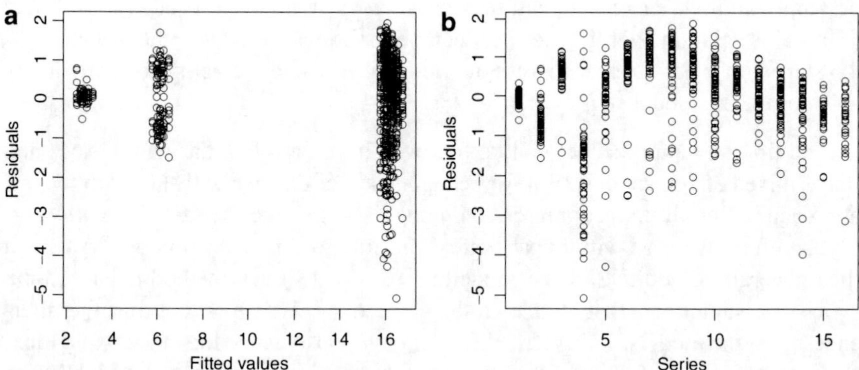

Fig. 8.5 (**a**) Residuals versus fitted values. Violation of homogeneity is detected because the spread is increasing with increased fitted values. (**b**) Residuals versus series. Note that there is violation of independence, as can be clearly seen from the pattern in the residuals

The question is then *why* we have this. The first thing we did was plotting the residuals versus each drug (these are simple boxplots, and are not shown here), and see whether any of these drugs cause higher or lower variation in the data. However, no indication of heterogeneity could be seen in residuals versus *Ethanol*, residuals versus *Propofol* or residuals versus *Amino_FZ*. A graph of the residuals versus *Larval stage* shows a clear difference in spread.

As to dependence, we can distinguish two types of dependence. The first type of dependence is due to a missing covariate or modelling a covariate as linear, whereas in reality, it has a non-linear effect. You can easily detect this by plotting the residuals of the linear regression model versus each continuous explanatory variable. You should not see any patterns in these graphs. If you do see a pattern, then the model has to be extended. This type of violation of independence is due to a bad model, and can easily be solved by improving the model by adding more covariates, interactions or by using smoothing models. A more problematic type of dependence is if you repeatedly sample the same physical unit (e.g. the same patient, or animal). To include this type of dependence structure in a regression model or GAM, a residual correlation structure can be implemented. Alternatively, a random intercept can be used, which introduces the so-called symmetric compound correlation between observations from the same patient or animal. Here, we have multiple observations from the same batch. We did fit the model in Eq. 8.6, which uses a random intercept for batch, but the likelihood ratio test showed that it was not better than the linear regression model (details how to compare models with, and without random effects can be found in Pinheiro and Bates 2000; West et al. 2006; Zuur et al. 2009).

Figure 8.5b shows that we can indeed see a clear pattern. Notice that even though *Series* was not used in the model, we should still produce this graph as it helps to explain where this violation takes place. This dependence structure is likely to be due to a bad model; the effect of time is modelled via the categorical variable *Larval stage*. We then conclude that this model is not good due to heterogeneity and lack of independence. Our first attempt to try to improve the model is by adopting a non-linear model that uses *Series* instead of *Larval stage*. This implies either the use of polynomial models or GAM. We will go for the second option as it more easily allows one to model non-linear patterns.

8.6 Generalised Additive Modelling

In the previous section, we discovered two main issues to address: Independence and homogeneity. Different options can be used to solve the dependence problem, namely (i) extend the linear regression model in Eq. 8.7 with more interaction terms, (ii) add more explanatory variables, or (iii) apply a transformation to linearise the relationships. We already discussed arguments against applying transformations. And all possible interactions and explanatory variables were already used in Section 8.5, however no improvements could be seen. Therefore, we move from linear regression models to smoothing techniques.

Smoothing models allow for non-linear relationships and belong to the family of generalised additive models (GAM). A detailed explanation on GAMs can be found in Hastie and Tibshirani (1990), Ruppert et al. (2003), Wood (2006), Zuur et al. (2007), Keele (2008), among others. A GAM example in forensic entomology can be found in Tarone and Foran (2008). Smoothing is a scientific field on itself, and there are many different ways of smoothing (Wood 2006). The easiest smoothing method is the moving average smoother, which calculates the average at a target value using all the observations that are in a window around this target value, and then this window is moved along the gradient, and each time an average is calculated. The other two phrases frequently seen in the smoothing literature are LOESS smoother and smoothing splines. The LOESS smoother is just an extension of the moving average in the sense that weighted linear regression is applied on all the observations in the window around the target value. The smoothing splines are a bit more difficult to explain. Basically the covariate gradient is split up in multiple bins, and on each bin a cubic polynomial is fitted. These are then glued together at the intersection points (also called knots). First order and second order derivatives are used to ensure smooth connections. From here onwards, it becomes rather technical; the smoothing splines are estimated by using a penalised sum of squares criterion which consists of a measure of fit and a penalty for the amount of non-smoothness or wiggliness. Full details can be found in Wood (2006).

Smoothing techniques have one major problem; whichever method is chosen, somehow the software needs to know the amount of smoothing. For the moving average and LOESS smoother, this means the size of the window. This is a value that you have to choose, and its effect can be between a straight line, or a curve that connects each point. Although this may sound rather subjective and scary, in practise it is not that difficult, and various tools exist to guide you, for example:

1. Use the AIC to choose the optimal amount of smoothing.
2. Decrease the amount of smoothing if there are patterns in the residuals.
3. Apply automatic smoothing selection using a tool called cross-validation. In here, observations are omitted in turn, the smoother is applied on the remaining data, the omitted data points are predicted, and prediction residuals are calculated. The value of the amount of smoothing that produces the lowest sum of squared prediction residuals is deemed as the optimal amount of smoothing, and is expressed as a number larger or equal than 0. We also called it the effective degrees of freedom (edf). A value of 1 means a straight line and 4 gives a fit similar to a third order polynomial and an edf of 10 is a highly non-linear curve.

The main advantage of a smoother is that it captures non-linear patterns in the data. So far, we have not mentioned anything about software. Although many packages can do linear regression, only a few can do GAMs. Without no doubt, S-PLUS (http://www.insightful.com/products/splus/default.asp) and R (R Development Core Team 2008) are the best software options for GAM. We used R, which is a free statistical software package that can be downloaded from www.r-project.org. Within R, various packages (a collection of functions) exists for GAMs, and we used the mgcv package (Wood 2004, 2006). The problem with R is that it has a steep learning curve and the analyses carried in this chapter can also be applied via an graphical user interface like for example Brodgar (www.brodgar.com).

The GAM applied on our data has the form:

$$\begin{aligned} Length_i = \alpha + \beta_1 \times Amino\_FZ_i + \beta_2 \times Propofol_i + \beta_3 \\ \times Ethanol_i + \beta_4 \times Amino\_FZ_i \times Propofol_i + \beta_5 \\ \times Amino\_FZ_i \times Ethanol_i + \beta_6 \times Propofol_i \times Ethanol_i \\ + \beta_7 \times Amino\_FZ_i \times Propofol_i \times Ethanol_i + f\left(Series_i\right) + \varepsilon_i \end{aligned} \quad (8.8)$$

The only difference with the linear regression model in Eq. 8.7 is that the explanatory variable *Larval stage* is replaced by *Series*, and a smoothing function is used to model its effect. The explanatory variable *Series* has more unique values than *Larval stage*, and is therefore better able to capture the non-linear time effects. The smoother was estimated as a smoothing spline, and to estimate the optimal amount of smoothing, cross-validation was used.

The smoother represents the *Length – Series* relationship, alias the growth pattern. The GAM in Eq 8.8 contains one smoother for the time effect. This means that it assumes that all observations have the same shape for the growth pattern over time, what ever the drug treatment is. However, it may well be that a certain drug treatment, or a combination of drug treatment result in different growth rates. It is relatively easy to extend the model in (8.8) to allow for a different growth pattern for each of the eight drug treatment combinations (this would give eight smoothers, one for each treatment). Alternatively, perhaps the maggots that received the ethanol treatment have a different growth pattern compared to those who did not receive Ethanol. In this case, we should consider a GAM with two smoothers. This is the interaction equivalent in a GAM.

Hence, in this context, each treatment combination is allowed to have a different *length – series* relationship. Obviously, we raise the question whether we should indeed use one series smoother for each treatment combination, or whether we can replace them by one overall smoother, or try different combinations. We used the AIC to find the optimal model.

As judged by the AIC, the model with eight series smoothers (one per drug treatment combination) was not better than the model with one overall smoother, nor did any other combination of smoothers give a lower (better) AIC. Hence, drug treatment does not change the shape of growth rate, just the absolute values.

8.6.1 Model Validation of the GAM

Instead of presenting the numerical and graphical output of the optimal GAM model, we go immediately to the model validation and show that there are still some serious problems even though we are increasing the complexity of the model. As in linear regression, with additive models we need to verify the underlying assumptions.

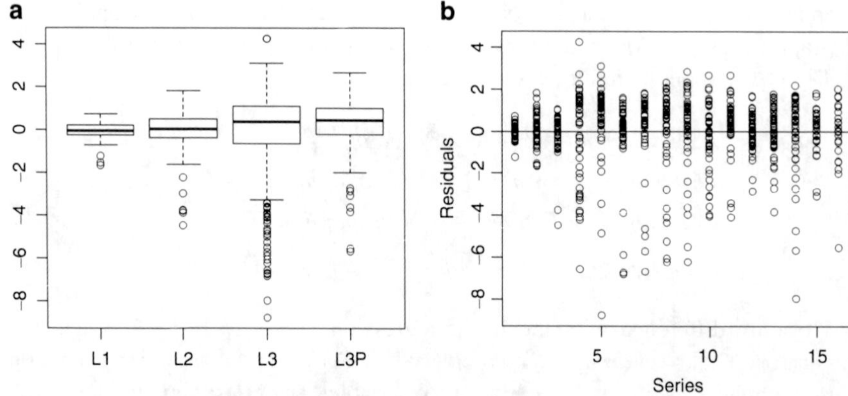

Fig. 8.6 (**a**) Residuals versus Larvae stage. Note the differences in residual spread. Larvae stage (L3) has the most variation. (**b**) Residuals versus Series. Independence is a valid assumption as the points are equally distributed above and bellow the zero line, indicating the same number of positive and negative residuals but some discrepancies can be expected

There is no doubt that much progress has been made in terms of the independence assumption as the smoother of Series seems to capture the patterns over time quite well (Fig. 8.6b). The smoother had 8.3 degrees of freedom, and was highly significant ($F = 698.4$, $p < 0.001$).

Although the GAM showed an improvement in terms of independence, the heterogeneity problems are still there, see Fig. 8.6a; it shows a boxplot of the residuals versus larvae stage. Hence, we are still violating the homogeneity assumption, and therefore further model improvement is required.

Because the GAM has less residual patterns than the linear regression model, it turned out to be an improvement. The good news is that there is life beyond GAMs and we will extend it to generalised additive mixed modelling (GAMM) to allow for the heterogeneity.

8.7 Generalised Additive Mixed Modelling

In the additive model in Eq. 8.8, we also assume independence and homogeneity. In the previous section, we aimed to improve the lack of fit due to the time effect (*Series*), but the model still assumes that the variation is the same everywhere; for different treatment effects, at the beginning of the experiment and at the end of the experiment, and also for different *Larval stages*. However, in Fig. 8.6a we saw heterogeneity in the residuals obtained from the GAM, and this is something we have to improve. One option is to transform length, but we prefer to avoid transformations. Another possible solution is generalised additive *mixed* modelling. In the same way as linear mixed effects modelling is an extension of linear regression, so is generalised additive mixed modelling an extension of generalised additive model. Pinheiro and Bates (2000) discuss a wide range of options to model heterogeneity. If, for example, the length variation increases over time, we can use:

$$\varepsilon_i \sim N(0, \sigma^2 \mid Series_i \mid^{2\delta}) \tag{8.9a}$$

The unknown variance parameters are σ and δ. For positive δ, the variance in the residuals increases for larger *Series* values, and vice versa. If $\delta = 0$, we obtain our familiar variance from Eq 8.4 again. It is also possible to assume that one particular drug treatment, say Propofol, causes more variation. This can be modelled with:

$$\varepsilon_i \sim N(0, \sigma_j^2) \tag{8.9b}$$

where $j = 1$ if an observation has the propofol treatment, and $j = 2$ if the observation does not have propofol treatment. The two unknown variance parameters are σ_1 and σ_2. An explanatory variable that is used in the variance is called a variance covariate. With three drug treatment variables there is a lot of choice as we can also have combinations of them (e.g. $j = 1$ if an observation received propofol and ethanol treatment, and 2 else). All these variance structures can be compared with each other using the AIC. However, instead of applying every possible combination, it is better to try and understand why we have heterogeneity, and for this we need to plot the residuals of the GAM in Eq. 8.8 against every drug treatment, combination of drug treatment, *Series* and *Larval stage*. The explanatory variable that shows the strongest heterogeneity can be selected as variance covariate. In the previous section, we only showed the graph with residuals plotted versus *Larval stage*, as it showed the strongest violation of heterogeneity. This means that *Larval stage* is the prime candidate to be the first variance covariate. Therefore, we can use the variance structure in Eq. 8.9b, where $j = 1, \ldots, 4$. Hence, we have 4 variances, and each variance corresponds to a *Larval stage*. The observations from a particular stage all have the same residual spread, but the different stages are allowed to have different spread.

The first thing we have to do is investigating whether this is actually a significant improvement. This is also part of the model selection process. Unfortunately, for a (additive) linear mixed effects model, the model selection process is slightly more complicated and consists of the following steps.

1. Start with a linear regression model or additive model from which you think that it contains all the important explanatory variables in the fixed part. If this model shows homogeneity and independence, you are finished.
2. If the model shows heterogeneity or dependence, then refit the model using restricted maximum likelihood (REML).
3. Inspect the residuals of the model fitted, and add a correlation or heterogeneity structure (or random intercepts).
4. Compare the model from step 2 with that in step 3. This can be done with the AIC or likelihood ratio test. Both models must contain the same fixed structure).
5. Inspect the residuals of the new model and if the residuals are homogenous and independence, go to step 6, else go to step 3 and add another heterogeneity or dependence structure.

6. Now that we have the optimal random structure it is time to find the optimal fixed structure. This process is similar to a model selection in linear regression; drop terms in turn and apply a likelihood ratio test. To compare nested models via a likelihood ratio test, use maximum likelihood estimation.
7. Present the final model estimated with REML and give an interpretation.

The justification of this protocol can be found in Diggle et al. (2002), and applications can be found in West et al (2006,), and Zuur et al. (2007, 2009). It contains phrases like REML estimation and nested models, but an explanation of these terms is outside the scope of this chapter.

We believe that the model in Eq. 8.8 is a reasonable starting point for step 1 of the protocol. We then added the variance structure in Eq. 8.9b, where *Larval stage* was the variance covariate. The likelihood ratio test shows that adding the multiple variances gives a significant improvement ($L = 176.16$, $df = 3$, $p < 0.001$). In step 5 of the protocol, we need to assess whether adding a variance covariate results in a model that complies with the assumptions. For this, you need to use the standardised residuals, which are defined as observed minus fitted values, divided by the square root of the variance in Eq. 8.9b. We plotted the standardised residuals versus *Larval stage* (Fig. 8.7a), fitted values (Fig. 8.7b) and *Series* (Fig. 8.7c) and none of these graphs showed any clear violation of homogeneity or independence, hence we can proceed to step 6 of the protocol.

In the second part of the protocol, we need to find the optimal fixed structure for the selected random structure. This means that we need to assess the significance of the three-way drug interaction term and the smoother of *Series*. Using a likelihood ratio

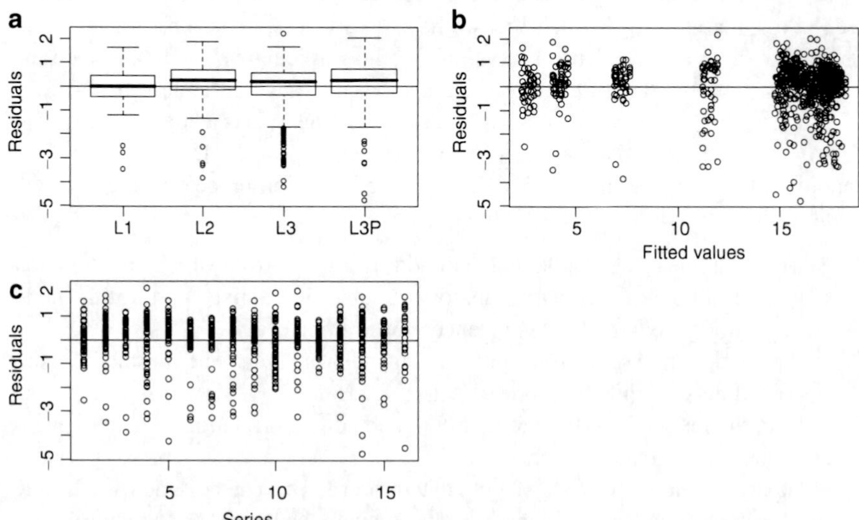

Fig. 8.7 (**a**) Standardised residuals versus fitted Larval stage. (**b**) Standardised residuals versus fitted values. C: Standardised residuals versus time. Note that there is no clear violation of homogeneity or independence in these graphs

Table 8.3 Estimated regression parameters, smoothers, t-values, F-value and p-values

	Estimate	S.E	t value	Pr(>|t|)
(Intercept)	13.87033	0.09789	141.697	<2e–16
Amino_FZ1 with	0.52095	0.11529	4.519	7.3e06
Propofol with	–0.29017	0.08106	–3.580	0.000368
Ethanol with	0.11874	0.11413	1.040	0.298516
Amino_FZ1&Ethanol with	–0.56055	0.16221	–3.456	0.000582
Approximate significance of smooth terms				
	edf	Est.rank	F	p-value
s(Series)	8.56	9	2,737	<2e–16

test, the three-way interaction was not significant. Using the same procedure, two two-way drug interaction terms were dropped and the final model is as follows.

$$Length = \alpha + \beta_1 \times Amino\_FZ_i + \beta_2 \times Propofol_i + \beta_3 \times Ethanol_i + \beta_4 \times Amino\_FZ_i \times Ethanol_i + f(Series_i) + \varepsilon_i \quad (8.10)$$

Hence, there are three main terms, and a two-way interaction between Amino_FZ and Ethanol, and a smoother representing the growth curve. So far, we have not presented any numeral output, as we first wanted to go over the underlying assumptions that need to be fulfilled. As this seems to be the best model, we now concentrate on main terms and interactions (fixed part of the model), and we provide information on p-values and fitted curves.

The estimated values for the regression parameters are given below. Note that the two interaction between Amino_FZ and Ethanol is highly significant (see Table 8.3).

When confronted for the first time with a GAM, one tends to be confused with the output and its interpretation. We therefore present again the GAM that we are fitting, but this time we substitute the estimated parameters.

$$Length_i = 13.9 + 0.5 \times Amino\_FZ_i - 0.3 Propofol_i + 0.11 \times Ethanol_i - 0.56 \times Amino\_FZ_i \times Ethanol_i + f(Series_i) + \varepsilon_i \quad (8.11a)$$

The coding in the drug variables in the statistical software is such that if a drug treatment is equal to "without" then the corresponding variable is 0, and 1 else. Therefore, an observation that was without any drug treatment (Amino_FZ = Propofol = Ethanol = "without", we get the following equation:

$$Length_i = 13.9 + f(Series_i) + \varepsilon_i \quad 8.11b$$

If an observation has for example only the Amino_FZ treatment, then we have to add 0.52 to the intercept. Hence, the estimated regression parameters in Table 8.3 show the difference between "without" and "with" for each drug treatment and the interaction. The last thing we need to discuss is the role of the smoother, which is given in Fig. 8.8. Suppose that we have an observation that did not receive any drug

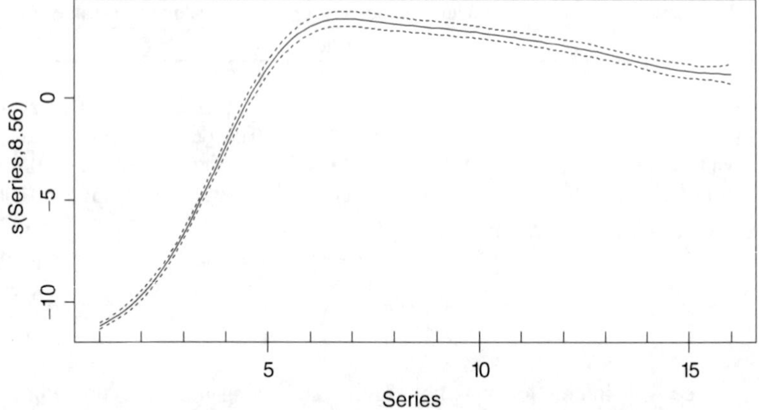

Fig. 8.8 Estimated smoother for the GAM in Eq. 8.11a. The solid line is the smoother and the dotted lines are the 95% point-wise confidence bands. The horizontal axis shows the values of Series, and the vertical axis is the contribution from the smoother to the fitted values. The smoother has 8.56 degrees of freedom

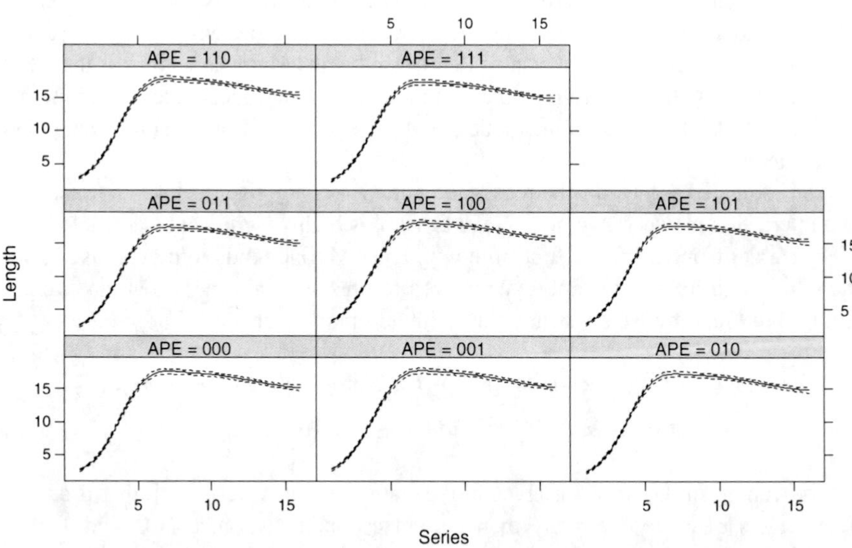

Fig. 8.9 Fitted values and 95% confidence intervals for the mean. Bootstrapping was used to obtain 95% confidence intervals. The solid lines are the predicted values and the dotted lines the 95% confidence intervals for the mean. The notation APE stand for the drug treatments Amino_FZ, Propofol and Ethanol and 0 for "without" and 1 for "with"

treatment, hence its underlying Equation is given in Eq. 8.11b. In order to get the fitted values for all such observations, we need to add the value of 13.9 to the smoother in Fig. 8.8. This means that the smoother is shifted vertically with a value of 13.9. Similar vertical shifts can be calculated for other drug treatment combinations, and the resulting eight curves are given in Fig. 8.9. Note that the confidence intervals

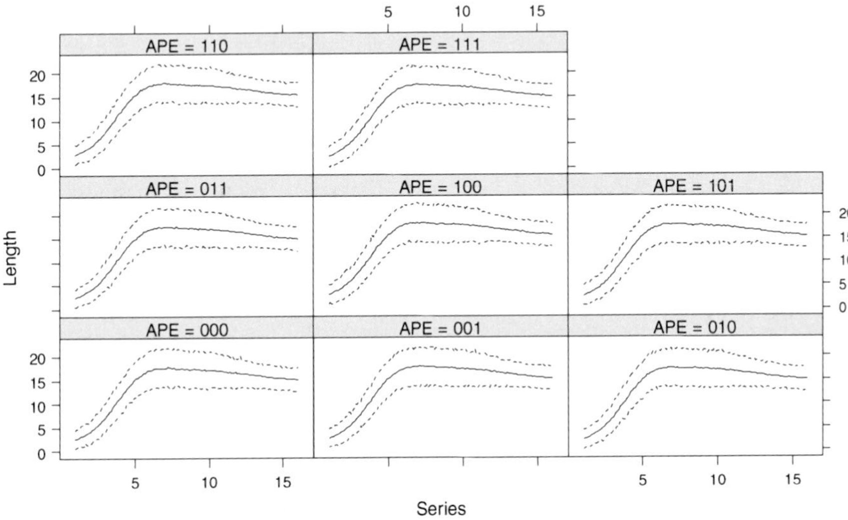

Fig. 8.10 Fitted values and 95% confidence intervals for the population. Bootstrapping was used to obtain 95% confidence intervals

are for the mean relationship. This means that if you repeat this experiment a large number of times, in 95% of the cases the real mean relationship is within the confidence bands. It is also possible to add confidence intervals for the population, and these show the range of possible length values at a certain time and drug treatment combination (Fig. 8.10).

For ordinary linear regression models and GAM, it is trivial to get the confidence intervals for the population, see for example Montgomery and Peck (1992) or Draper and Smith (1998). Suppose that the variance of the fitted values is given by σ_{fit}^2 and for the residuals σ^2. A confidence interval for the mean around a linear regression line or smoother is based on σ_{fit}^2. For the population confidence intervals, you need to add these two variances, and the square root of this can be used to obtain confidence intervals for the population values. Hence, the confidence intervals for the population are always larger. A problem arises because we use multiple variance σ_j^2 for the residuals, where j stands for *Larval stage*. Therefore, we cannot use standard statistical equations to obtain a confidence interval for the population values around the smoother. There are multiple ways of still getting a confidence interval for the population, and whichever approach we used, the results were similar. The easiest approach is bootstrapping.

8.7.1 Bootstrapping

To obtain 95% confidence intervals in Fig. 8.9 and 8.10, bootstrapping was used, following Keele (2008). In a bootstrapping procedure, we create a large number (say 1,000) of similar data sets, and on each of these data sets we fit the GAMM.

The 1,000 estimated smoothers can be used to obtain confidence intervals. We implemented the following algorithm.

1. Fit the GAMM and store the fitted values and residuals. Denote these by F_i and E_i respectively.
2. For j is 1 to 1,000, carry out the following steps.
   a. Permute the residuals, and call these E_i^b, where i is the observation index. Because the variance structure in Eq. 8.9b is used, we cannot close our eyes and randomly permute the residuals. Instead, we need to permute the residuals that have the same σ_j. Formulated differently, we permute the residuals from the same larvae stage.
   b. Add the residuals E_i^b to the fitted values F_i and apply the GAM in Eq. 8.10 on the bootstrapped data $Y_i^b = F_i + E_i^b$.
   c. Predict the fitted values for each drug combination and store these in a matrix B_{ji}.
3. Once the loop for the bootstrap has finished, sort the 1,000 bootstrapped values for each observation i, and take the median, 25th and 975th values. The latter two form the lower and upper bands for the 95% quantile confidence interval.

The method above gives the bootstrapped 95% quantile confidence interval for the mean of a GAM (Keele 2008). To obtain the 95% quantile confidence interval for the population, there are various options. In the first option that we tried, we added a random value ε_i from a Normal distribution with mean 0 and variance σ_j^2 to the predicted values in step 2c. The only problem is that we don't know the exact *Larval stage* at a certain time. We also sampled larvae stages from a multinomial logistic regression model in which drug treatment and *Series* were used as explanatory variables. This was done in each bootstrap iteration. As an alternative larvae stages can be drawn from a distribution based on the observed frequencies in the measured data. Another option is to store the 1,000 predicted standard errors for the population as well, and use the median standard error in Fig. 8.10.

This all sounds overly complicated, but all that it does is to generate 1,000 similar data sets for each of the 8 drug treatments, based on our (hopefully valid) model. The complicating factor is that for the confidence interval for the population, we need to know the *Larval stage* at a certain time. Because there is considerable variation in the distribution of the larvae stages along time, it takes a bit bootstrapping effort to create 1,000 realistic data sets. The different bootstrap approaches give very similar results.

8.8 Inverse Modelling

The second underlying question in this chapter is how we can estimate the time of oviposition for a maggot of a given length, and known larval stage and drug treatment. This is called inverse prediction, and for linear regression models, existing formulae can be used to estimate the value of the explanatory variables (see also Wells and LaMotte 1995). However, for GAM and models with multiple variances, there

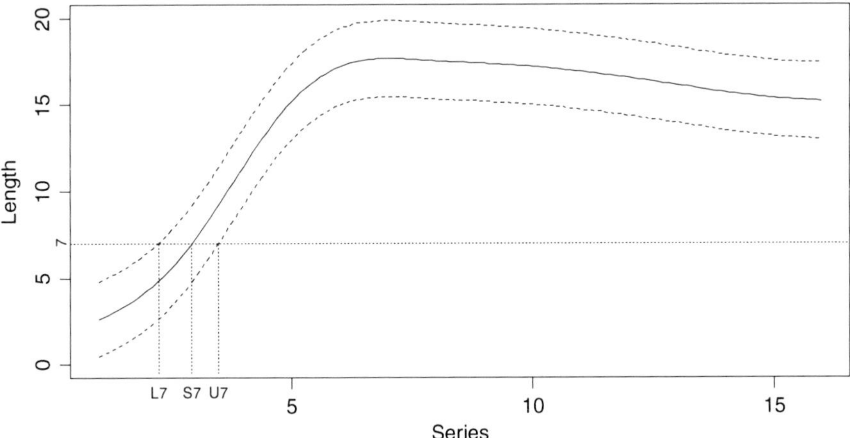

Fig. 8.11 Estimated model for the observations that did not receive any drug combination. The solid line represents the estimated fitted values and was obtained by bootstrapping. The dotted lines around the smoother are 95% confidence intervals for the population. For a maggot of length 7 and larval stage 2, inverse prediction using bootstrapping gives a value of Series = 2.92, with a lower confidence interval of 2.22 and upper confidence interval of 3.47

are no existing equations (to the best of our knowledge). Therefore, we use our bootstrapping scheme of the previous section. The solid line in Fig. 8.11 is the predicted values for the observations that did not receive a drug treatment (this is the panel labelled APE = 000 in Fig. 8.10). The question we ask ourselves is as follows. Suppose we have a 7 mm long maggot, and it has reached larval stage 2, what was the time of oviposition? Fig. 8.11 contains a dotted horizontal line which intersects the vertical axis at length = 7. Intuitively, we look up where this vertical line intersects with the smoother, and the corresponding *Series* value gives us the answer. This is the value of *Series* labelled as S7 in the graph. We would also like to have a confidence interval around S7. These are taken from intersection the 95% confidence bands for the population with the horizontal dotted line. In the graph, these are denoted by L7 (lower confidence band) and U7 (upper confidence band). Draper and Smith called these the "fiducial limits". They advise to consider the interval as the inverse confidence limits for X, given a Y.

There are a couple of potential problems. First of all, if the length is chosen too large (say 17), we end with an inverse confidence interval between approximately 6 and infinity. In terms of biology, this means that you cannot accurately determine the time of oviposition. The second problem is how to determine the lower (L7) and upper (U7) confidence bands. We are still working with the model in Eq. 8.11a.

The difference with the situation in the previous section is that we now know the value of the larval stage, and we can therefore more easily get the population confidence interval. We again created 1,000 similar data sets for each drug treatment using the bootstrap approach. For each of the 1,000 data sets, we fitted the GAM and predicted the smoother along the entire time axis. Because we know that larval stage is 2, we can easily create the 95% population confidence bands. For each of these models, we determined L7, S7 and U7. This gives us 1,000 realisations of L7, S7 and U7 for each drug treatment. The median L7, S7 and U7 for this group are 2.227, 2.926, and 3.476 respectively. For the observations that received the three drug treatments, we have 2.312, 2.985, and 3.530. We can use these median values as estimators of L7, S7 and U7.

A serious problem is that the L7–U7 interval is not a real confidence interval. For a given length and larval stage, it just provides a range of plausible time values. What we really want is a probability distribution that would allow us to make statements along the lines of:

P(Time of oviposition < 2 days) = 0.3

Or something like: In 95% of the cases the oviposition was between 2 and 3 days ago. For this, we need to derive a probability distribution for Series, given length and stage. This is illustrated in Fig. 8.12. The curve shows the fitted values for the observations without any drug treatment. The dotted lines are population confidence intervals, and the density curves on top of the smoother show the probability of other length values at a certain time. Hence, a maggot of length 7 may be from Series = 2.92, but it is also possible that it is from Series = 2, or even from Series = 1, albeit with a small probability. This probability is visualised as vertical lines along the *x*-axis. Note that it is not a discrete but continuous distribution.

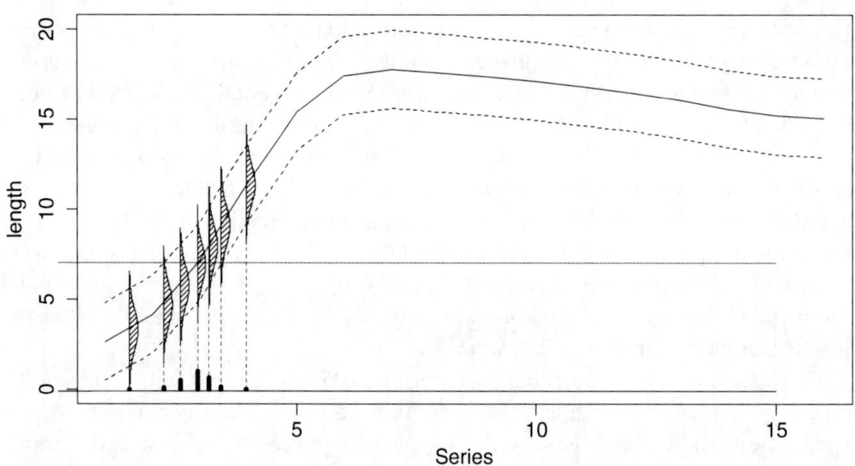

Fig. 8.12 Illustration of probability distribution for time. The solid line shows the fitted values for observations with no drug treatment, and the dotted line are its population confidence intervals. On top of the fitted line, we have sketched Gaussian density curves, and these show the range of likely length values at different time points. Hence, a maggot of length 7 is most likely from Series = 2.9, but there is also a small probability that it is from Series = 4. The vertical thick lines show the probability distribution for the Series values, for a maggot of length 7

To obtain the probability distribution, we need Bayes theorem:

$$P(Series \mid Length, \ Stage) = \frac{P(Length, Stage \mid Series) \cdot \mathrm{P(Series)}}{\mathrm{P(Length,Stage)}}$$

To calculate these probabilities for a range of series values, various arguable choices were made. Discussing these choices requires concepts like MCMC and Bayesian statistics, and is outside the scope of this chapter. Further research is required. It is interesting to know that the 95% confidence interval obtained by this approach is given by 2.19–3.43, which is similar to our bootstrapping approach.

8.9 Discussion

In this chapter, we discussed a range of techniques that can be used to analyse certain types of forensic entomology data. By no means is this the only way to analyse these data, and one can easily imagine more complicated models using random effects of temporal correlation structures.

The experiment presented in this case study was carried out at only one temperature value. It needs to be repeated under different temperature conditions, and this will lead to more complicated GAMM models that may include temperature interactions.

Another question often asked by forensic entomologists is how many samples to take. Obviously, this question can be split in two questions: How many samples do we need for an experiment as described in Section 8.2, and how many samples should the police take at the crime scene. The first question may be answered using data and models presented in this chapter. It brings us in the world of power analysis (Zar 1999). The only problem is that conventional power analysis equations are based on linear regression, and not for models with heterogeneity, random effects, auto-correlation and smoothers. Bootstrapping may be an option here. Zuur et al. (2007) applied some simulation tools to see what happens with estimated patterns if 5%, 10% or 25% of the data are removed. A similar approach may be adopted here. As to the second question, how many samples should the police take, we cannot address this question without having access to such data and investigate the variation in such samples. This will be the next step.

References

Akaike H (1973) Information theory as an extension of the maximum likelihood principle. In: Petrov BN, Csaki F (eds) Second international symposium on information theory. Akadémiai Kiadó, Budapest, Hungary, pp 267–281

Amendt J, Krettek R, Zehner R (2004) Forensic entomology. Naturwissenschaften 91:51–65

Burnham KP, Anderson DR (2002) Model selection and multimodel inference: a practical information-theoretic approach, 2nd edn. Springer, New York, USA

Cleveland WS (1993) Visualizing data. Hobart, Summit, NJ, USA, p 360

Crawley MJ (2005) Statistics. An introduction using R. Wiley, New York

R Development Core Team (2008) R: A language and environment for statistical computing. R Foundation for Statistical Computing, Vienna, Austria. ISBN 3-900051-07-0, URL http://www.R-project.org

Diggle PJ, Heagerty P, Liang KY, Zeger SL (2002) The analysis of longitudinal data, 2nd edn. Oxford University Press, Oxford, England

Donovan SE, Hall MJR, Turner BD, Moncrieff CB (2006) Larval growth rates of the blowfly Calliphora vicina over a range of temperatures. Med Vet Entomol 20:106–114

Draper N, Smith H (1998) Applied regression analysis, 3rd edn. Wiley, New York

Fremdt H (2008) Der Einfluss von Rohypnol® und Ethanol auf Sukzession und Entwicklung nekrophager Insekten. Unpublished Master thesis, University of Kiel

Grassberger M and C Reiter (2001) Effect of temperature on Lucilia sericata (Diptera: Calliphoridae) development with special reference to the isomegalen-and isomorphen-diagram. Forensic Sci. Int. 120, 32–36

Grassberger M and C Reiter (2002) Effect of temperature on development of the forensically important holoarctic blow fly Protophormia terraenovae (Robineau-Desvoidy) (Diptera Calliphoridae). Forensic. Sci. Int.128, 177–182

Hastie T, Tibshirani R (1990) Generalized additive models. Chapman and Hall, London

Jacoby WG (2006) The dot plot: a graphical display for labeled quantitative values. The Pol Methodologist 14(1):6–14

Keele LJ (2008) Semiparametric regression for the social sciences, Wiley, New York

LaMotte LR and JD Wells (2000) P-values for postmortem intervals from arthropod succession data. Journal of Agricultural, Biological and Environmental Statistics 5:58–68

Marchenko MJ (2001) Medicolegal relevance of cadaver entomofauna for the determination of time since death. Forensic Sci Int 120:89–109

Montgomery DC, Peck EA (1992) Introduction to linear regression analysis. Wiley, New York 504

Pinheiro J, Bates D (2000) Mixed effects models in S and S-Plus. Springer, New York, USA

Quinn GP, Keough MJ (2002) Experimental design and data analysis for biologists. Cambridge University Press, Cambridge

Ruppert D, Wand MP, Carroll RJ (2003) Semiparametric Regression. Cambridge Univerity Press, New York

Schabenberger O, Pierce FJ (2002) Contemporary statistical models for the plant and soil sciences. CRC, Boca Raton, FL

Schwarz G (1978) Estimating the dimension of a model. Ann Stat 6(2):461–464

Smith KGV (1986) A Manual of Forensic Entomology. Cornell University Press, Ithaca

Tarone AM, Foran DR (2008) Generalized additive models and *Lucilia sericata* growth: assesing confidence intervals and error rates in forensic entomology. J Forensic Sci 53(4)

Van Laerhoven SL (2008) Blind validation of postmortem interval estimates using developmental rates of blow flies. Forensic Sci Int. 180:76–80

Wells JD, Lamotte LR (1995) Estimating maggot age from weight using inverse prediction. J Forensic Sci 40(4):585–90

West B, Welch KB, Galecki AT (2006) Linear mixed models: a practical guide using statistical software. Chapman and Hall/CRC, Boca Raton, FL

Wood SN (2004) Stable and efficient multiple smoothing parameter estimation for generalized additive models. J Am Stat Assoc 99:673–686

Wood SN (2006) Generalized additive models: an introduction. Chapman and Hall/CRC,

Zar JH (1999) Biostatistical analysis, 4th edn. Prentice-Hall, Upper Saddle River, USA

Zuur AF, Ieno EN, Smith GM (2007) Analysing ecological data. Springer, Berlin p 680

Zuur AF, Ieno EN, Walker NJ, Saveliev AA, Smith GM (2009) Mixed effects models in ecology with R (2008-In press). Springer, Berlin p 650

Zuur, AF, Ieno, EN, Walker, N, Saveliev, AA, and Smith, GM (2009) Mixed effects models and extensions in ecology with R. Springer

Zuur AF Ieno EN Meesters EHWG (2009b) A Beginner's Guide to R Springer

Chapter 9
Toxicology and Forensic Entomology

Lucila Maria Lopes de Carvalho

9.1 Introduction

Toxicology is a scientific field with many distinct branches, and as such, its various specialists share distinct interests. For example, for forensic investigations of deaths caused by poisoning, the combined effort of pathology, biology, chemistry and pharmaceutics experts and others is needed. Every detail must be based on sampled evidences of the highest quality and best conditions possible, to be coupled with other information for a conclusive picture of the circumstances of death.

Identifying illicit substances or their metabolic sub-products during the quantitative and qualitative analyses of every sample is of crucial relevance to determine if these are of natural or synthetic origin. For example, the presence of many synthetic drugs can be determined through detection and quantification of secondary contaminants, such as adulterants and other substances or solvents mixed to it (Tanaka et al. 1994).

Deaths related to drugs have increased in several countries over the recent years. Frequently the body of the victim remains undiscovered for a period of time ranging from a few days to months (Goff and Lord 1994).

There are many toxicological analysis techniques and methodologies available today, but it is often difficult obtaining proper samples for the following reasons (Kintz et al. 1990a; Pounder 1991):

The cadaver may be greatly decomposed or contaminated.
There is no blood or urine available to be sampled.
Some biological traits may have been altered according to the death circumstances, e.g. death by intoxication with carbon monoxide.
The relatives of the dead person have religious beliefs that forbid collecting samples from the body.

In this context, insects (mainly maggots) have frequently been used in the absence of body samples for toxicological analyses and they, in fact, present some advantages. It is possible to detect drugs and illicit substances in insects directly feeding from

L.M.L. de Carvalho (✉)
Department of Zoology, Paulista State University, São Paulo, Brazil

J. Amendt et al. (eds.), *Current Concepts in Forensic Entomology*,
DOI 10.1007/978-1-4020-9684-6_9, © Springer Science+Business Media B.V. 2010

the body or carcass tissues, provided that they were not exposed to other sources of drug contamination (Beyer et al. 1980; Kintz et al. 1990b). The analyses of insects found on decomposed bodies are useful not only as criminal evidences and tools to aid estimating postmortem intervals (PMI), but also qualitatively and quantitatively identify the presence of drugs (Beyer et al. 1980).

Entomotoxicology is a rather recent forensic entomology branch that deals with the use of toxicological analyses of carrion-feeding insects in order to identify drugs and toxins present in tissues from dead bodies. It also investigates the effects caused by these drugs on insect lifecycles, thus aiding correctly estimate PMI (Introna 2001).

9.2 Detection of Substances on Insects

Successful detection of substances has been accomplished by several extraction methods from maggots, pupae and adults of Diptera and even from the feces of beetles (Miller et al. 1994).

Insects may also be reliable animal models for immunohistochemical studies for detecting and surveying metabolism of some substances. Morphine was detected on third larval instar maggots from *Calliphora vicina* Linnaeus (Diptera: Calliphoridae) fed with an artificial diet mixed with the drug. Drug-positive maggots presented a characteristic haemolymph colour and intense immune reaction between the exocuticle and the endocuticle. These results indicated that morphine was stored inside the cuticle of the maggots during their development (Bourel et al. 2001a). Cocaine was detected on sections of second and third stage larvae by immunohistochemistry using the peroxidase-complex method. Positive specimens showed a more intense immunoreaction in an area located at the limit between exocuticle and endocuticle. These results constitute an evidence of cocaine accumulation in the cuticle of the larvae during their development. Neglecting the importance of this information may cause errors of up to12 h in estimating the PMI (Alves Jr et al. 2007).On the other hand, diethylpropion (Inibex®) was not detected in larvae of *Chrysomya megacephala* and *Chrysomya putoria* using this methodology, suggesting the rapid excretion of drug (Alves Jr et al. 2008).

Beyer et al. (1980) was the first to purpose analysing insects collected from decomposed bodies to detect substances. Ever since, there were researchers testing various drugs and techniques on maggots, frequently aimed at using them as toxicological samples for detecting substances like bromazepam, levomepromazine (Kintz 1990a), malathion (Gunatilake et al. 1989), fenobarbital, brazotam, oxazepam (Kintz 1990b), cocaine (Goff et al. 1989), morphine (Bourel et al. 2001b), diazepam (Carvalho et al. 2001), and such in cases of criminal nature like suicide, homicide, kidnapping, etc (Bourel et al. 1999; Gunatilake and Goff 1989; Goff et al. 1991, 1992; Sadler et al. 1997).

The insects most frequently used to estimate PMIs and as toxicological analysis samples belong to the families Calliphoridae (blowflies) and Sarcophagidae (flesh flies). These flies are expert agile flyers that feed on corpses, and are usually the first to arrive to the corpse or death scenario (Goff and Lord 1994).

Homogenized insect samples can be quite easily analyzed through toxicological procedures like radio-immune-analysis (RIA), gas chromatography (GC), thin layer chromatography (TLC), high pressure liquid chromatography (HPLC) and gas mass analysis (GC-MS). The most indicated procedure is fractioning the samples by chromatography followed by mass spectrometry analysis, because their combined precision permits separating, identifying and quantifying chemical substances and contaminants from illicit substances at minute amounts (Moore and Casale 1994; Collins et al. 1997). For example, chromatography is commonly applied to at least three types of investigation involving *Cannabis* (marijuana) (Gudzinowicz et al. 1980):

Identification of *Cannabis* in obtained plant samples
Determining the origin of the obtained *Cannabis* samples
Designing profiles of *Cannabis* samples according to their geographic origins

Gas chromatography is remarkable for its very high resolution and sensitiveness. With it, it is possible analyzing samples composed of multiple substances at minute quantities, a fact that would limit the use of other techniques. On the other hand, sample preparation must follow several steps and the results may take considerable time (Collins et al. 1997).

The advantages of using maggots instead of body tissues can be seen in chromatography. Chromatograms obtained from larval extracts present less endogenous contaminant peaks than chromatograms from human body extracts (Kintz et al. 1990b). This is particularly useful if the sampled material is decomposed, and insects are rather easily collected and reared in laboratory.

Larvae from flies (maggots) feeding from contaminated human tissues ingest drugs and toxins recently taken by the dead person when alive. These substances are transferred through the food chain, usually bioaccumulating in beetles that predate on blowfly larvae. Consequently, the beetles can also be used for toxicological analysis in some cases (Introna et al. 2001).

Nuorteva and Nuorteva (1982) detected mercury in larvae, pupae and adult insects reared on mercury-contaminated fish. Similar procedures were employed in a forensic investigation in where the body of a woman was found in putrefactive state and covered with maggots in a rural region of Finland. The distinctly low mercury levels present in the maggots indicated a probable geographic origin for the victim: an area almost free of mercury contamination. They also detected mercury in adult beetles of *Creophilus maxillosus* (Staphylinidae) and *Tenebrio molitor* (Tenebrionidae) that were feeding from necrophagous maggots reared on tissues containing mercury. No adverse effects on the staphilinids development were observed, whereas, adult tenebrionids displayed a significant reduction of their activities.

In the absence of maggots, especially in bodies on advanced putrefaction, necrophagous and some predator coleopterans that feed on necrophagous insects may be used as toxicological samples. *Dermestes frischi* Kugelann (Diptera:

Dermestidae) and *Thanatophilus sinuatus* Fabricius (Diptera: Silphidae) beetles were reared on food containing morphine and it was possible to detect the drug in all development stages of *D. frischi* and in the larval stages of *T. sinuatus* (Bourel et al. 2001c).

Nolte et al. (1992) recorded the first forensic case where it was possible detecting cocaine and benzoylecgonine in fly larvae collected from decomposing humans. The body was practically bones and no injuries were apparent. What could be the cause of death? The victim's girlfriend told the authorities he could have been on cocaine when he disappeared. Samples of muscles and insect larvae were submitted for toxicological analysis and results were positive to benzoylecgonine, a cocaine metabolite. Cocaine is toxic and can cause fatal effects in the heart, intestines, brain and lungs. In this investigation, the association of this knowledge with the absence of injuries and the drug metabolites on larvae and human muscle allowed the establishment of cocaine overdose as the cause of death.

In 1985 Leclercq and Brahy detected the presence of arsenic through toxicological analysis of species of flies of the families Piophilidae, Psychodidae and Fanniidae in a case of accidental death in France. Other researchers, Gunatilake and Goff (1989) detected traces of the organophosphate insecticide Malathion in calliphorid maggots from a suicidal, through GC analysis.

Parathion is an insecticide and acaricide widely used in agriculture. It is rapidly absorbed by the soil and it is degraded by photolysis. Cases of intoxication by inappropriate use, inadequate use of protection gear and suicides through direct ingestion are fairly frequent, especially in agricultural regions. Wolff et al. (2004) employed high performance liquid chromatography columns to separate and quantify parathion in arthropods collected from decomposing drug-treated rabbits. It was detected in different development stages of ten arthropod species of the orders Diptera, Coleoptera, Hymenoptera, Isopoda and Acari, which were collected during all decomposition phases (from fresh stage to dry remains). Besides their small size, isopods and acarids accumulated parathion in quantities enough to be detected. The chitinous/proteic matrix of the cuticle contributes in fixing the drug onto the arthropods (Introna et al. 2001). The drug presented deterrent and insecticidal activity at the mouth entrance of the drug-treated rabbits. This is relevant since most necrophagous insects lay eggs on body cavities. No significant alterations on insect succession were observed, in contrast to results from a similar experiment using malathion (Gunatilake and Goff 1989), wherein a delay in insect oviposition and different taxa were found on carcasses containing (or not) insecticidal contaminants. Further research is warranted to understand the various effects of insecticides over carrion-feeding entomofauna.

Kintz et al. (1990c) successfully demonstrated the potentialities of entomotoxicology by describing many cases. An interesting case was that of a body found a couple of months after death. Liquid chromatography was employed to analyse tissue samples of his heart, liver, lung, spleen and kidneys as well as calliphorid maggots collected from his body. The results from both materials indicated the presence of triazolam, oxazepam, phenobarbitol, alimemazine and clomipramine.

Positive identification of morphine was achieved by Introna et al. (1996) from empty puparia of calliphorid flies reared on a diet containg the drug. Likewise, Goff et al. (1993) detected amitriptyline and nortriptyline in fly larvae developing on rabbit carcasses containing different concentrations of these substances. Drugs have also been detected in fly puparia and beetle exuvia found around the mummified body of dead person missing for about 2 years (Miller et al. 1994).

Manhoff et al. (1991) used gas chromatography and mass spectrometry to detect the presence of cocaine in mummified tissues, blood samples, maggots and beetle faeces from a corpse. Hair samples from a person can also be used to detect cocaine and other substances several years after his death (Kidwell 1993). Substances may deposit on puparia and drugs may be released from insect chitin.

Carvalho et al. (2001) reared *C. albiceps* and *C. putoria* (Diptera: Calliphoridae) larvae on livers from rabbits given twice the lethal dosage (LD) of diazepam. Drug was detected in all analysed tissues of rabbit and in almost all stages of development of both fly species showing insects can be used as an alternative sample for toxicological analysis (Table 9.1) although it was not observed a correlation between tissue and larvae concentrations.

Comparisons of drug concentrations between those in human tissues and blowfly larvae use to show different patterns of distribution that may be related to differences in species physiology as well to the chemical properties of the drugs (Campobasso et al. 2004). Several previous studies have confirmed the reliability of entomological specimens for qualitative analyses, although quantitative extrapolations are still unreliable. However, even if only as qualitative specimens, carrion feeding immature and adult stages of insects may serve as alternative toxicological evidence or sources of toxicological analysis, supporting the final diagnosis of narcotic intoxication (Campobasso et al. 2004).

Table 9.1 Diazepam quantification (μg/mg) in samples of rabbit tissues and fly specimens through of GC-MS analysis

Sample	Quantification(μg/mg)
Heart	9.531 (a)
Liver	5.943 (a)
Blood	3.228 (b)
Lung	3.341 (b)
Urine	4.029 (b)
Larvae of *C.albiceps*	0.342 (b)
Larvae of *C.putoria*	0.479 (b)
Puparia of *C.albiceps*	0.233 (b)
Puparia of *C.putoria*	3.022 (b)
Adult of *C.albiceps*	0.051 (b)
Adult of *C.putoria*	0.000 (b)

Means in a column followed by the same letters are not significantly different ($P < 0.05$)

9.3 Toxicological Analyses of Entomological Specimens and a Protocol for Collecting and Preserving Arthropods

It is recommended some cares to participate on the investigation and on collection of entomological evidence especially at the death scene. FE experts should wear protective clothing, gloves and shoe covers or boots mainly to avoid any contamination of the scene with fibres or other material (Amendt et al. 2007).

All material or equipment used to collect sample and for toxicological analysis must be clean and, if possible washed in hot water. A plastic film (magi pack®) or aluminium paper can be put in a lid of glass to avoid contamination of the sample.

Tissue, fluid and insects must be *in natura* and collected in vials free of preservatives or any external contamination. In most cases a refrigerated storage before the analysis is recommended. If possible, should be also stored a part of individuals of each sample in different vials under refrigerated condition to keep as reference sample.

Samples can be collect on the corpse (natural orifices, eyes, traumatic wounds or everywhere on the body), clothes, shoes and around the corpse (Amendt et al. 2007).

Though many samples for analysis are best sent in their original state, others require additives to maintain them in optimum conditions until they reach the laboratory (Knight 1991). Entomological samples can be collected by using forceps and spoons for collecting immature stages of flies, fine paintbrush for collecting eggs or handheld insect capture net for flying adults (Amendt et al. 2007). The toxicological analysis should be done as soon as possible to guarantee the integrity of samples.

When insects were collected from a carcass or corpse, they must be sampled in individual glass vials labelled with proper information (see protocol sheet for the collection of entomological evidence illustrated in Table 9.2). They must be taken to the laboratory, where larvae (if alive can be killed by freezing prior the analysis) must be rinsed in distilled water and pupae/puparia must be rinsed with methanol prior to extraction to avoid contaminants. Then, the sample must be prepared for analysis. The methods for toxicological analysis take into account the substances to be analysed, inorganic (metals) or organic (drugs and pesticides) as well as major and minor affinities to organic solvents, which depend on lipid, protein and cartilaginous components (Gagliano-Candela and Aventaggiato 2001). It is important to emphasize that all material used for analysis must be of glass to avoid contamination of phthalate present on plastic vials. Adult flies also can be sent to analysis although the drug concentration is less than immature stages.

Contaminations of entomological samples used in forensic investigations to determine the PMI or to detect toxic compounds can occur when arthropods collected from a body or body discovery site originate from a source other than the deceased (Archer et al. 2006). Some contaminating insects can be collected in the mortuary due accidental transfer of insects between corpses or infestation of bodies by insects living in the mortuary. For example, in a case of drug related death, if contaminating maggots are collected, the result can be negative once these maggots are free of drug and the toxicological research is affected.

Besides, some errors caused by post-mortem changes before the sample is removed cannot be avoided. These are variable, depending partly on the conditions

Table 9.2 Protocol of collecting samples for toxicological analysis

Case no:———			Collection date:	———	
Collector Name:	———		Date found:	———	
Body details:	———		Date reported missing:	———	
Age:	———		Gender:	———	
Height:	———		Weight (kg)	———	
City:	———	Region	Urban	Country side	Wildland
Death scenario:	Indoors		Outdoors	Field	River
	Ocean		Road	Dense vegetation	Burned
Temperature:	Ambient			Relative humidity	Precipitation
Decomposition stage:	Fresh	Bloated	Early decomposition	Advanced decomposition	Skeletonized
Number of sp/body part:	———			Evidence of Scavengers:	———
Insects					
Imatures:	Living		Egg	Maggot	Pupae
	Dead		Egg	Maggot	Pupae
Adults:	Living				
	Dead				
Site of collection:	Skin		Head,neck	Thorax, abdomen	Hands, foot
	Internal organs		Brain	Liver, kidneys, intestines	Heart, lungs
	Soil, coffin				
Samples for analysis:	Immatures		Adults		
Others:	Drug vials		Syringes	Prescriptions	
Probable substance:	———				

of post-mortem environment and partly on the microbiological population on corpse (Knight 1991).

There are two alternatives to obtain samples for toxicological investigations in animal experimentation. The first method is used an animal model to administrate the drug and then their samples are offered to immature insects to complete the development. Experimentation on animal models is usually necessary when trying to detect, quantify and evaluate drug effects on arthropod development. The domestic rabbit is the most indicated animal for this kind of research; for it is of easily handled and satisfactory results can be obtained with fair drug dosages. We would recommend using ten 3.0–4.0 kg rabbits of same sex for each experiment (five treatment replicates and five controls), minding that the animals should be kept in separate cages provisioned with water and food for as long as 5 days before the procedures, to avoid excessive stress. The drugs must be in contaminant-free saline solution for intravenous administration. As such, lethal, sub-lethal or LD50 dosages of each drug can be tested and determined. Past 30 min from the administration of the drugs, any surviving rabbits must be killed without inflicting external injuries, to avoid exposing blood and other liquids. Samples of the blood, liver, spleen, brain, urine, heart are then obtained through immediate necropsy and stored in labelled vials for posterior analysis. After that, additional samples from the rabbit's tissues must be taken for directly exposure to larvae of insects of forensic importance.

Each insect species must be separately placed with a tissue sample. Larval development time, pupation period until emergence, longevity length and mortality rates are to be recorded from both control and treatment conditions, and then compared to evaluate any drug alterations over development. Toxicological analysis may be performed on samples from these insects, usually by the same manner as is done with samples from mammals, as explained below.

For preparing the samples for toxin extraction, about 1.0 g of tissue, 1.0 mL for fluids and 10 insect samples (maggot, pupae or adult) are mixed separately with a solvent with homogenizing equipment. There are several techniques for extracting each different drug or chemical, and the methods employed may vary according to the laboratory or Institution of Legal Medicine standards and the available materials.

The extracted material can then be injected into GC-MS equipment to obtain a chromatogram (Fig. 9.1). A curve of calibration is prepared to quantify the drug and compare it to other concentrations. An appropriate standard solution is mixed to the sample, if necessary. Several substances involved in drug-related deaths have been already detected in previous studies by researchers of different countries (Table 9.3). The identity of each substance must be verified by comparison of their retention time and mass spectra with authentic standards, when possible (Casale and Moore 1994).

The second method, immature insects can be directly reared on artificial diets containing known concentrations of the drug. However, this method is restricted for lacking the metabolic processes on the drug that would have occurred after the previous ingestion and incorporation by a living organism.

Herein is suggested a protocol for collecting arthropod necrofauna aiming to facilitate and standardize procedures in investigations and toxicological analyses.

9.4 Drug Effects over Arthropod Development

Insects have been demonstrated to be effective alternative samples for toxicological analyses; especially in cases where the dead body is mostly bone remains.

Drugs and toxins can be detected in larvae whenever their metabolic accumulation rates exceed the excretion rates. It is not fully understood exactly how these processes occur in maggots neither how they affect their development. Nevertheless these alterations should be determined for accurately estimating PMIs.

Some experiments have described a direct correlation between the drug concentration in larvae and in human tissues, but other studies have found no correlation or have even demonstrated that drug concentrations found in larvae may be significantly lower than in body tissues. Sadler et al. (1997) concluded in a research of barbiturates and analgesics in *Calliphora vicina* larvae that it is impossible to predict on the basis of chemical structure which drugs are likely to be detectable. By this reason the absence of drug in insect sample may not indicate the substance was not present in the corpse.

```
File:           E:\MSFILES\ARICIO\COCA1.D
Operator:       trigo
Date Acquired:   3 Dec 97   8:14 am
Method File:    COCAINA1.M
Sample Name:                      1 ug cocaina tratada pelo metodo 1
Misc Info:                        1/50
ALS vial:       1
```

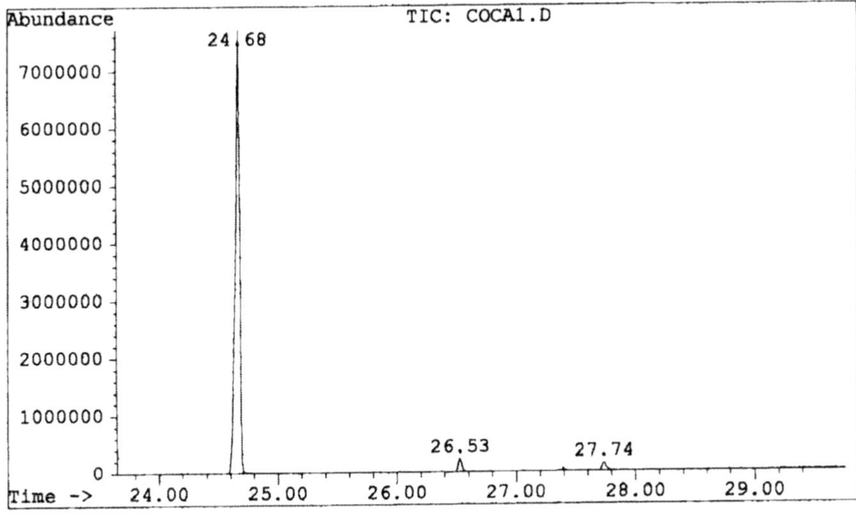

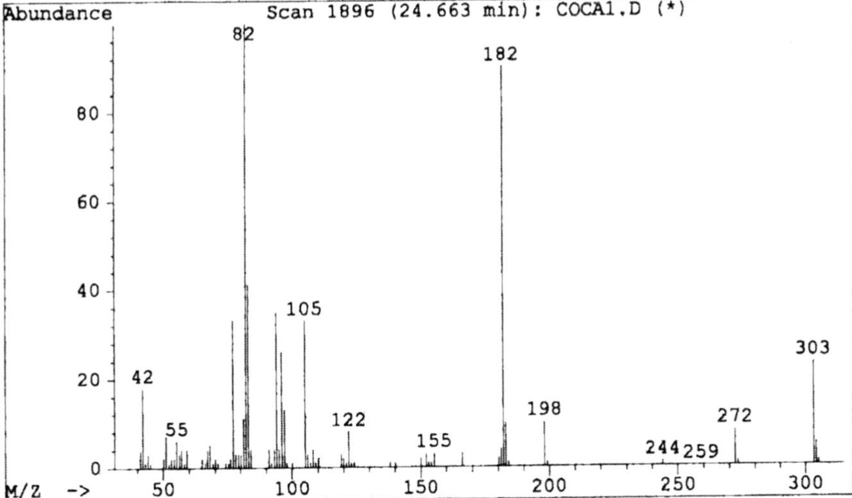

Fig. 9.1 Example of positive chromatogram to cocaine

Some experimental findings seem to demonstrate qualitative and quantitative correlations between substances found in tissues, larvae, puparia and insect feces.

However, Pounder (1991) did not detect any correlations between larval and tissue drug concentrations, suggesting that larvae may have only qualitative relevance since the quantifications based on these are not accurate. This is easily observed in actual

Table 9.3 Substances detected in arthropods

Substance	Sample	Reference
Copper, iron, zinc	Housefly (adult)	Sohal and Lamb (1979)
Phenobarbital	Blowfly (larvae)	Beyer et al. (1980)
Mercury	Blowfly (larvae, puparium, adult) and beetles	NuortevaandNuorteva (1982)
Arsenic	Piophilidae,Psychodidae and Muscidae	Leclerq and Brahy (1985)
Selenium	Adult house fly	Simmons et al. (1988)
Malathion	Fly larvae	Gunatilake and Goff (1989)
Cocaine	Fly larvae	Goff et al.(1989)
Bromazepam, levomepromazine, morphine, phenobarbital, triazolam,oxazepam, phenobarbital, alimemazine, clomipramine	Fly larvae	Kintz et al.(1990a,b,c)
Opiates	Fly larvae	Introna et al.(1990)
Heroin	Fly larvae	Goff et al.(1991)
Cocaine	Fly larvae and beetle fecal material	Manhoff et al.(1991)
Cocaine	Fly larvae and puparia	Nolte et al.(1992)
Amitriptyline, propoxyphene and acetaminophen	Fly larvae	Wilson et al.(1993)
Opiates	Fly larvae	Kintz et al.(1994)
Amitriptyline and nortriptyline	Empty fly puparia and beetle exuvie	Miller et al.(1994)
Phenobarbital, paracetamol	Fly larvae	Sadler et al.(1997)
3,4 methylenedioxy methamphetamine	Fleshfly (larvae and puparium)	Goff et al.(1997)
Morphine	Fly larvae	Hedouin et al.(1999)
Secobarbital	Fly larvae	Levine et al.(2000)
Diazepam	Blowf ly (larvae, puparia and adult)	Carvalho et al.(2001)
Morphine	Fly larvae	Bourel et al.(2001)
Cocaine, opiates, phenobarbital, levomepromezine, amitriptyline, nortriptyline, tioridazine and clomipramine	Blowfly(larvae) and human tissues	Campobasso et al.(2004)
Parathion	Diptera,Coleoptera, Hymenoptera, Isopoda and Acari	Wolff et al.(2004)
Nordiazepam and oxazepam	Fly larvae and puparium	Pien et al.(2004)
Cannabis sativa, cocaine and dietilpropione (amphepramone)	Blowfly(larvae, puparium and adult)	Carvalho (2004)

cases of criminal investigations: there are differences between the concentrations of substance found in human fluids shortly after death and in sampled larvae, several days or months after death. This may result from some internal redistribution of the drug through the body after death, molecular alterations on the drug structure (especially at the body surface area), and pharmacokinetics variations between arthropod species or different development stages or also in cases of concomitant use of drug (Tracqui et al. 2004). Besides, drug concentration varies depending on the tissue and site-to-site variability on the same organ. Williams and Pounder (1997) observed that skeletal muscle is not homogeneous with respect to concentrations of different drugs in cases of fatal overdose. Such findings suggest that entomological results can be affected by this variability related especially to quantitative analysis.

In this way, significant variation can be found in the concentration of many substances depending on the place from which sampling was carried out. In life there may be variations when arterial compared to venous blood is used. Similarly, portal blood may have a substantially higher concentration of a substance that is being absorbed from the intestine, before it is extracted by passage through the liver. After death, most variation is caused by uneven destruction by enzymatic and microbiological activity and by diffusion from sites of higher concentrations once the barriers formed by living cell membranes were broken down. Applying these facts to toxic compounds, the concentrations may vary considerably according to the sampling site (Knight 1991). For example, diazepam contains benzodiazepine that produces sedative and tranquilizing effects on humans, being widely prescribed as sedative, antispasmodic and antiepileptic. It is one of the sedatives that cause most fatal incidents by accidental overdose in Brazil (Bortolleto 1993). A human fatal concentration in blood is 5–18 mg/kg and in liver is 3 mg/kg (Knight 1991). The effects of diazepam over the development and lifecycle of blowflies were evaluated, and it was found to increase insect development rates when compared to negative controls, although there was no direct correlation between the concentrations measured from insects and the liver substrate from which maggots were feeding. Then, larvae feeding from tissues containing the drug developed faster than unexposed larvae: their required time for pupation and emergence of the adults was significantly shorter, suggesting that the drug is capable of accelerating the development of these flies, affecting their lifecycle (Fig. 9.2). It was also observed that diazepam affected the size and shape of puparium (Fig. 9.3) (Carvalho et al. 2001).

In an experiment developed by Musvasva et al. (2001) with chicken liver containing different concentrations of hydrocortisone and sodium methohexital with *Sarcophaga tibialis* (Diptera: Sarcophagidae), maggots exposed to either drugs (especially to small dosages) took significantly longer to pupate than unexposed ones, while those larvae that had ingested sodium methohexital also displayed shorter pupation periods than unexposed ones. This experiment illustrated that the drug effects are likely to be dependent on the insect development stage and the taken dosage.

Flesh fly *Boettcherisca peregrina* (Diptera: Sarcophagidae) maggots were reared on rabbit tissues containing cocaine and benzoylecgonine to evaluate the influence over their development rates. It was found that larvae developed more rapidly 36 h after egg eclosion when reared from tissues containing cocaine than from the negative controls, This increment in development speed lasted until 76 h after exposition to the drug and the period required to pupation and emergence of

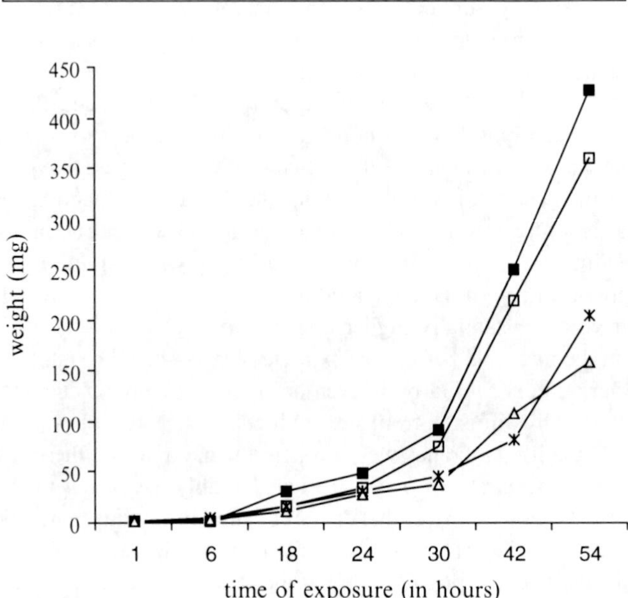

Fig. 9.2 Rates of development of maggots of *Chrysomya albiceps* and *Chrysomya putoria* reared on liver tissues from rabbits containing diazepam (Ca = C. albiceps, Cp = C. putoria)

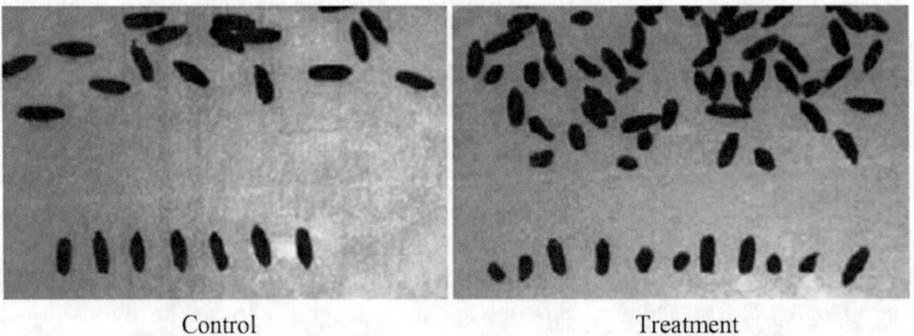

Fig. 9.3 Puparium of *Chrysomya putoria* larvae not exposed to drug (control) and exposed to drug (diazepam treatment)

the adults was shorter than in negative controls (Goff et al. 1989). Maggots from other flesh flies were reared on heroin-containing tissues, where they were found to develop faster than unexposed maggots between 18–96 h after egg eclosion. These alterations on development rate would be sufficient to cause PMI mistaken estimates if these heroin effects were overlooked during an actual investigation (Goff et al. 1991).

Wilson et al. (1993) analyzed maggots reared on human skeleton muscles obtained from suicides by overdose with co-proxamol and amitriptyline. Amitriptyline, nortriptyline and propoxyphene were detectable in third instar larvae in concentrations less than those detected in the muscles. Results were negative on puparia, indicating that the drugs vanished during the metamorphosis between prepupae and puparium formation. Thus, no drugs were detected in adult flies. These results indicate that the drugs were not accumulated during the lifecycle of these calliphorids, suggesting efficient excretion by the Malphigian tubules, in spite of the different chemical properties of these drugs (e.g. solubility in lipid/water).

In an experiment dealing with different concentrations of paracetamol in pig liver, there was no significant concentration variation during larval development of *C. vicina*, but there an increase on the development rates especially between the second and fourth days of drug exposition (O'brien and Turner 2004).

Sadler et al. (1995) measured drug bioaccumulation and excretion by *C. vicina* maggots reared on the muscles of three suicides containing amitriptyline, temazepam, trazodone and trimipramine. High concentrations of the drugs were found in larvae, with a peak concentration on the seventh and eighth days of exposition, followed by an abrupt decrease during pupation. The drugs were not detected in pupae through low-sensitivity analyses. Since drug concentration varied considerably from immature stages to adult, the authors concluded that the larvae were metabolizing and eliminating drugs at different rates. Later, Pien et al. (2004) detected nordiazepam and sub-products in empty puparia, suggesting that bioaccumulation could be occurring without excretion of the substances.

It has been already observed that effects from substances may vary depending on the dosage administered (if sub-lethal, lethal, twice the lethal, etc.), sometimes accelerating, delaying or not affecting development rates. The tissue type used in the experiment has also to be considered, because larval development length in maggots on different human tissues. Kaneshrajah and Turner (2004) have verified that the development rate of *C. vicina* maggots on pork liver is faster (up to 2 days shorter) than the rates of maggots reared on spleen, kidney, heart or brains. This difference is bound to influence PMI estimates. Probably it seems to be related to site-to-site variability of drug concentration or to the larvae feeding preference for a tissue. Necrophagous arthropods feeding from human tissues can be considered a valuable source of clues in investigations on criminal deaths helping estimate PMI and serving as reliable samples for toxicological analysis, in the absence of appropriate human tissues and fluids. If the body is in advanced decomposition state (e.g. dry remains), analyses of insect samples may be more accurate than analyses of the tissues. In a case study of a decomposed and skeletonized body, samples of calf muscle and maggots were sent for toxicological analysis. The result was negative to muscle and positive to maggots (Levine et al. 2000). Then, the absence of a drug from tissue or fluid does not imply its absence from feeding maggots.

These biological variations are vast and difficult to access, being further experimentation involving interactions between drugs and arthropod species necessary. It is important bearing in mind that some substances, if present on tissues may produce changes in the lifecycle of insects.

The many mentioned studies illustrate that insects found on corpses can be used for toxicological analysis. They demonstrate the occurrence of developmental alterations capable of causing mistakes on PMI estimates and draw attention to the phenomenon of secondary bioaccumulation.

References

Alves JR MJ, Thyssen PJ, Giorgio S, Mello, MMF, Linhares AX (2007) Detection of cocaine in *Chrysomya albiceps* (Diptera: Calliphoridae) larvae reared from a human corpse: Report of a forensic entomology case in southeastern Brazil. Annals of the Entomological Society of America - ESA, 55th ESA Annual Meeting 06 to 12 of December, 2007 Denver, USA. Available in World Wide Web http://esa.confex.com/esa/2007/techprogram/paper_29565.htm. Accessed 22 july 2008

Alves JR MJ, Esteban CS, Lima CGP, Thyssen PJ, Linhares AX, Giorgio S (2008) Uso da imuno-histoquímica para detecção da dinâmica e conversão metabólica de anfepramona em imaturos de califorídeos (Diptera). Proceedings of XXVII Brazilian Congress of Zoology, Curitiba-Pr, 17–21 de fevereiro de 2008

Amendt J, Campobasso CP, Gaudry E, Reiter C, Lê Blanc HN, Hall MJR (2007) Best practice in forensic entomology – standards and guidelines. Int J legal Med 121:90–104

Archer MS, Elgar MA, Briggs CA, Ranson DL (2006) Fly pupae and puparia as potential contaminants of forensic entomology samples from sites of body discovery. Int J Legal Med 120:364–368

Beyer JC, Enos WF, Stajic M (1980) Drug identification through analysis of maggots. JFSCA 25:411–412

Bortolleto ME (1993) Tóxicos, civilização e saúde. Contribuição à análise dos sistemas de informações tóxico-farmacológicas no Brasil, série política de Saúde, nº. 12, Rio de janeiro, pp 61–64

Bourel B, Hedouin V, Martin-Bouyer L, Becart A, Tournel G, Deveaux M, Gosset D (1999) Effects of morphine in decomposing bodies on the development of *Lucilia sericata* (Diptera: Calliphoridae). J Forensic Sci 44:354–358

Bourel B, Fleurisse L, Hedouin V, Cailliez JC, Creusy C, Goff ML, Gosset D (2001a) Immunohistochemical contribution to the study of morphine metabolism in Calliphoridae larvae and implications in forensic entomotoxicology. J Forensic Sci 46:596–599

Bourel B, Tournel G, Hedouin V, Deveaux M, Goff ML, Gosset D (2001b) Morphine extraction in necrophagous insects remains for determining ante-mortem opiate intoxication. Forensic Sci Int 120:127–131

Bourel B, Tournel G, Hedouin V, Goff ML, Gosset D (2001c) Determination of drug levels in two species of necrophagous coleoptera reared on substrates containing morphine. J Forensic Sci 46:600–603

Campobasso CP, Gherardi M, Caligara M, Sironi L, Introna F (2004) Drug analysis in blowfly larvae and in human tissues: a comparative study. Int J Legal Med 118:210–214

Carvalho LML (2004) Detecção e efeito de drogas no crescimento e desenvolvimento de formas imaturas e adultas de *Chrysomya albiceps* (Wiedemann) e *Chrysomya putoria* (Wiedemann) (Diptera: Calliphoridae), duas moscas varejeiras de interesse forense.Tese de Doutorado, 104p. Universidade Estadual de Campinas, Instituto de Biología. Avaiable *on line*: http://libdigi.unicamp.br/document/?code=vtls000347096

Carvalho LML, Linhares AX, Trigo JR (2001) Determination of drug levels and the effect of diazepam on the growth of necrophagous flies of forensic importance in southeastern Brazil. FSI 120:140–144

Casale JF, Moore JM (1994) 3, 4, 5-Trimethoxy-substituted analogs of cocaine, cis-trans-cynnamoylcocaine and tropacocaine: characterization and quantitation of new alkaloids in coca leaf, coca paste and refined illicit cocaine. JFSCA 39:462–472

Collins CA, Braga GL, Bonato PS (1997) Introdução a métodos cromatográficos, 7 ed. Editora Unicamp

Gagliano-Candela R, Aventaggiato L (2001) The detection of toxic substances in entomological specimens. Int J Legal Med 114:197–203

Goff ML, Lord WD (1994) Entomotoxicology: a new area for forensic investigation. Am J Forensic Med Pathol 15:51–57

Goff ML, Omori AI, Goodbrod JR (1989) Effect of cocaine in tissues on the development rate of *Boettcherisca peregrina* (Diptera: Sarcophagidae). J Med Entomol 26(2):91–93

Goff ML, Brown WA, Hewadikaram KA, Omori AI (1991) Effect of heroin in decoposing tissues on the development of *Boettcherisca peregrina* (Diptera: Sarcophagidae) and implications of this effect on estimation of postmortem intervals using arthropod development patterns. JFSCA 36(2):537–542

Goff ML, Brown WA, Omori AI (1992) Preliminary observations of the effect of methanphetamine in decoposing tissues on the development rate of *Parasarcophaga ruficornis* (Diptera: Sarcophagidae) and implications of this effect on the estimations of postmortem intervals. JFSCA 37(3):867–872

Goff ML, Brown WA, Omori AI, LaPointe DA (1993) Preliminary observations of the effects of amitriptyline in decomposing tissues on the development of *Parasarcophaga ruficornis* (Diptera: Sarcophagidae) and implications of this effect to estimation of postmortem interval. JFSCA 38:316–322

Goff ML, Miller ML, Paulsson JD, Lord WD, Richards E, Omori AI (1997) Effects of 3, 4-methylenedioxymethamphetamine in decomposing tissues on the development of *Parasarcophaga ruficornis* (Diptera:Sarcophagidae) and detection of the drug in postmortem blood, liver tissue, larvae and puparia. J Forensic Sci 42:276–280

Gudzinowicz BJ, Gudzinowics MJ, Hologgitas J, Driscoll JL (1980) Advances in chromatography, vol 18, chapter 5: The analysis of Marijuana cannabinoids and their metabolites in biological media by GC and/or GC-MS techniques.

Gunatilake K, Goff ML (1989) Detection of organophosphate poisoning in a putrefying body by analysing arthropod larvae. JFSCA 34:714–716

Hedouin V, Bourel B, Martin-Bouyer L, Becart A, Tournel G, Deveaux M, Gosset D (1999) Determination of drug levels in larvae of *Lucilia sericata* (Diptera: Calliphoridae) reared on rabbit carcasse containing morphine. J Forensic Sci 44:351–353

Introna F Jr, Lo Dico C, Caplan YH, Smialek JE (1990) Opiate analysis of cadaveric blow fly larvae as an indicator of narcotic intoxication. JFSCA 35:118–122

Introna F, Gagliano-Candela R., Di Vella G (1996) Opiate analysis on empty puparia-positive results. In: Proceedings of XX international congress of entomology. Firenze, Italy, 25–31 August 1996, p 755

Introna F Jr, Campobasso CP, Goff ML (2001) Entomotoxicology. FSI 120:42–47

Kaneshrajah G, Turner B (2004) *Calliphora vicina* larvae at different rates on different body tissues. Int J Legal Med 118:242–244

Kidwell DA (1993) Analysis of phencyclidine and cocaine in human hair by tandem mass spectrometry. J Forensic Sci CA 38:272–284

Kintz P, Tracqui A, Mangin P (1990a) Toxicology and fly larvae on a putrefied cadaver. J Forensic Sci Soc 30:243–246

Kintz P, Tracqui A, Ludes B, Waller J, Boukhabza A, Mangin P, Lugnier AA, Chaumont AJ (1990b) Fly larvae and their relevance in forensic toxicology. Am J Forensic Med Pathol 11:63–65

Kintz P, Godelar A, Tracqui A, Mangin P, Lugnier AA, Chaumont AJ (1990c) Fly larvae: a new toxicological method of investigation in forensic medicine. J Forensic Sci Soc 35:243–246

Kintz P, Tracqui A, Mangin P (1994) Analysis of opiate in fly larvae sampled on putrefied cadaver. J Forensic Sci Soc 34:95–97

Knight B (1991) Forensic pathology, chapter 31–33. Oxford University Press, New York

Leclerq M, Brahy G (1985) Entomologie et médicine legale: datation de la mort. J Med Leg 28:271–278

Levine B, Golle M, Smialek JE (2000) An unusual death involving maggots. Am J Forensic Med Pathol 21:59–61

Manhoff DT, Hood I, Caputo F, Perry J, Rosen S, Mirchandani HG (1991) Cocaine in decomposed human remains. J Forensic Sci 36:1732–1735

Miller ML, Lord WD, Goff ML, Donnelly B, McDonough ET, Alexis JC (1994) Isolation of amitriptyline and nortriptyline from fly puparia (Phoridae) and beetle exuvia (Dermestidae) associated with mummified human remains. J Forensic Sci 39:1305–1213

Moore JM, Casale JF (1994) In depth chromatography analyses of illicit cocaine and its precursor, cocas leaves. J Chromatogr A 674:165–205

Musvasva E, Williams KA, Muller WJ, Villet MH (2001) Preliminary observations on the effects of hydrocortisone and sodium methohexital on development of *Sarcophaga* (Curranea) *tibialis* Macquart (Diptera: Sarcophagidae), and implications for estimating post mortem interval. Forensic Sci Int 120:37–41

Nolte KB, Pinder RD, Lord WD (1992) Insect larvae used to detect cocaine poisoning in a decomposed body. J Forensic Sci 4:179–185

Nuorteva P, Nuorteva SL (1982) The fate of mercury in sarcosaprophagous flies and in insects eating them. Ambio 11:34–37

O'Brien C, Turner B (2004) Impact of paracetamol on *Calliphora vicina* larval development. Int J Legal Med 118:188–189

Pien K, Laloup M, Pipeleers-Marichal M, Grootaert P, De Boeck G, Samyn N, Boonen T, Vits K, Wood M (2004) Toxicological data and growth characteristics of single post-feeding larvae and puparia of *Calliphora vicina* (Diptera: Calliphoridae) obtained from a controlled nordiazepam study. Int J Legal Med 118:190–193

Pounder DJ (1991) Forensic entomo-toxicology. JFSS 31:469–472

Sadler DW, Fuke C, Court F, Pounder DJ (1995) Drug accumulation and elimination in *Calliphora vicina* larvae. Forensic Sci Int 71:191–197

Sadler DW, Patl MR, Robertson L, Brown G, Fuke E, Pounder DJ (1997) Barbiturates and analgesics in *Calliphora vicina* larvae. J Forensic Sci 42(3):481–448

Simmons TW, Jamall IS, Lockshin RA (1988) Accumulation, distribution and toxicity of selenium in the adult house fly, *Musca domestica*. Comp Biochem Physiol 91:559–563

Sohal RS, Lamb RE (1979) Storage excretion of metalic cations in the adult housefly *Musca domestica*. J Insect Physiol 25:119–124

Tanaka K, Ohmori T, Inoue T, Seta S (1994) Impurity profiling analysis of illicit methamphetamine by capillary gas chromatography. JFSCA 39:500–511

Tracqui A, Tracqui CK, Kintz P, Ludes B (2004) Entomotoxicology for the forensic toxicologist: much ado about nothing? Int J Legal Med 118:194–196

Williams KR, Pounder DJ (1997) Site-to-site variability of drug concentrations in skeletal muscle. Am J Forensic Med Pathol 18:246–250

Wilson Z, Hubbard S, Pounder DJ (1993) Drug analysis in fly larvae. Am J Forensic Med Pathol 14:118–120

Wolff M, Builes A, Zapata G, Morales G, Benecke M (2004) Detection of parathion (0,0-diethyl 0-(4-nitrophenyl) phosphorothioate) by HPLC in insects of forensic importance in Medellín, Colombia. Aggrawal's Internet. J Forensic Med Toxicol 5:6–11

Chapter 10
Cuticular Hydrocarbons: A New Tool in Forensic Entomology?

Falko P. Drijfhout

10.1 Introduction

The cuticle of all insects is covered with a very thin epicuticular layer of wax. This layer consists of free lipids, a class of compounds that includes hydrocarbons, alcohols, fatty acids, waxes, acylglycerides, phospholipids and glycolipids (Gibbs and Crockett 1998), although the presence of the last three groups may reflect aggressive extraction techniques which remove both internal as well as external lipids. This waxy layer prevents desiccation and penetration of micro-organisms (Gullan and Cranston 1994) as well as encoding various chemical signals. In the majority of the insects, and nearly all social insects so far studied, the free lipids are dominated by hydrocarbons (Lockey 1988). Cuticular hydrocarbons are found in all life stages of insects and are biologically very stable. Their biosynthesis is genetically based and modulated by factors such as reproductive status (Monnin 2006), developmental stage (Martin et al. 2001), diet (Buczkowski et al. 2005) or temperature (Toolson 1982; Savarit and Ferveur 2002; Rouault et al. 2004). The link found between diet and hydrocarbon production in social insects by Liang and Silverman (2000, 2001) is disputable, since behavioural changes occur in <2 min suggesting direct hydrocarbon transfer via contact with the prey rather than via diet, which explains why the host acquires the entire cuticular hydrocarbon profile of the prey (Liang and Silverman 2000).

Necrophagous insects are important insects for the determination of the postmortem interval. In establishing the postmortem interval a forensic entomologist will try to establish the age of the oldest colonising species; for which various methods exists (see review by Amendt et al. 2004). One of these methods could involve the identification of hydrocarbons present on the cuticle of the insects or even on their pupae and puparia as these contain hydrocarbons as well (Gilby and McKellar 1970) as some recent studies (Zhu et al. 2006, 2007; Ye et al. 2007; Roux et al. 2008) have shown that the composition of the hydrocarbon profile found on either the cuticle

F.P. Drijfhout
School of Physical and Geographical Sciences, Keele University, United Kingdom

J. Amendt et al. (eds.), *Current Concepts in Forensic Entomology*,
DOI 10.1007/978-1-4020-9684-6_10,

of the larvae or the pupae is not static but changes over time. If these changes occur as part of the development of larvae into adults and if this can be incorporated into a model, these hydrocarbons could be a very useful tool in estimating the age of a larvae or pupae and hence could increase the accuracy of the PMI.

The chapter aims to discuss the possible use of hydrocarbons in forensic entomology through a general overview of the structure of hydrocarbons as well as how they can be chemically analysed and identified. The second part focuses on the physiological function of hydrocarbons in insects and possible changes that may occur either related to their developmental stage or different climate conditions that may exist.

10.2 Hydrocarbon Structure

Cuticular hydrocarbons are compounds consisting only of carbon and hydrogen atoms, and all have the same basic structure consisting of a long carbon chain, which in social insects appears to be between 19 and 35 carbon atoms long. However, the apparent abundance of C_{19}–C_{35} cuticular hydrocarbons may be a reflection of the limitations of the detection techniques, since new high temperature GC columns (Akino 2006) and MALDI-MS techniques (Cvačka et al. 2006) are revealing cuticular hydrocarbons with chain lengths of up to 70 carbon atoms. Of particular interest is the discovery that *Formica truncorum*, and probably all other wood-ants (*Formica* s.str.), have 56% of their cuticular hydrocarbons with chain-lengths $>C_{34}$ (Akino 2006).

Hydrocarbon chains occur in their saturated or unsaturated form and may have one or more methyl groups (CH_3) attached (Fig. 10.1). In the saturated form (alkanes or

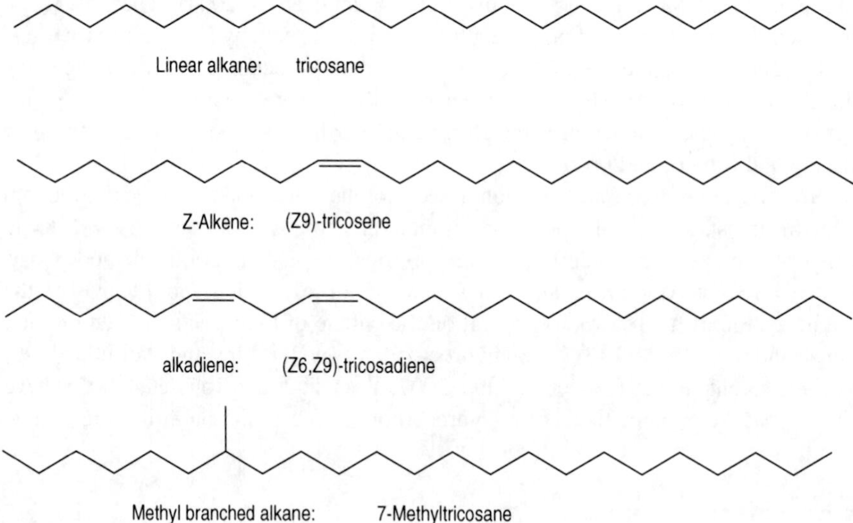

Fig. 10.1 Structures of some of the main classes of hydrocarbons

sometimes referred to as paraffins) all of the carbon atoms are joined by single bonds, while the unsaturated compounds (olefins) have either one (alkenes or monoenes), two (alkadienes or dienes) or three (alkatrienes or trienes) double bonds at various positions along the chain. In addition, olefins can take the form of one of two isomeric forms between carbon atoms at the double bond in the hydrocarbon chain. These are referred to as *cis*-alkenes (or Z-alkenes) and *trans*-alkenes (or *E*-alkenes), however, all known insect cuticular alkenes have the '*Z*' configuration.

The chain length, in addition to the presence and position of methyl groups and/or double bonds, has a large impact on the physical properties of the compound, such as structure and volatility (Gibbs 1995), which in turn underlies its suitability for any particular function. The shape of saturated hydrocarbons (alkanes) allows close packing of the molecules and they are therefore ideally suited to function as waterproofing molecules. Species like *Drosophila pseudoobscura* and *D. mojavensis*, appear to be incapable of producing alkanes in reasonable quantities and therefore suffer from high cuticular permeability and are susceptible to desiccation stress (Blomquist et al. 1985; Toolson et al. 1990). Also, as chain-length increases, alkanes become less volatile and are better at producing films with low water permeability. Therefore, a C_{33} alkane is better at reducing *trans*-cuticular water flux than C_{23} alkane. This is because of the stabilizing effect of weak intermolecular forces (van der Waals forces), which increase in strength as the molecular size increases. It is these forces that are disrupted by methyl-branches and double bonds consequently lowering the melting point (Gibbs 1998). For example, alkanes larger than C_{18} are wax-like solids at room temperature; however, the introduction of double bonds (Toolson and Kuper-Simbron 1989; Gibbs and Pomonis 1995) or methyl groups (Morgan 2004) into the molecule will drastically decrease their melting points. It can be assumed that boiling point and thus volatility is affected in a similar way but as yet little is known about the boiling points of alkenes and methyl branched hydrocarbons. The rich mixes of different groups of hydrocarbons e.g. alkanes and alkenes, ensures that the melting temperature range is low and broad due to intermolecular interactions (Morgan 2004), which is needed to regulate the permeability of the cuticle in a highly variable terrestrial environment. Gibbs (1998) suggested that the cuticular lipid layer is a mixture of microscopic solid areas (alkanes) and liquid areas (alkenes). The role of melting point in determining the action of compounds has already been suggested for ant repellents used by wasps, where it appears that its state (liquid or solid) is crucial to its function (Dani et al. 2003). The state of cuticular hydrocarbons will also affect their mobility and hence how accessible they are to chemoreceptors in insects.

10.3 Collecting Hydrocarbons

Before cuticular hydrocarbons can be analysed they need to be removed from the cuticle of the insect. Extracting the hydrocarbons can be done in two different ways; liquid extracting or solid phase micro extraction (SPME). Extraction needs to be done carefully as if the method is too aggressive, many of the internal compounds from the insect will be extracted as well. Insects that are collected from plants or within their nests usually give clean extracts, but if material (larvae, eggs, maggots, larvae) are collected from corpses, this potentially could lead to some more dirty extracts as some of the corpse's tissue may have accumulated on the material. Millar and Sims (1998) give a very good review on the preparation, cleanup and fractionation of extractions.

10.3.1 Liquid Extraction

In liquid extraction the hydrocarbons are extracted from the cuticle by dissolving them in an organic solvent such as hexane or dichloromethane. The insect is usually placed in a glass vial and a minimum of solvent is added to the vial. After circa 15 min the insect can be removed from the vial and the extract can be either analysed directly or dried down to a smaller volume to increase the concentration of compounds present. For the extraction of hydrocarbons a non-polar solvent is preferable as the hydrocarbons themselves are non-polar. Using too polar solvents, such as methanol or ethanol, will only extract more polar compounds or even internal compounds.

10.3.2 Solid Phase Micro Extraction

Solid phase micro extraction is a technique invented in 1990 and ever since then it has been used very extensively for many different purposes. The extraction is based on the affinity for compounds to absorb on an absorbent (a liquid stationary phase) (Pawliszyn 1997). It was first introduced by Pawliszyn and co-workers (Arthur and Pawliszyn 1990) for the extraction of organic compounds from an aqueous medium. However, many people have seen the benefits of this technique and now it is used for the analysis of headspace samples to identify volatiles emitted by living plants (Vercammen et al. 2000), identification of sex pheromones (BorgKarlson and Mozuraitis 1996; Rochat et al. 2000) as well as sampling solid samples (Lommelen et al. 2006). The major benefit of SPME is that it is very quick and can be used in connection with a gas chromatograph (GC, see below) or a liquid chromatograph. During sampling the compounds of interest are absorbed on the stationary phase and after a certain amount of time the compounds can be thermally desorbed within the injector of the GC; see sampling diagram in Fig. 10.2. With solid sampling it is even possible to rub the SPME needle over the body part of an insect, in order to absorb only those compounds found on the specific body part (Lommelen et al. 2006).

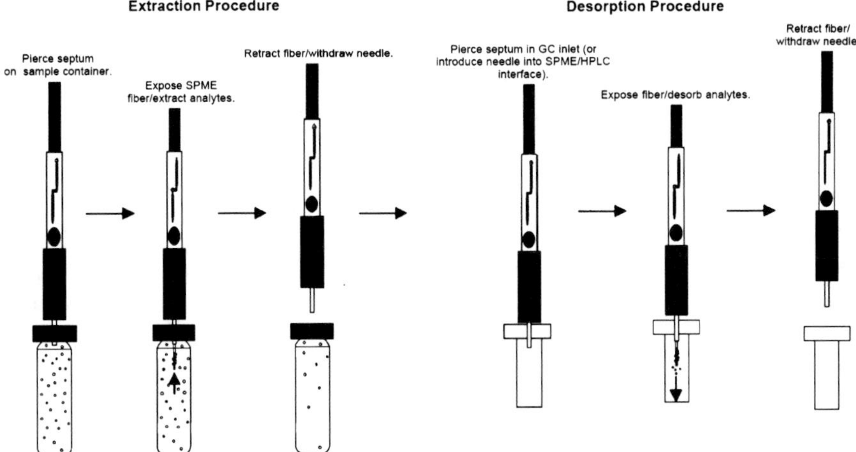

Fig. 10.2 Sampling by solid phase microextraction (SPME) and thermal desorption of the analytes from the coated fiber into a gas chromatograph (Adapted with permission from Supelco, Bellefonte, PH)

10.4 Chemical Analysis of Hydrocarbons

When extracts are prepared from insects, it can either reveal clean or 'dirty' extracts. When an insect is left too long in the solvent in a liquid-liquid extraction, changes are that many internal (polar) molecules will be extracted as well, yielding 'dirty' extracts. In this case the extract will not only contain hydrocarbons, but also fatty acids, phospholipids and other large polar molecules. Hence, before the extract can then be analysed to determine the hydrocarbons present, these hydrocarbons need to be separated from the more polar molecules. This can be done on a small column with silica gel, or on a TLC plate. Millar (1998) describes in detail how these can be carried out. Separation of hydrocarbons from lipids is a rather simple procedure. Separation of hydrocarbons among themselves in their different classes is more complicated. Saturated and unsaturated hydrocarbons can be separated via a silica gel column impregnated with silver nitrate (Millar 1998). Separation of linear alkanes and branched alkanes is more complicated although possible (Xu and Sun 2005), but quite often this is not needed for the identification of the hydrocarbons found on the cuticle.

Due to the non-polar nature of hydrocarbons and the fact that the majority of the cuticular hydrocarbons are volatile, the chemical analysis of extracts in order to determine the hydrocarbons is mainly carried out by gas chromatography (GC). Gas chromatography has been used since the 1960s and is still one of the standard methods of analysis of hydrocarbons. One important reason for this is that GC is a simple, fast, flexible and relative inexpensive method that can be used on a routine basis in almost every lab. It is also a technique that requires minimal training. Another important feature of GC is its ability to separate up to 100 compounds in a single run (Heath and Dueben 1998). When compounds are injected in the heated injector port of the GC, they are vaporized and are carried with a carrier gas on to the column. Injection can be done either in split or in splitless mode. In split mode, part of the mixture that is injected is split off through a splitvent. This is done to prevent the column being overloaded with analytes as the column has a limited capacity to separated compounds efficiently. However, if the concentration of the analytes within a sample is very low, the injection is carried out in splitless mode. In that case all of sample is carried with the carrier gas onto the analytical column. As they move through the column the molecules will interact with the stationary phase in the column. These interactions as well as the volatility of the compound determine the retention of the compound on the column. The less volatile the molecule is and the more interaction it has, the more it will be retained on the column, i.e. leaving the column at the later time, resulting in a higher retention time. In GC the driving forces for separation compounds as they move through the column are their affinity for the liquid coating the walls of the column and the temperature of the oven; as the temperature increases the compounds move into the vapor phase and move more quickly through the column (Heath and Dueben 1998). Therefore, in general the oven temperature is not constant throughout the run, but started at a low temperature (40–60°C), and consequently varied slowly to a high temperature (280–320°C) at 10–20°C/min to ensure higher boiling compounds will elute from the column. Special high temperature columns are now available to elute compounds in the temperature range of 420–450°C. The plotted result from a GC analysis is called a chromatogram: a plot of the retention time versus the concentration of a compound or several compounds.

Detection of the compounds eluting from the column can be done by several detectors. The standard detector used in GC is a Flame Ionisation Detector (FID) or more commonly nowadays, a mass selective detector (MSD or mass spectrometer). With an FID the compound eluting from the column is burnt within the flame of the detector. This produces ions in the detector resulting in an ion current that can be measured. The amount of ions produced (and consequently the ion current) is linear with the concentration of the chemical. FID is a general detector that detects all organic compounds, it is very sensitive and can detect up to 10–100 pg. However, one of the major disadvantages of the FID is that it is a very non-selective detector. A peak in the chromatogram only reveals the presence of a chemical, but it doesn't tell us anything about the nature or structure of that chemical. The time at which it elutes from the column (retention time) does however reveal information about the volatility of the chemical. The higher the retention time, the less volatile the chemical will be. Due to the non-selectivity of the FID, analysis of hydrocarbons with a gas chromatograph with an FID (GC-FID) can lead to false results. Two compounds, especially two isomeric hydrocarbons, can have very similar retention times, but still be different compounds (see also Fig. 10.6).

With the rapid development of mass spectrometry in the 1940s and 1950s and more important the ability to couple a mass spectrometer to the gas chromatograph (GC-MS) in 1956, much more information on the chemical in question can be obtained. In mass spectrometry the compound that enters the instrument (in high vacuum) through the inlet system, is hereafter ionized within the so-called ion source.

Ionisation can be carried out either by hitting the compound with accelerated electrons (electron ionisation, EI) or with an ionised gas (chemical ionisation, CI). This process leads to ionized molecules which then fragment (molecular ion or parent ion) into smaller fragments (daughter ions). These ions (molecular ions and daughter ions) are then separated according to their mass/charge ratio in the mass analyser and subsequently detected through an electron multiplier. Figure 10.3 shows these processes in a schematic diagram. The result is a mass spectrum that is a plot of the mass of the fragments and its intensity that is often given as a percentage of the most abundant peak (see Fig. 10.4).

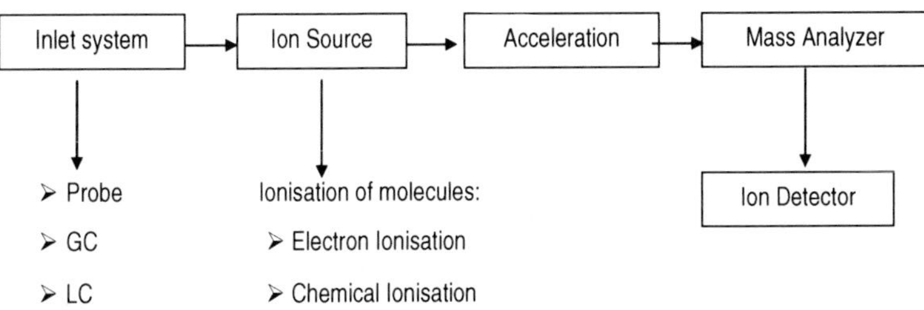

Fig. 10.3 Schematic diagram of the most important processes within mass spectrometry

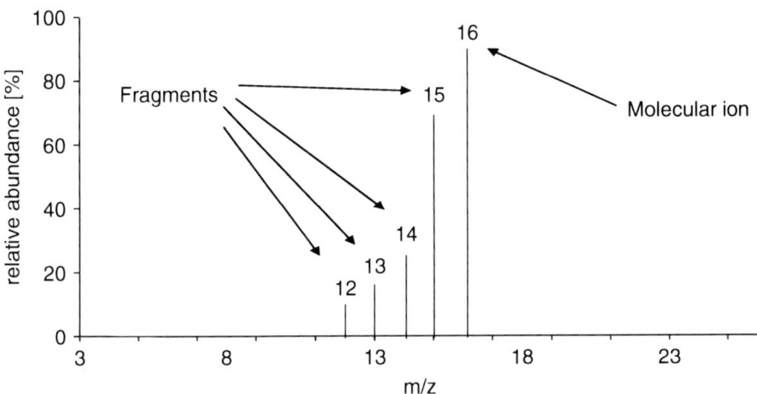

Fig. 10.4 Model mass spectrum of methane (CH_4; molecular weight = 16)

The current field of mass spectrometry is very wide and is mainly determined by two important stages in mass spectrometry; either the method of ionization or the method of ion separation. These two methods are independent from each other and also more or less independent of the class of compound to be analysed. This chapter is not intended to cover all these different methods used in mass spectrometry. The reader is referred to other textbooks such as Introduction to Mass Spectrometry (Watson 1997) and Mass spectrometry – Principles and Applications (De Hoffmann and Stroobant 2001) for more detailed information regarding ionization methods and ion separation methods in mass spectrometry. For more background information of the interpretation of mass spectral data the reader is referred to the standard reference book of McLafferty and Turcek (1993).

The manner in which a compound fragments in the ion source is very characteristic for each compound. The obtained mass spectrum (plot of the mass-to-charge (*m/z*) versus the abundance of each mass) can therefore be compared to those found in databases and unknown compounds may be identified based on their mass spectrum. When a gas chromatogram is coupled to a mass spectrometer, both quantitative and qualitative data can be obtained from compounds present in a sample. In GC-MS compounds within a sample are first separated in the GC and one by one each peak (compound) enters the mass spectrometer. This means that a mass spectrum can be obtained for each peak in the chromatogram. The area under a peak corresponds to the amount of material, whereas the mass spectrum of each peak gives structural information about the compound or compounds.

The different classes of hydrocarbons found on the cuticle of insects give different mass spectra. Figure 10.5 gives examples of mass spectra of an alkane, alkene, methyl branched alkane and dimethyl branched alkane respectively. Alkanes are characterised by high intensity *m/z* at 41, 55, 71 and decreasing intensities of ions of the series 85, 99, 113, etc. These ions are similar for all alkanes, so different alkanes (with different chain lengths) can only be identified through their molecular ions that correspond to the molecular weight of the alkane. These ions are usually more abundant then the fragment ions that can be found in the high mass region, see Fig. 10.5a. Alkenes have a similar pattern, although the masses are different. Values for *m/z* = 53, 69, 83 and 97 are present in high abundance followed by a decrease of *m/z* after the series 111, 125, 139, etc. Alkenes give a slightly more distinct molecular ion (see Fig. 10.5b), two mass units less then the corresponding alkanes.

Monomethyl alkanes have a very similar pattern to the alkanes, except that the characteristic pattern is interrupted in the centre of the mass spectrum if the methyl group is located more or less in the middle of the alkane. Ions (or rather pair of ions) with increased abundance indicate the position of the methyl group on the linear chain. An example is given in Fig. 10.5c, the mass spectrum of 9-methylpentacosene (9MeC25) where two ion pairs are present. The pair of ions at *m/z* 140:141 and 280:281 are formed after the compound fragments around the methyl group. In general within these ion pairs, the even mass ion is slightly higher in abundance than the odd mass ion. These more intense ions can be used to locate the position of the methyl group. If the methyl group is located at the beginning of the carbon chain, a slightly changed pattern is observed. There are no increased ions in the

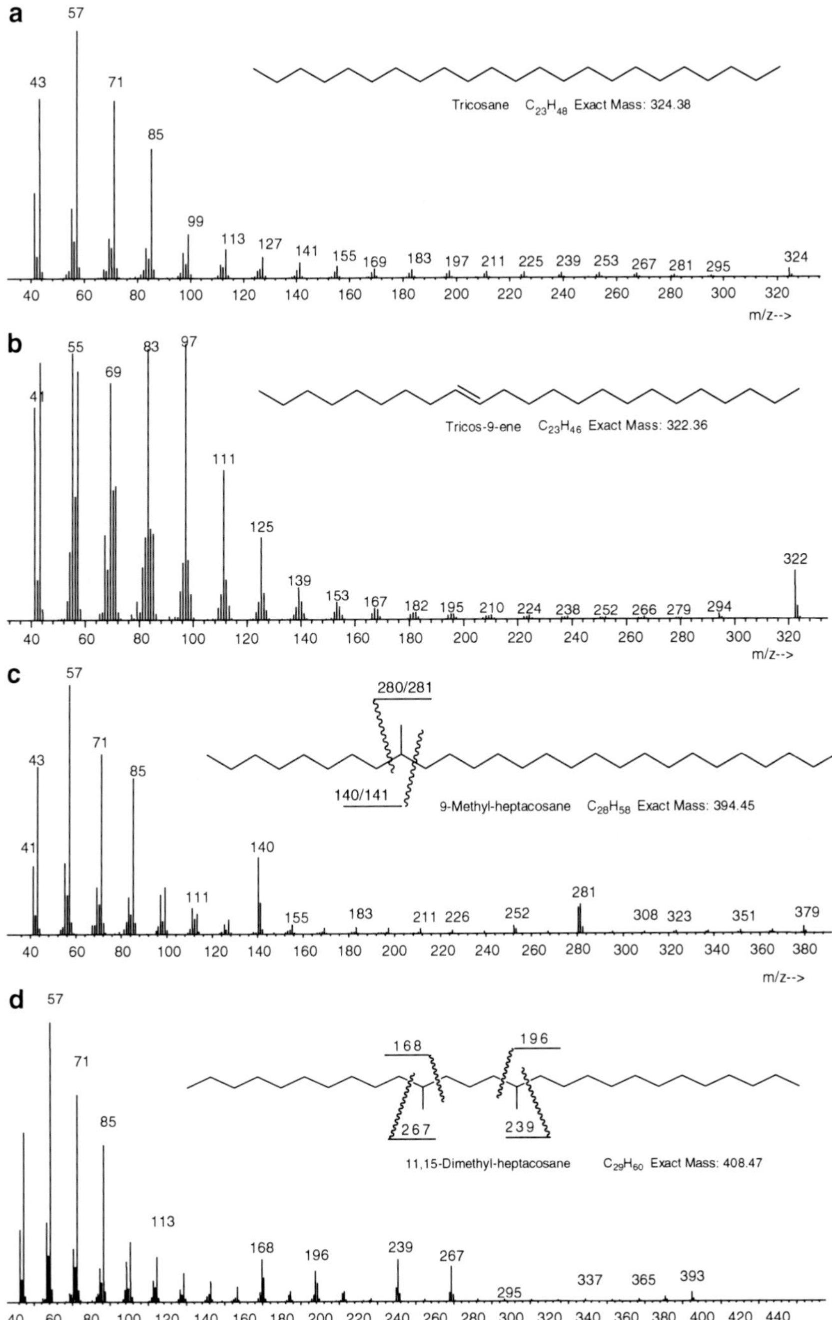

Fig. 10.5 Examples of mass spectra of respectively an alkane (**a**), alkene (**b**), monomethylalkane (**c**) and dimethylalkane (**d**)

centre of the mass spectrum, but two more intense ions; one in the lower mass region and one at the higher mass region, near the molecular ion.

A similar process and pattern arises with dimethylalkanes, but instead of each compound yielding two additional characteristic ion pairs, now each compound will yield four characteristic ion pairs. Whereas in monomethylalkanes the even ions are more abundant, in dimethylalkanes, two of the ion pairs have the odd ion in higher abundance, while the other two have the even ion in higher abundance (see Fig. 10.5d).

One of the main problems in identifying branched hydrocarbons is that often two isomeric compounds, e.g. 11MeC27 and 13MeC27 elute as a single peak. The mass spectra of this 'single peak' will therefore be a mixture of two or sometimes even more branched hydrocarbons that cannot be separated on the GC column. The mass spectrum however will have the characteristic ions for both monomethylalkanes, making identification more difficult. In general monomethylalkanes with the methyl group at the beginning of the chain can be separated out, e.g. 3MeC27 and 5MeC27 are easily separated. The nearer the methyl groups are towards the middle of the molecule, the more difficult it is to separate them. Dimethylalkanes with only one methyl group on a different carbon in the chain (e.g. 5,11-diMeC27 and 5,15-diMeC27) are even more difficult to separate.

Mass spectrometry has the major advantage over the FID that it gives information (masses of molecule or its fragments) about the peak shown on the chromatogram. Therefore it enables the researcher to investigate the mass spectrum of the peak and determine whether the mass spectrum is of a pure compound (hence the peak consists of a single compound) or whether the mass spectrum is a mixture of different compounds (hence the peak consist of more than one compound, e.g. 11MeC27 and 13MeC27). This is illustrated in Fig. 10.6, which shows the chromatographic analysis of the contents of the Dufour glands of two bumblebee species. The chromatograms (top of the figure) could represent the analysis carried out by GC-FID and if only the chromatogram is observed, peaks A and B appear to be the same compound, due to identical retention times. However, further investigation of the mass spectra of peak A and B clearly shows that they are not the same compounds. In fact peak A is a single peak while peak B consist of four different compounds eluting at the same time.

10.5 Production and Physiological Role of Cuticular Hydrocarbons

Hydrocarbons are found on the epicuticle of almost all insects. It is however not entirely sure where the hydrocarbons are produced and or where they are stored within the insect. In flies cuticular hydrocarbon production is believed to be under direct control of ecdysteroid hormones that are indirectly influenced by the juvenile hormone (Trabalon et al. 1994; Wicker and Jallon 1995). Genetic feminization studies in *Drosophila melanogaster* have shown that cuticular hydrocarbon production

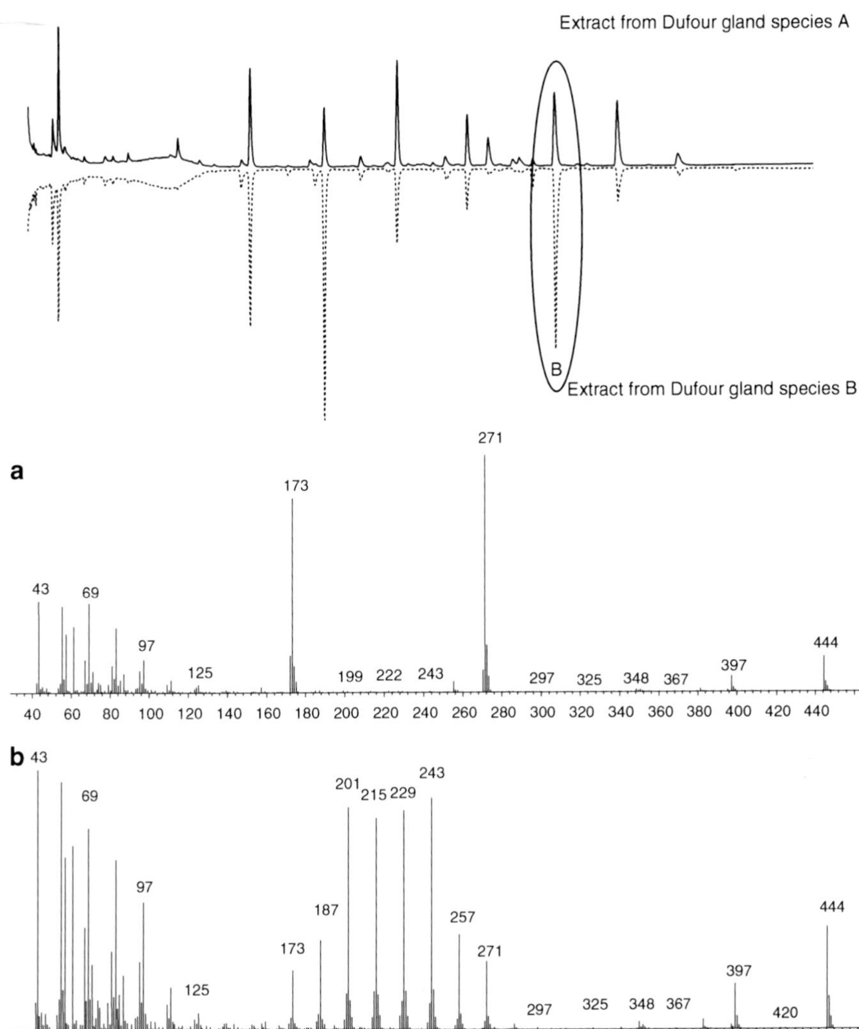

Fig. 10.6 Chromatographic analysis of the contents of the Dufour glands from two bumblebee species. The top part shows the chromatograms (mirrored), the bottom part the mass spectra of peak A (from bumblebee species A) and B (from bumblebee species B) respectively

occurs in the sub-cuticular abdominal cells (oenocytes) of adult flies (Ferveur et al. 1997) and since these cells also produce cuticular hydrocarbons in mosquitos, locusts and cockroaches (Schal et al. 1998) they, along with the epidermal cells, are probably the site of cuticular hydrocarbon production in most insects. The hydrocarbons are then transported from the oenocytes by a high-density haemolymph lipoprotein to various target tissues, including the ovaries (since insect eggs also require protection from desiccation), and pheromone or cuticular hydrocarbon-emitting glands (Gobin et al. 2003). It is possible that hydrocarbons are produced in several different glands depending on the function of produced hydrocarbon in insects (Soroker and Hefetz 2000; Thompson et al. 1981). Undecane for example is frequently found in the Dufour gland of Formicine ants, and is thought to have a function as an alarm pheromone (Fujiwara-Tsujii et al. 2006).

The principle role of cuticular hydrocarbons is controlling the *trans*-cuticular water flux in insects due to their high surface: volume ratio. Although cuticular hydrocarbons account for about 0.1% of total mass of a typical insect, their presence can reduce the insect's permeability to water by up to 1,300% (Edney 1977). Hydrocarbons are very non-polar molecules and hence prevent water from passing through a layer of hydrocarbons. Due to their structure alkanes can be nicely layered and hence they are more suitable for waterproofing then alkenes or methyl branched alkanes. The hydrophobicity of cuticular hydrocarbons also prevents the wetting of insects, since a single drop of water would increase the body weight of an insect enormously. Having evolved a wide range of cuticular hydrocarbons to protect against dehydration, the insects had the potential to develop them as signalling molecules. Communicating with chemicals is a fundamental process, which drives speciation and evolution of social structures, in a similar way that communicating by sound, such as language or song, has been a key factor in the evolution of vertebrates. Howard and Blomquist (2005) identified up to six different areas of signalling where cuticular hydrocarbons may play an important role, these are: species and gender recognition, nest-mate recognition, task-specific cues, dominance and fertility cues, chemical mimicry, and acting as primer pheromones. Hydrocarbons have also been shown to be used as deterrents in ants, although only one example exists of this (Martin et al. 2007),

The rise in the number of studies on cuticular hydrocarbons has been driven by the increased availability of gas chromatography coupled to mass spectrometry (GC-MS). Despite this, the number of cuticular hydrocarbons that have been shown to have a pheromonal effect remains relatively low, especially in relation to the number of compounds identified. Direct evidence is lacking (Dani 2006) in all but a few cases that are highlighted below. In the next paragraph the four main classes of hydrocarbons will be discussed in relation to their physiological role within insects.

10.5.1 Alkanes

Linear alkanes or *n*-alkanes (straight chains) normally dominate the cuticular hydrocarbon profile of insects (Blomquist and Dillwith 1985). Alkanes occur as a continuous homologous series of either odd numbered linear alkanes or even numbered linear alkenes across narrow (e.g. C_{19}–C_{21} in *Gnantogenys striatula*, Lommelen et al. 2006) or broad (e.g. C_{21}–C_{37} in *Dinoponera quadriceps*, Monnin et al. 2002) ranges. Odd linear *n*-alkanes are always more abundant than the even linear *n*-alkanes.

Despite being well studied there is little conclusive evidence yet that alkanes are key signalling compounds. Pickett et al. (1982) demonstrated that honeybee olfactory neurons did not respond to C_{23}–C_{27} alkanes which was supported by calcium imaging studies, where no glomerular (neural) responses in the antennal lobe of honeybees were obtained for alkanes $>C_{10}$, whereas a wide range of alcohols, aldehydes and ketones elicited strong neural responses (Sachse et al. 1999). This explains why honeybees were unable to learn to discriminate well between alkanes (C_{27}–C_{31}), whereas they learnt and discriminated well between most alkenes (Châline et al. 2005). These findings are supported by the observation that only altering the alkene and not the alkane profile results in a clear behavioural response in honeybees (Dani et al. 2005), the paper wasp *Polistes dominulus* (Dani et al. 2001) and ant *Formica exsecta* (Martin et al. 2008c). However, alkanes can be learnt by honeybees (Châline et al. 2005) just not very well. They may play a role in nest-mate recognition in *Formica japonica* (Akino et al. 2004), but the overall lack of behavioural response to alkanes suggests that their primary function is water-proofing.

10.5.2 Alkenes

Whereas alkanes have one feature, i.e. their chain lengths, alkenes have two added features; the angle formed by the double bond and the lengths of the chains either side of the double bond. This means that a single alkene can have different functions, depending on the position of the double bond. For example, the sex pheromone of the housefly (*Musca domestica*) is (*Z*9)-$C_{23:1}$ with the double bond at the nineth position, whereas (*Z*7)-$C_{23:1}$ with the double bond at the seventh position, induces a dose-dependent inhibition of male-male courtship in *D. melanogaster* (Scott 1986). Therefore, it is essential that the position of all the double bonds in the alkenes is determined e.g. through dimethyldisulfide (DMDS) derivatization (Carlson et al. 1989).

Many other alkenes are found to be involved in sexual communication (Howard and Blomquist 1982), for example, the stable fly, *Stomoxys calcitrans* uses (*Z*9)-$C_{31:1}$ and (*Z*9)-$C_{33:1}$ as sex pheromones (Sonnet et al. 1979). Among the social insects, honeybees and wasps are much more responsive to alkenes than alkanes (Dani et al. 2001; Dani et al. 2005; Châline et al. 2005). Furthermore, the proportions of alkenes in three termite (*Macrotermes falciger*) phenotypes were associated with inter-group aggression levels (Kaib et al. 2002). Alkenes are often found on the cuticle of ants and correlation studies have shown that they may be used as a fertility signal (Monnin 2006). However, there is now direct evidence using bioassays with synthetic compounds that the (*Z*9)-$C_{23:1}$–(*Z*9)-$C_{29:1}$ and (*Z*9)-$C_{25:1}$–(*Z*9)-$C_{33:1}$ alkenes found in *Formica japonica* (Akino et al. 2004) and *F. exsecta* (Martin et al. 2008c) ants respectively, act as nest-mate recognition cues in these species.

10.5.3 Alkadienes and Alkatrienes

Relative to alkanes and alkenes, alkadienes and alkatrienes are much less common especially within the social insects. The introduction of more double bonds increases the structural complexity of the molecule; hence it is not surprising to find these compounds being involved in communication. For example, 9,19-alkadienes are oxidized and cleaved then act as sex pheromones of the yellow-headed spruce sawfly, *Pikonema alaskensis* (Bartelt et al. 1982) and two 7,11-alkadienes from *D. melanogaster* are involved in stimulating male courtship of females (Antony and Jallon 1982). The major female sex pheromone from the almond seed wasp, *Eurytoma amygdali*, was identified as a mixture of two dienes, 6,9-tricosadiene and 6,9-pentacosadiene (Krokos et al. 2001; Mazomenos et al. 2004). An example of an alkatriene is found in the arctiid moth, *Utethesia ornatrix*, which uses 3,6,9-heneicosatriene as short range orientation cues for males (Conner et al. 1980). Additional examples can be found in the extensive review by Lockey (1988).

10.5.4 Methyl-Branched Alkanes

Twenty-one of the twenty-four cuticular hydrocarbons with known biological activity are methyl-branched alkanes (Nelson 1993). These were functioning as sex pheromones, kairomones and anti-aphrodisiacs. In the last 10 years the number of methyl-branched cuticular hydrocarbons suspected to be involved in chemical communication continues to grow e.g. colony recognition in *Polistes* wasps (Dani et al. 2001) and fertility signal in ants (Endler et al. 2004). Although numerous studies identified methyl-branched alkanes thought to be involved in various aspects of communication, it has been very difficult to prove this with bioassays. Synthesising these compounds is currently difficult and laborious. Furthermore, establishing the chirality of an isomer and subsequent synthesis of the correct isomer is difficult. However, the diversity of identified methyl-branched alkanes is far greater than any other group of cuticular hydrocarbons (Martin and Drijfhout, in press), since not only can the chain length vary, but also the position and number of the methyl groups along the chain, all of which alter the molecular structure. This diversity makes them prime candidates as signalling compounds.

10.5.5 Methyl-Branched Alkenes

In contrast to the abundance of saturated methyl-branched hydrocarbons the unsaturated versions are less common. Information about methyl-branched alkenes in communication is sparse, although 13-methyl-Z6-heneicosene has been identified as the sex pheromone of the herald moth, *Scoliopteryx libatrix* (Francke et al. 2000), while 13, 17-dimethyl-1-tritriacontene and 13, 17-dimethyl-1-pentatriacontene have been identified as the contact sex pheromone components of the female tsetse fly, *Glossina austeni* (Kimura et al. 2001). Recent studies on the cuticular hydrocarbons of the invasive garden ant (*Lasius neglectus*) and some of its related species indicate that several methyl branched alkenes are present in relatively large quantities (Ugelvig et al. 2008), but their functions are not yet known.

10.6 Do Hydrocarbon Profiles Change?

Although it is pretty clear that all insects will have some hydrocarbons on their cuticle, and we can assign a function to some of these hydrocarbons (although this is frequently only via indirect evidence), it is still not known or fully understood if these hydrocarbons change over time in the life of the insect. A phenomenon that will be crucial if these hydrocarbons are to be used in forensic entomology. In a preliminary study, the author with co-worker Dr. Martin studied the hydrocarbon profile of bees of different ages. In this study it became clear that bees just emerging from the cells have a different hydrocarbon profile than worker bees within the colony. This change however is very rapid. Other studies have indicated that the hydrocarbon profile of ants could change with a different diet (Buczkowski et al. 2005). Therefore is seems plausible to assume that the hydrocarbon profile of an insect will fluctuate over time, hence indicating a certain status (age, gender, caste) of the insects. However, this fluctuation should not interfere with biological signalling and should therefore not involve those signalling compounds. However within a hydrocarbon profile only some hydrocarbons are used for communication, while others often have different functions (Martin et al. 2008a,b,c).

Several factors are important for changing the hydrocarbon profile of an insect. In the above section some have already been mentioned. These are genetic factors such as the age of a bee or the gender of a fly. In general the age or gender of an insect can influence the profile of the hydrocarbons present. This genetical component of hydrocarbon production has been well studied in *Drosophila* spp. (Ferveur 2005). Several genes are known to either induce or reduce hydrocarbon production. Much of this work has been carried out in relation to the sex pheromones in *D. domestica* that is (Z9)-tricosene. The (*Z7*) isomer which is produced by the males can either be up-regulated or down-regulated depending on the enzyme. However, the amount of saturated versus unsaturated hydrocarbons can also be altered, as when the PGa14 transposon was inserted within desat1 in *D. melanogaster* production of unsaturated hydrocarbons was reduced and saturated hydrocarbon production increased (Marcillac et al. 2005). Savarit and co-workers have found that overexpression of the *UAS-tra* transgene in *D. melanogaster* after a heat shock resulted in the complete elimination of their cuticular hydrocarbons (Savarit et al. 1999). Unfortunately this level of detail is not available for all insects, and in many species the production and regulation of hydrocarbons is unknown.

A factor related to the genetic aspect is the developmental stage in which an insect is. In bees it has been shown that the profile can change over time. In ants several studies have confirmed the fact that ants of different casts have different HC profiles, e.g. foragers have more saturated hydrocarbons than ants that work within the nest (Greene and Gordon 2003; Martin and Drijfhout 2009).

Environmental factors such as diet, or temperature could have an effect on the hydrocarbons present. Diet has already been mentioned but this effect is debatable. Geoclimate or temperature can also have an effect of the hydrocarbons present (Rouault et al. 2001; Savarit and Ferveur 2002). In *D. pseudoobscura*, when they are found in the Mojave Desert, mainly saturated hydrocarbons are found, but when the same ants are reared in the lab mostly unsaturated hydrocarbons are produced (Toolson and Kuper-Simbron 1989). Similarly, the ratio of C_{35}:C_{37} alkadienes is decreased when *D. mojavensis* is reared in conditions where the temperature is raised from 17°C to 34°C (Markow and Toolson 1990). All these observations can be explained by the fact that either alkanes (or longer alkanes) are more suitable for waterproofing compared to alkenes (or shorter alkanes).

For living organisms, there is another factor that has been shown to be important, and that is the micro-organisms and fungal pathogens living on, e.g., insects (Crespo et al. 2000). The reader is referred to an excellent review by Pedrini et al. (2007) on entomopathogenic fungi invading their hosts through the cuticle. Napolitano and Juarez (1997) were the first to actually show that entomopathogenic fungi can use hydrocarbons as an energy source. Hydrocarbons extracted from the blood sucking bug, *Triatoma infestans* appear to be a much better energy source for both *Metarrhizium anisopliae* and *Beauveria bassiana* than synthetic hydrocarbons that contained only linear alkanes. Epicuticular hydrocarbon changes were also measured on both *Ostrinia nubilalis* and *Melolontha melolontha* due to infections of two entomophatogenous strains, *Beauveria bassiana* and *B. brongniartii*. In *O. nubilalis* the total amount of hydrocarbons on the cuticle was reduced from 6.88 ± 1.6 ng to 0.84 ± 0.08 ng in just 96 h after the application of *B. bassiana* (Lecuona et al. 1991). A similar effect was observed in *M. melolontha* where the amount of extractable hydrocarbons dropped from 7.71 ± 2.53 ng to 2 ± 0.8 ng in 24 h. After 96 h the amount increased again to 15.2 ± 4.8 ng indicating a restoration of the hydrocarbon profile or deposition of hydrocarbons from the spores on the insect's cuticle (Lecuona et al. 1991). Of the hydrocarbons disappearing, the monomethylalkanes and dimethylalkanes were the first molecules to decrease or disappear from the cuticle (Lecuona et al. 1991). This is in accordance with earlier results from Napolitano and Juarez (1997) who showed that linear alkanes was a lesser energy source compared to a hydrocarbon extract that contained some branched hydrocarbons.

10.7 Weathering of Hydrocarbons

Hydrocarbons on living organisms such as insects will change over time as new compounds are synthesised each time through the insects' biosynthesis. Furthermore, as compounds are emitted, either as signalling compounds or used in nestmate recognition, or during trophilaxis some compounds will decrease over time. However, the question to be asked is, whether the hydrocarbon composition will change over time *independently of any biosynthesis*, a situation that might occur on the puparia. In this case, no new compounds can be added in time and hence the stability of hydrocarbons is important. The question is how likely will they degrade or change over time? A large number of papers have been published on this topic, but not in relation to insects but in relation to oil spills and gasoline releases (for reviews see Christensen and Tomasi 2007; Medeiros and Simoneit 2007; Morrison 2000).

However these studies can give valuable information to a forensic entomologist on the behaviour of hydrocarbons in more natural conditions. Oils and gasoline consist mainly of hydrocarbons (linear and cyclic, saturated and unsaturated) and depending on the origin of an oil sample the hydrocarbon profile or composition will be different. When an oil spill has been observed it is very important to establish the time when the actual spill or release occurred. The area of environmental forensics deals with this type of research and already several models have been investigated in order to determine the origin of an oil/gasoline (oil spill fingerprinting, e.g. Christensen and Tomasi 2007) and to estimate when certain oil has been released. An excellent review in the textbook Environmental Forensics presents these different models or attempts to establish the time of release (Morrison 1999).

This change of the hydrocarbon composition is referred to as *weathering*, which can be further divided into (1) physical weathering (e.g. evaporation, dissolution), (2) biological weathering (microbial degradation, see above) and (3) chemical weathering (e.g. photodegradation under influence of sunlight) (Malmquist et al. 2007). Of these three processes the physical weathering will transport compounds from one area to the other, whereas the latter two will degrade and/or chemically alter the individual compounds. Very stable and unreactive compounds will survive and will see a much smaller change over time than less stable compounds. In general the amount of straight chain alkanes (especially those with low molecular masses) will decrease first followed by branched chain alkanes, aromatics and lastly by cycloalkanes. Within these models (e.g. to estimate the time of release) quite often the ratio of certain compounds are used, as this ratio seem to change over time. In an attempt by Schmidt et al. (2002), the ratio of toluene (a cyclic aromatic hydrocarbon) against octane (a linear alkane) was used to identify the origin and time of a certain gasoline release. Studying several GC chromatograms over time, there seems to be a trend in this ratio over the last 30 years, with the ratio of toluene/octane increasing and interestingly a jump in 1994. However as the authors state in the article, this is only useful if less than 50% of the gasoline release has evaporated. This model of using the ratio between several important compounds (such as biomarkers, see review by Medeiros and Simoneit 2007) has been used by other researchers as well. Frequently a compound that is very stable (low volatility and low water solubility) will be used as a (bio)marker and ratios of this marker to other less stable hydrocarbons are calculated. Examples that have been used are the ratio of heptadecane/pristine (Christensen and Larsen 1993) or octadecane/phytane (Sauer et al. 1998).

10.8 How Important Are Hydrocarbons in the Area of Forensic Entomology?

From the information given above, it is quite clear that hydrocarbons are very important, and as such the time and research spend on these compounds can easily be justified. Hydrocarbons can give us some crucial information with regards to understanding communication system in insects, such as in ants. Furthermore, due to the many different hydrocarbons that exists on the cuticle of insects, each insect generally has a very distinctive hydrocarbon profile. This profile, in combination with powerful statistics, such as Principle Component Analysis (PCA), is a valuable tool in identifying species, a research area referred to as chemo-taxonomy. This use of chemicals to identify a species is used in plant taxonomy as well as insect taxonomy and hydrocarbons have been proven to be very useful within chemo-taxonomy (Page et al. 1997; Urech et al. 2005).

Within the context of forensic entomology, there are two possible areas where hydrocarbons could be very useful if not essential. The first area would be in the species identification of either larvae or eggs. Currently any larvae or eggs found on a body, when it is not obvious to which species they belong, needs to be reared to full adulthood before a positive identification can be obtained. DNA-Barcoding can be used to identify the species when it concerns eggs, larvae or adults, however it can be a time consuming process and will be very difficult in the case of empty puparia. Yet, due to the species specificity of a hydrocarbon profile, the hydrocarbons can be extracted from eggs or larvae, and the obtained hydrocarbon profile can be compared to a database with known hydrocarbon profiles from different species. If a match is found the egg or larvae can be identified to its species within a couple of hours. It is possible to extract hydrocarbons from eggs, larvae, adults and even exuviae. The latter was shown by Ye and co-workers (Ye et al. 2007) where the authors extracted hydrocarbons from the pupal exuviae of six necrophagous flies. After analysing the extracts with GC-MS and performing discriminant analysis on all the observed peaks, all exuviae from the six different flies could be separated from each other. Both linear alkanes (such as tricosane and octacosane) and branched alkanes (such as methylpentacosane, dimethyl hexacosane) were important hydrocarbons in order to discriminate between the six different extracts. This study emphasised the forensic importance of hydrocarbons and how these can be used in conjunction with DNA-Barcoding. Up till now no other papers have been published using hydrocarbons in the context of combining forensic entomology and chemotaxonomy, but this may only be due to the fact that this is a very new research area. In addition, there is evidence (Rouault et al. 2001; Ugelvig et al. 2008; Cremer et al. 2008) that insects could have varying cuticular hydrocarbon profiles, depending on the geographical region they are found in. If this variation is significant in those insects frequently found on corpses, the hydrocarbon profile could reveal potential transport activities of the body if the profile of the hydrocarbons is different from those insects found in a certain region.

The second area in which cuticular hydrocarbons can be a valuable tool in aiding forensic entomology is in establishing the post-mortem interval (PMI). Two recent papers have been published whereby the authors state that a significant change of the hydrocarbon profile on the blowfly, *Chrysomya rufifacies* has been observed. This changed was correlated either with larval age or with weathering conditions created in the lab. In the first study the hydrocarbon composition of *C. rufifacies* larvae were correlated with age (Zhu et al. 2006). It was found that the ratio of the peak area of a linear alkane, nonacosane (n-C_{29}) divided by the combined peak areas of another eight selected peaks increased with larval age. In the second study Zhu et al. (2007) used the puparial cases of *C. megacephala* and placed them in incubators to simulate weathering conditions. The relative abundance of several linear alkanes (especially the low molecular weight even numbered *n*-alkanes) increased over time, whereas almost all branched alkanes seem to decrease over time. In contrast the abundance of the high molecular weight hydrocarbons with chain length of more than 31 carbons all increased over time. Up till now this is the only paper dealing with the change of cuticular hydrocarbons on puparia over time, yet it showed that there is a real potential to use the puparia in the PMI estimation. However, there are also still some unexplained results e.g. why the abundance of 5MeC29 decreased whereas the abundance of 7MeC31 increased? Is this observation related to the fact that the abundance of most hydrocarbons $>C_{31}$ increased as well? And is therefore chain length more important than branching? All this indicates that further investigation is necessary, but if these results are consisted, they could very well be used in estimating the PMI even though we do not quite fully understand the process behind these changes. In the most recent study, Roux et al. (2008) provided even further evidence for the importance of hydrocarbons as an alternative method for evaluating the postmortem interval through a complete ontogenetical study revealing the changes in hydrocarbon profiles in three calliphorid Diptera of forensic interest, *Calliphora vomitoria*, *C. vicina* and *Protophormia terraenovae*. Results enabled them to obtain good resolution of larvae within 1 day's precision. Some early results indicate that there is ontogenetic variation of the hydrocarbon profiles in *Lucilia sericata* as well (Drijfhout to be published). Although mosquitoes may not be directly of forensic importance, research showed that female *Aedes aegypti* of different ages also have different hydrocarbon profiles (Desena et al. 1999a). As with *C. rufifacies* it was the linear alkanes (pentacosane and nonacosane) that corresponded to different ages. After refining the method, it is now even possible to reveal information about the age of a female *A. aegypti* only using the cuticular hydrocarbons found on the leg of mosquitoes (Desena et al. 1999b). Results form all these studies showed that there is a big difference in the hydrocarbon profiles from eggs, to larvae to adults. Adults seem to have an increase in hydrocarbons with a higher molecular weight (Roux et al. 2008; Drijfhout to be published). Furthermore, the ratio of odd linear alkanes versus even linear alkanes also changed from eggs to larvae to pupae. The rational behind these changes are still not fully understood, although events such as moulting and pupation can in some cases be linked to a change in the hydrocarbon profile.

10.9 Conclusion and Final Remarks

In conclusion, there is ample evidence that hydrocarbons in general, but also cuticular hydrocarbons, can assist many different research areas. Many hydrocarbons have been identified as compounds to be essential in insect development and that these compounds are under strict genetic control. The main problem right now is the lack of basic research in how these compounds are regulated and how they will change over time. If these processes are known and the relation between the cuticular hydrocarbon compositions and e.g. larval age is known, this would greatly enhance the precision in determining the larval age of larvae found on corpses. Extracting the hydrocarbon content on eggs, larvae and or pupae is fairly simple and has been carried in many other disciplines. The consequent analysis of these extracts with either GC-FID or GC-MS is a quick and sensitive method of obtaining information on the composition of the hydrocarbons present. This method is quick and requires minimal work, certainly if this is compared to rearing eggs or larvae to adulthood. Added to this is that morphology is sometimes not clear enough to distinguish either the species or age of larvae.

Although the process of weathering of hydrocarbons on the cuticle of insects is less well understood, it has certainly the potential to become a powerful tool in establishing the time of death, if the changes that will occur under certain conditions are known. This research area is fairly new and much more is needed to see whether the change in hydrocarbon composition due to e.g. larval development and weathering is independent from each other or not. Can we treat each of these processes independently, or do we need to develop a more complex model in which all of these factors are taken into account? All studies on the change of hydrocarbons in necrophagous flies and on their empty puparia (Zhu et al. 2006, 2007; Roux et al. 2008) were carried out under strict climate conditions. However a change in temperature or humidity in nature will affect the weathering process of hydrocarbons on empty puparia as well as the development of larvae. A change in temperature could e.g. result in the insect entering the diapause stage. Will this have any effect on the hydrocarbon profile? To my knowledge no research has been done on what effect the diapause stage has on the production and maintenance of hydrocarbons. Nelson and Lee (2004) reported that overwintering larvae of the goldenrod gall fly, *Eurosta solidaginis*, increase their total hydrocarbons from 122 ng/larva collected in autumn (September) to 4,900 ng/larva collected in winter (January). Although there was no indication that certain individual hydrocarbons increased more than others, the percent composition of the hydrocarbons changed from larvae collected in autumn compare to those collected in winter. It is however important to note that larvae collected in autumn were young larvae and these collected in winter were old larvae, hence this change could be due to larval development (and hence useful in age determination of larvae) rather than due to a decrease in temperature. From this it is clear that there are at least two classes of factors that may influence the hydrocarbon patterns of larvae or adult flies (Zhu et al. 2007). The first class is related to the larvae/pupae/puparia in unweathered conditions; factors include e.g. age, sex, and geographic population. The second class of factors are those related to the weathering of hydrocarbons; factors include sunlight wind or rainfall. Therefore further research is needed to determine if the results obtained so far in the lab will be similar to those obtained in the field. The ultimate experiment will be to have eggs hatch in the field and compare their change in hydrocarbon patterns with those that have hatched and grown in the lab.

These early studies clearly indicate the forensic importance of these cuticular hydrocarbons. It is clear that cuticular hydrocarbons can be used to identify insect species found on bodies if the cuticular hydrocarbons composition are known for the more well known species found on corpses. In addition these compounds have a great potential in establishing the time of death, as their cuticular hydrocarbons change with age. However, factors other than age may influence the hydrocarbon profiles and therefore much more research is needed to exploit the full potential of hydrocarbons in establishing PMI.

Acknowledgements The author thanks Dr S. Martin and Prof. E.D. Morgan for their help during the preparation of the manuscript.

References

Akino T (2006) Cuticular hydrocarbons of *Formica truncorum* (Hymenoptera: Formicidae): description of new very long chained hydrocarbon components. Appl Entomol Zool 41:667–677

Akino T, Yamamura K, Wakamura S et al (2004) Direct behavioural evidence for hydrocarbons as nest mate recognition cues in *Formica japonica* (Hymenoptera: Formicidae). Appl Entomol Zool 39:381–387

Amendt J, Krettek R, Zehner R (2004) Forensic entomology. Naturwissenschaften 91:51–65

Antony C, Jallon J-M (1982) The chemical basis for sex recognition in *Drosophila melanogaster*. J Ins Physiol 28:873–880

Arthur CL, Pawliszyn J (1990) Solid-Phase Microextraction with thermal-desorption using fused-silica optical fibers. Anal Chem 62:2145–2148

Bartelt RJ, Jones RL, Kulman HM (1982) Evidence for a multicomponent sex-pheromone in the yellowheaded spruce sawfly. J Chem Ecol 8:83–94

Blomquist GJ, Dillwith JW (1985) Cuticular lipids. In: Kerkut GA, Gilbert LI (eds) Comprehensive Insect Physiology Biochemistry and Pharmacology, 1st edn. Pergamon, Oxford, pp 117–154

Blomquist GJ, Toolson EC, Nelson DR (1985) Epicuticular hydrocarbons of *Drosophila pseudoobscura* (Diptera: Drosophilidae): identification of unusual alkadiene and alkatriene positional isomers. Insect Biochem 15:25–34

BorgKarlson AK, Mozuraitis R (1996) Solid phase micro extraction technique used for collecting semiochemicals. Identification of volatiles released by individual signalling *Phyllonorycter sylvella* moths. Zeitschrift Fur Naturforschung C-a Journal of Biosciences 51:599–602

Buczkowski G, Kumar R, Suib SL, Silverman J (2005) Diet-related modification of cuticular hydrocarbon profiles of the Argentine ant *Linepithema humile* diminishes intercolony aggression. J Chem Ecol 31:829–843

Carlson DA, Roan C-S, Yost RA, Hector J (1989) Dimethyl disulfide derivatives of long chain alkenes alkadienes and alkatrienes for gas chromatography/mass spectrometry. Anal Chem 61:1564–1571

Châline N, Sandoz J-C, Martin SJ, Ratnieks FLW, Jones GR (2005) Learning and discrimination of individual cuticular hydrocarbons by honeybees (*Apis mellifera*). Chem Senses 30:327–335

Christensen LB, Larsen TH (1993) Method for determining the age of diesel oil spills in the soil. Ground Water Monit Remediation 13:142–149

Christensen JH, Tomasi G (2007) Practical aspects of chemometrics for oil spill fingerprinting. J Chromatogr A 1169:1–22

Conner WE, Eisner T, Vander Meer RK, Guerrero A, Chiringelli D, Meinwald J (1980) Sex attractant of an arctiid moth (*Utethesia ornatrix*): a pulsed chemical signal. Behav Ecol Sociobiol 7:55–63

Cremer S, Ugelvig LV, Drijfhout FP, Schlick-Steiner BC, Steiner FM et al (2008) The evolution of invasiveness in garden ants. PLoS ONE 3(12):e3838. doi:101371/journal.pone.0003838

Crespo R, Juarez MP, Cafferata LFR (2000) Biochemical interaction between entomopathogenous fungi and their insect-host-like hydrocarbons. Mycologia 92:528–536

Cvačka J, Jiroš P, Šobotník J, Hanus R, Svatoš A (2006) Analysis of insect cuticular hydrocarbons using matrix-assisted laser desorption/ionization mass spectrometry. J Chem Ecol 32:409–434

Dani FR (2006) Cuticular lipids as semiochemicals in paper wasps and other social insects. Ann Zool Fennici 43:500–514

Dani FR, Jones GR, Destri S, Spencer SH, Turillazzi S (2001) Deciphering the recognition signature within the cuticular chemical profile of paper wasps. Anim Behav 62:165–171

Dani FR, Jones GR, Morgan ED, Turillazzi S (2003) Re-evaluation of the chemical secretion of the sternal glands of Polistes social wasps (Hymenoptera Vespidae). Ethol Ecol Evol 15:73–82

Dani FR, Jones GR, Corsi S, Beard R, Pradella D, Turillazzi S (2005) Nestmate recognition cues in the honey bee: differential importance of cuticular alkanes and alkenes. Chem Senses 30:477–489

De Hoffmann E, Stroobant V (2001) Mass spectrometry – principles and applications, 2nd edn. Wiley, Chichester

Desena ML, Clark JM, Edman JD, Symington SB, Scott TW, Clark GG, Peters TM (1999a) Potential for aging female *Aedes aegypti* (Diptera: Culicidae) by gas chromatographic analysis of cuticular hydrocarbons including a field evaluation. J Med Entomol 36:811–823

Desena ML, Edman JD, Clark JM, Symington SB, Scott TW (1999b) *Aedes aegypti* (Diptera: Culicidae) age determination by cuticular hydrocarbon analysis of female legs. J Med Entomol 36:824–830

Edney EB (1977) Water balance in land arthropods. Springer, New York

Endler A, Liebig J, Schmitt T, Parker JE, Jones GR, Schreier P, Hölldobler B (2004) Surface hydrocarbons of queen eggs regulate worker reproduction in a social insect. Proc Natl Acad Sci USA 101:2945–2950

Ferveur J-F (2005) Cuticular hydrocarbons: Their evolution and roles in *Drosophila* pheromonal communication. Behav Genet 35:279–295

Ferveur J-F, Savarit F, O'Kane CJ, Sureau G, Greenspan RJ, Jallon J-M (1997) Genetic feminization of pheromones and its behavioral consequences in *Drosophila* males. Science 276:1555–1558

Francke W, Plass E, Zimmermann N, Tietgen H, Tolasch T, Franke S, Subchev M, Toshova T, Pickett JA, Wadhams LJ, Woodcock CM (2000) Major sex pheromone component of female herald moth *Scoliopteryx libatrix* is the novel branched alkene (6Z 13)-Methylheneicosene. J Chem Ecol 26:1135–1149

Fujiwara-Tsujii N, Yamagata N, Takeda T, Mizunami M, Yamaoka R (2006) Behavioral responses to the alarm pheromone of the ant *Camponotus obscuripes* (Hymenoptera: Formicidae). Zool Sci (Tokyo) 23:353–358

Gibbs AG (1995) Physical properties of insect cuticular hydrocarbons: model mixtures and interactions. Comp Biochem Physiol B 112:667–672

Gibbs AG (1998) The role of lipid physical properties in lipid barriers. Am Zool 38:268–279

Gibbs AG, Crockett EL (1998) The biology of lipids: integrative and comparative perspectives. Am Zool 38:265–267

Gibbs AG, Pomonis JG (1995) Physical properties of insect cuticular hydrocarbons: the effects of chain length methyl-branching and unsaturation. Comp Biochem Physiol 112B:243–249

Gilby AR, McKellar JW (1970) Composition of empty puparia of a blowfly. J Insect Physiol 16: 1517–1529

Gobin B, Ito F, Billen J (2003) The subepithelial gland in ants: a novel exocrine gland closely associated with the cuticle. Acta Zoologica (Stockholm) 84:285–291

Greene MJ, Gordon DM (2003) Social insects – cuticular hydrocarbons inform task decisions. Nature 423:32–32

Gullan PJ, Cranston PS (1994) The insects: an outline of entomology. Chapman and Hall, London

Heath RR, Dueben D (1998) Analytical and preparative gas chromatography. In: Millar JG, Haynes KF (eds) Methods in chemical ecology. Kluwer, Dordrecht, pp 85–126

Howard RW, Blomquist GJ (1982) Chemical ecology and biochemistry of insect hydrocarbons. Annu Rev Entomol 27:149–172

Howard RW, Blomquist GJ (2005) Ecological behavioural and biochemical aspects of insect hydrocarbons. Annu Rev Entomol 50:371–393

Kaib M, Franke S, Francke W, Brandl R (2002) Cuticular hydrocarbons in a termite: phenotypes and a neighbour – stranger effect. Physiol Entomol 27:189–198

Kimura T, Carlson AD, Mori K (2001) Synthesis of all the stereoisomers of 13, 17-dimethyl-1-tritriacontene and 13, 17-dimethyl-1-pentatriacontene the contact sex pheromone components of the female tsetse fly *Glossina austeni*. Eur J Org Chem 17:3385–3390

Krokos FD, Konstantopoulou MA, Mazomenos BE (2001) Alkadienes and alkenes mediating mating behaviour of the almond seed wasp *Eurytoma amygdali*. J Chem Ecol 27: 2169–2181

Lecuona R, Riba G, Cassier P, Clement JL (1991) Alterations of insect epicuticular hydrocarbons during infection with *Beauveria bassiana* or *B. brongniartii*. J Invertebr Pathol 58:10–18

Liang D, Silverman J (2000) 'You are what you eat': diet modifies cuticular hydrocarbons and nestmate recognition in the Argentine ant *Linepithema humile*. Naturwissenschaften 87:412–416

Liang D, Silverman J (2001) Colony disassociation following diet partitioning in a unicolonial ant. Naturwissenschaften 88:73–77

Lockey KH (1988) Lipids of the insect cuticle: origin composition and function. Comp Biochem Physiol 89B:595–645

Lommelen E, Johnson CA, Drijfhout FP, Billen J, Wenseleers T, Gobin B (2006) Cuticular hydrocarbons provide reliable cues of fertility in the ant *Gnamptogenys striatula*. J Chem Ecol 32:2023–2034

Malmquist LMV, Olsen RR, Hansen AB, Andersen O, Christensen JH (2007) Assessment of oil weathering by as chromatography-mass spectrometry time warping and principal component analysis. J Chromatogr A 1164:262–270

Marcillac F, Bousquet F, Alabouvette J, Savarit F, Ferveur JF (2005) A mutation with major effects on *Drosophila melanogaster* sex pheromones. Genetics 171:1617–1628

Markow TA, Toolson EC (1990) Temperature effects on epicuticular hydrocarbons and sexual isolation in *Drosophila mojavensis*. In: Barker JSF, Starmer WT, MacIntyre RJ (eds) Ecological and evolutionary genetics of *Drosophila*. Plenum, New York, pp 315–331

Martin C, Salvy M, Provost E, Bagnères A-G, Roux M, Crauser D, Clement J-L, Le Conte Y (2001) Variations in chemical mimicry by the ectoparasitic mite *Varroa jacobsoni* according to the developmental stage of the host honey-bee Apis mellifera. Ins Biochem Mol Biol 31:365–379

Martin SJ, Jenner EA, Drijfhout FP (2007) Chemical deterrent enables a socially parasitic ant to invade multiple hosts. Proc Roy Soc B Biol Sci 274:2717–2722

Martin SJ, Helantera H, Drijfhout FP (2008a) Colony-specific hydrocarbons identify nest mates in two species of Formica ant. J Chem Ecol 34:1072–1080

Martin SJ, Helantera H, Drijfhout FP (2008b) Evolution of species-specific cuticular hydrocarbon patterns in Formica ants. Biol J Linnean Soc 95:131–140

Martin SJ, Vitikainen E, Helantera H, Drijfhout FP (2008c) Chemical basis of nest-mate discrimination in the ant *Formica exsecta*. Proc Roy Soc B Biol Sci 275:1271–1278

Martin SJ, Drijfhout FP (2009) Nestmate and task cues are influenced and encoded differently within ant cuticular hydrocarbon profiles. J Chem Ecol 35:368–374

Martin SJ, Drijfhout FP (2009) A review of ant cuticular hydrocarbons. J Chem Ecol in press

Mazomenos BE, Athanassiou CG, Kavallieratos N, Milonas P (2004) Evaluation of the major female *Eurytoma amygdali* sex pheromone Components (Z, Z)-6, 9-tricosadiene and (Z, Z)-6, 9-pentacosadiene for male attraction in field tests. J Chem Ecol 30:1245–1255

McLafferty FW, Turceck F (1993) Interpretation of mass spectra, 4th edn. University Science Books, Mill Valley, CA

Medeiros PM, Simoneit BRT (2007) Gas chromatography coupled to mass spectrometry for analyses of organic compounds and biomarkers as tracers for geological environmental and forensic research. J Sep Sci 30:1516–1536

Millar JG (1998) Liquid chromatography. In: Millar JG, Haynes KF (eds) Methods in chemical ecology. Kluwer, Dordrecht, pp 85–126

Millar JG, Sims JJ (1998) Preparation cleanup and preliminary fractionation of extracts. In: Millar JG, Haynes KF (eds) Methods in chemical ecology. Kluwer, Dordrecht, pp 85–126

Monnin T (2006) Chemical recognition of reproductive status in social insects. Annales Zoologici Fennici 43:515–530

Monnin T, Ratnieks FLW, Jones GR, Beard R (2002) Pretender punishment induced by chemical signalling in a queenless ant. Nature 419:61–65

Morgan ED (2004) Biosynthesis in insects. The Royal Society of Chemistry Cambridge, UK

Morrison R (1999) Environmental forensics: principles and applications. CRC, Boca Raton

Morrison RD (2000) Application of forensic techniques for age dating and source identification in environmental litigation. Environ Forensics 1:131–153

Napolitano R, Juarez MP (1997) Entomopathogenous fungi degrade epicuticular hydrocarbons of *Triatoma infestans*. Arch Biochem Biophys 344:208–214

Nelson DR (1993) Methyl-branched lipids in insects. In: Stanley-Samuelson DW, Nelson DR (eds) Insect lipids: chemistry biochemistry and biology. Nebraska University Press, Lincoln, NE, pp 270–315

Nelson DR, Lee RE (2004) Cuticular lipids and desiccation resistance in overwintering larvae of the goldenrod gall fly *Eurosta solidaginis* (Diptera: Tephritidae). Comp Biochem Physiol 138B:313–320

Page M, Nelson LJ, Blomquist GJ, Seybold SJ (1997) Cuticular hydrocarbons as chemotaxonomic characters of pine engraver beetles (Ips spp.) in the grandicollis subgeneric group. J Chem Ecol 23:1053–1099

Pawliszyn J (1997) Solid phase microextraction: theory and practice. Wiley-VCH, New York

Pedrini N, Crespo R, Juarez MP (2007) Biochemistry of insect epicuticle degradation by entomopathogenic fungi. Comp Biochem Physiol 146C:124–137

Pickett JA, Williams IH, Martin AP (1982) Z)-11-eicosen-1-ol an important new pheromonal component from the sting of the honey bee *Apis mellifera* L. (Hymenoptera:Apidae. J Chem Ecol 8:163–175

Rochat D, Ramirez-Lucas P, Malosse C, Aldana R, Kakul T, Morin JP (2000) Role of solid-phase microextraction in the identification of highly volatile pheromones of two Rhinoceros beetles *Scapanes australis* and *Strategus aloeus* (Coleoptera Scarabaeidae Dynastinae). J Chromatogr A 885:433–444

Rouault J, Capy P, Jallon JM (2001) Variations of male cuticular hydrocarbons with geoclimatic variables: an adaptative mechanism in *Drosophila melanogaster*? Genetica 110:117–130

Rouault J-D, Marican C, Wicker-Thomas C, Jallon J-M (2004) Relations between cuticular hydrocarbons (HC) polymorphism resistance against desiccation and breeding temperature; a model for HC evolution in *D. melanogaster* and *D. stimulans*. Genetica 120:195–212

Roux O, Gers C, Legal L (2008) Ontogenetic study of three Calliphoridae of forensic importance through cuticular hydrocarbon analysis. Med Vet Entomol 22:309–317

Sachse S, Rappert A, Galizia CG (1999) The spatial representation of chemical structures in the antennal lobe of honeybees: steps towards the olfactory code. Eur J Neurosci 11:3970–3982

Sauer TC, Michel J, Hayes MO, Aurand DV (1998) Hydrocarbon characterization and weathering of oiled intertidal sediments along the Saudi Arabian Coast two years after the Gulf War oil spill. Environ Int 24:43–60

Savarit F, Ferveur J-F (2002) Temperature affects the ontogeny of sexually dimorphic cuticular hydrocarbons in *Drosophila melanogaster*. J Exp Biol 205:3241–3249

Savarit F, Sureau G, Cobb M, Ferveur J-F (1999) Genetic elimination of known pheromones reveals the fundamental chemical bases of mating and isolation in *Drosophila*. Proc Natl Acad Sci USA 96:9015–9020

Schal C, Sevala VL, Young HP, Bachmann JAS (1998) Sites of synthesis and transport pathways of insect hydrocarbons: cuticle and ovary as target tissues. Am Zool 38:382–393

Schmidt GW, Beckmann DD, Torkelson BE (2002) A technique for estimating the age of regular/mid-grade gasolines released to the subsurface since the early 1970's. Environ Forensics 3:145–162

Scott D (1986) Sexual mimicry regulates the attractiveness of mated *Drosophila melanogaster* females. Proc Natl Acad Sci USA 83:8429–8433

Sonnet PE, Uebel EC, Lusby WR, Schwarz M, Miller RW (1979) Sex pheromone of the stable fly. Identification, synthesis, and evaluation of alkenes from female stable flies. J Chem Ecol 5:353–361

Soroker V, Hefetz A (2000) Hydrocarbon site of synthesis and circulation in the desert ant *Cataglyphis niger*. J Insect Physiol 46:1097–1102

Thompson MJ, Glancey BM, Robbins WE, Lofgren CS, Dutky SR, Kochansky J, Vandermeer RK, Glover AR (1981) Major hydrocarbons of the post-pharyngeal gland of mated queens of the red imported fire ant *Solenopsis invicta*. Lipids 16:485–495

Toolson EC (1982) Effects of rearing temperature on cuticle permeability and epicuticular lipid composition in *Drosophila pseudoobscura*. J Exp Zool 222:249–253

Toolson EC, Kuper-Simbron R (1989) Laboratory evolution of epicuticular hydrocarbon composition and cuticular permeability in *Drosophila pseudoobscura*: effects on sexual dimorphism and thermal-acclimation ability. Evolution 43:468–473

Toolson EC, Markow TA, Jackson LL, Howard RW (1990) Epicuticular hydrocarbon composition of wild and laboratory-reared *Drosophila mojavensis* Patterson and Crow (Diptera: Drosophilidae). Ann Entomol Soc Am 83:1165–1176

Trabalon M, Campan M, Hartmann N, Baehr J-C, Porcheron P, Clément J-L (1994) Effects of allatectomy and ovariectomy on cuticular hydrocarbons in *Calliphora vomitoria* (Diptera). Arch Insect Biochem Physiol 25:375–391

Ugelvig LV, Drijfhout FP, Kronauer DJC, Boomsma JJ, Pedersen JS, Cremer S (2008) The introduction history of invasive garden ants in Europe: integrating genetic chemical and behavioural approaches. BMC Biology 6:11

Urech R, Brown GW, Moore CJ, Green PE (2005) Cuticular hydrocarbons of buffalo fly *Haematobia exigua* and chemotaxonomic differentiation from horn fly *H-Irritans*. J Chem Ecol 31:2451–2461

Vercammen J, Sandra P, Baltussen E, Sandra T, David F (2000) Considerations on static and dynamic sorptive and adsorptive sampling to monitor volatiles emitted by living plants. Hrc-J High Res Chromatogr 23:547–553

Watson JT (1997) Introduction to mass spectrometry, 3rd edn. Lippincott-Raven, New York

Wicker C, Jallon J-M (1995) Influence of ovary and ecdysteroids on pheromone biosynthesis in *Drosophila melanogaster* (Diptera: Drosophilidae). Eur J Entomol 92:197–202

Xu SP, Sun YG (2005) An improved method for the micro-separation of straight chain and branched/cyclic alkanes: Urea inclusion paper layer chromatography. Org Geochem 36:1334–1338

Ye GY, Li K, Zhu JY, Zhu GH, Hu C (2007) Cuticular hydrocarbon composition in pupal exuviae for taxonomic differentiation of six necrophagous flies. J Med Entomol 44:450–456

Zhu GH, Ye GY, Hu C, Xu XH, Li K (2006) Development changes of cuticular hydrocarbons in *Chrysomya rufifacies* larvae: potential for determining larval age. Med Vet Entomol 20:438–444

Zhu GH, Xu XH, Yu XJ, Zhang Y, Wang JR (2007) Puparial case hydrocarbons of *Chrysomya megacephala* as an indicator of the postmortem interval. Forensic Sci Int 169:1–5

Chapter 11
Exploiting Insect Olfaction in Forensic Entomology

Hélène N. LeBlanc and James G. Logan

11.1 Introduction

Insects, specifically blowflies (Diptera: Calliphoridae), are often the first to arrive at the scene of a crime and provide crucial information including post mortem interval and whether the body has been moved from its original location, amongst other useful information. History tells us that insects' association with death was recognised as early as documentation of events could be made (Greenberg and Kunich 2005; Benecke 2001). As we continue to understand this link dramatic advances, such as those mentioned throughout this book, are continually being made in the field of Forensic Entomology in relation to different situations, environments, as well as the incorporation of new approaches. While the methods used to determine the postmortem interval (PMI), such as larval age determination and arthropod succession, are continually being used and further investigated the mechanism which attracts the flies to the body has not been fully explored. It is well documented that female flies will lay eggs near wounds or natural orifices soon after death so that the larvae may develop in a moist area (Smith 1986; Anderson 2001). However, determining exactly what attracts insects to a decomposing body and cause behavioural responses such as mating and laying eggs (oviposition), has still not been identified.

As humans, we primarily sense our world using vision, sound and touch (Cadré and Millar 2004). It is therefore understandable that we, at times, underestimate the importance of olfaction, the sense of smell. Insects perceive the world differently to humans and their ecology relies, sometimes almost exclusively, on chemicals they detect from their environment. Carrion insects are no exception to this.

It has been widely accepted that female carrion flies are attracted to volatile chemical cues emitted by a decomposing body (their host) in order to establish a

H.N. LeBlanc
Faculty of Science, University of Ontario Institute of Technology, Oshawa, Canada

J.G. Logan
Biological Chemistry Department Hertfordshire, Centre for Sustainable Pest and Disease Management, UK

J. Amendt et al. (eds.), *Current Concepts in Forensic Entomology*,
DOI 10.1007/978-1-4020-9684-6_11,

suitable site for oviposition (Ignell and Hansson 2005). Although we know that body-derived odours are likely to attract the carrion insect to the body, we have not yet identified the chemicals responsible for the attraction and whether the chemicals covey additional information to the insects. The aim of this chapter is to describe recent advances in forensic entomology research and state-of-the-art techniques used to investigate insect responses to volatile chemicals from a decomposing body. The identification of such chemicals could aid in the development of new tools to estimate a more accurate PMI.

11.2 Insect Olfaction and Decomposition

11.2.1 Insect Olfaction

Insects must locate food sources in order to sustain life, obtain energy and gain nutrients required for the production of offspring. This is achieved by means of efficient sensory processes and behavioural mechanisms that are mediated by external and internal stimuli (Agelopoulos and Pickett 1998). Insects use chemical signals to navigate through their environment. They are able to quickly process the information within an odour plume coming from a source, such as a decomposing body. Generally, insects use different sensory perceptions to locate food, a mate, an oviposition site, and detect danger (Cragg and Cole 1956; Borror et al. 1989; Castner 2001). The cues can be visual, auditory, olfactory, gustatory, and physical and each is likely to play a role in the induction of a series of behaviours, leading to the successful location of the source of interest. The most dominant cues used by insects is considered to be olfactory stimuli. Most insects have a highly developed olfactory system and use it to detect volatile chemicals.

The capacity to detect and respond to volatile chemicals present in the environment exists in nearly all living creatures; however, this ability is particularly important in insects (Vickers 2000). Insect olfactory organs involved in the response to volatile chemicals are located primarily on the antennae (Borror et al. 1989). Often in nature the morphology and position of chemosensory appendages, such as the antenna, may help determine its importance and efficiency in capturing chemical cues. For example, most insects possess long, movable antennae which provide greater capacity to detect volatiles without requiring the insect to re-position its body frequently to detect an odour (Vickers 2000). These evolutionary features indicate that chemical cues play an important role in insect behaviour and survival.

Insect antennae are covered with a large number of sensillae (Castner 2001; Shields and Hildebrand 2001). Each sensillum houses olfactory receptor neurones (ORN) which detect volatile chemicals (Shields and Hildebrand 2001). The chemicals enter through pores on the sensilla where they are transported across the sensillum lymph by odorant binding proteins to the dendrites of the olfactory neurones (McIver 1982) (Fig. 11.1). The sensory neurones input information directly into the central nervous system and this induces a behavioural response in the insect

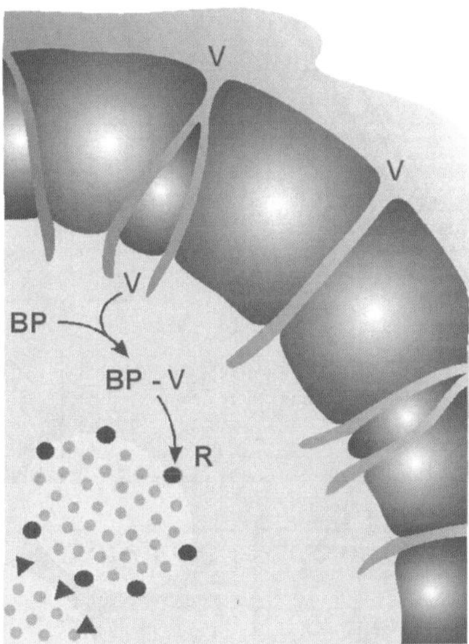

Fig. 11.1 Sensillum. V = volatile compound, BP = binding protein, R = receptor

(Hansson 2002; Zhou et al. 2004). Most insects respond not only to single compounds, but also to mixtures of compounds. With the correct combination of sensory inputs, animals or plants are recognised as hosts (Bruce et al. 2005). Insects also detect chemical gradients, giving them vital information about the location of an odour source (Vickers 2000). They detect volatile chemicals that indicate host suitability and also the presence of potential predators or competitors (Pickett et al 1998; Shields and Hildebrand 2001). As this chapter explains, even the state of decomposition of a body is revealed through the volatiles released. These volatile "signals" are also called semiochemicals.

11.2.2 Semiochemicals

The word "semiochemical" is derived from the Greek word *simeon*, which means 'sign' or 'signal' (Agelopoulos et al. 1999). Semiochemicals are volatile in nature and when airborne, they can be detected from long distances and potentially perceived by a number of other organisms of the same or different species (Agelopoulos and Pickett 1998; Selby 2003). Semiochemicals convey information between organisms and can be classified into two groups, pheromones and allelochemicals, according to the effect produced on the receiver or emitter (Nordlund and Lewis 1976; Blight 1990). Pheromones are chemicals which cause interactions between individuals of the same species (intra-specific) such as those that initiate behaviours

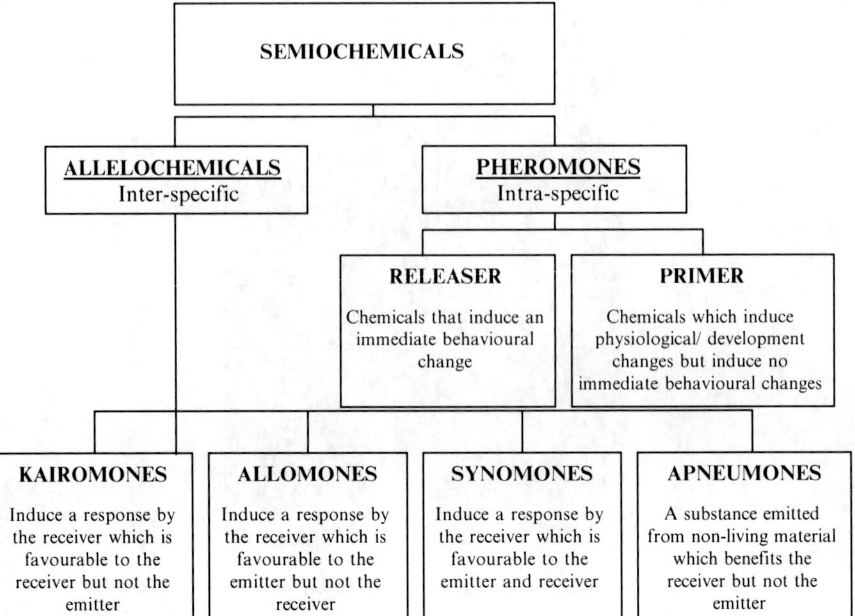

Fig. 11.2 Classification of semiochemicals (Nordlund and Lewis 1976; Howse et al. 1998)

such as mating; while allelochemicals create interactions between different species (inter-specific) (Agelopoulos et al. 1999) (Fig. 11.2). Semiochemicals are often perceived by the receiver beyond its visual range (Gikonyo et al. 2003) and a behavioural response can be triggered with only very small quantities of chemicals (Cork et al. 1990). Some volatiles are released in such small quantities, in fact, that they are barely detectable by the most advanced analytical techniques (Zumwalt et al. 1982); however, these can still be detected by insects.

The successful location of a plant or animal host by an insect is reliant on its ability to detect semiochemicals that give information about host suitability or physiological state (Pickett et al. 1998). For example, female mosquitoes (Diptera: Culicidae) detect odours such as carbon dioxide, ammonia and lactic acid from their animal or human hosts and these chemicals are of major importance in the successful location of an appropriate host in order to obtain a blood meal (Blackwell et al. 1992; Takken and Knols 1999). Semiochemicals may also be used to locate a suitable oviposition site or mate. For example, volatiles released during decomposition of a body allow blowflies to find the carcass, thereby increasing its chance of finding a suitable oviposition site, a mate, and food for their offspring (Smith 1986). Additionally, olfactory stimuli may function in combination with other stimuli, for example, oriental fruit moths, *Grapholita molesta* (Lepidoptera: Tortricidae), are unable to remain orientated when placed in a visually diminished "blank" environment containing a chemical attractant, implying that a visual cue is vital for site location (Vickers 2000). Similarly, the blackfly, *Simulium arcticum* (Diptera: Simuliidae), relies heavily on visual cues, such as shape and colour, to locate a host at close range and, therefore, responds not only to the CO_2 being released by the living host (Sutcliffe et al. 1991).

11.2.3 The Decomposition Process

Decomposition commences almost immediately after death and it is believed that the semiochemicals utilised by carrion insects are produced at the onset of decomposition (Vass et al. 1992; Dix and Graham 2000; Vass et al. 2002; Dent et al. 2004). The cells begin to die and enzymes digest the cells from the inside out, a process called autolysis or self-digestion. This action causes the cells to rupture and release nutrient-rich fluid (Dix and Graham 2000; Vass 2001). Tissues containing more digestive enzymes, such as the liver, are digested at a faster rate than those containing fewer enzymes. Putrefaction occurs as the bacteria, already present in the large intestine, destroy the soft tissues resulting in the production of liquids and various gasses (hydrogen sulfide, carbon dioxide, methane, ammonia, sulfur dioxide, and hydrogen) (Vass 2001). These bacteria gain access to the vascular system and spread throughout the body. There is often a green discolouration associated with these changes (Williams et al. 2001). As the blood begins to break down within the blood vessels and the skin loses pigmentation, the dark stained blood vessels can be observed through the skin producing an effect called marbling. The outer layers of skin begin to slip off the body while fluid under the slipping skin form blisters. Trapped gasses cause the body to become 'bloated'. The body swells, primarily within the abdomen, and decomposed blood or faecal matter may be 'purged' from the lungs, airways, or rectum. Once the trapped gasses have escaped, a more active stage of decomposition can be observed. Volatile compounds derived from the decomposition of materials, such as proteins and fats, are subsequently produced (Vass et al. 2002; Dent et al. 2004). The greatest physical changes to the cadaver occur at this time. The organs degrade and become unrecognisable forming a grey exudate within and beneath the body.

The odour associated with putrefaction is mainly caused by the release of sulfur-containing compounds and various inorganic gases that are produced in the bowel. However, bacteria, fungi, protozoa and even insects aid in the breakdown of soft tissues of the body during putrefaction and this results in the production of various gases including CO_2, H_2S, CH_4, NH_3,SO_2, H_2 and a variety of volatile organic compounds (Statheropoulos et al. 2005). Following this active decomposition the body may become skeletonised leaving behind just dry leathery skin and bones, depending on the environment in which the body has been resting (Clark et al. 1997; Dix and Graham 2000). The body can go through several phases and rates of decomposition and these are highly dependent on weather, temperature, humidity, and the environment, i.e. indoors, outdoors, buried, under water, wrapped or concealed (Clark et al. 1997; Dix and Graham 2000).

In the initial stages of decomposition, there are no visual or odour effects obvious to humans at this time, however, some insects are able to detect the decomposition immediately (Anderson 2001). Blowflies are most often the first insects to oviposit on a carcass and it is likely that a number of factors initially attract these insects to the body, including volatile semiochemicals. Throughout decomposition, bodies constantly change and emit hundreds of chemicals (Vass et al. 2002). It is currently unclear which semiochemicals are detected by the different carrion insects that are found on a decomposing body at the different stages of decomposition. However, chemical ecology research is unravelling this complex interaction between insects and decomposing corpses. This could potentially be exploited to develop a more accurate time of death, alongside the insect identifications relating to succession.

For the purposes of forensic entomology, the process of decomposition is divided into five visually distinct stages. These are the fresh, bloated, active decay, advanced decay, and dry stages, originally described by Payne (1965) and Anderson and VanLaerhoven (1996). The descriptions of these stages are based on physical condition, odour, and at times varying insect activity (Payne 1965). The "fresh" stage refers to the period immediately after death and continues until the body is bloated. Chemical breakdown occurs during this stage; however, few morphological changes are observed. There is no obvious odour to humans. The "bloated" stage becomes evident when an accumulation of gasses from the activity of anaerobic bacteria produce a swollen, bloated appearance. There is an obvious odour present at this time. The "active decay" stage is recognisable by the deflation of the carcass due to the gases escaping from the body, often due to the insect activity occurring on the body. There is a very strong putrid odour that can be detected – this is when the strongest odours are detected. During the "advanced decay" stage, a large amount of the flesh has been removed; however, there is still some moist tissue present. The odour is less obtrusive than in the previous stage, but it is still quite noticeable. The "dry" stage, also at times refered to as "skeletonisation", has been reached when the carcass has been reduced to bones, cartilage, and dry skin. At this stage, only a slight odour is present.

11.2.4 Carrion Insects and Semiochemicals

Different species of carrion insect have developed different seasonal life-cycle development times, variations in habitat, and preferences in carcass size, host species, and decomposition stage due to the high levels of interspecific competition (Fisher et al. 1998). Some, like the blowfly *Lucilia sericata* (Diptera: Calliphoridae), are able to develop on live sheep and other warm-blooded vertebrates as well as in carrion (Fisher et al. 1998). Such strategies greatly increase the chance of survival of their offspring.

Carrion odour contains a wide range of chemicals. Volatile molecules appear almost immediately after death but human olfactory perception is too insensitive to detect such short-term degradation. However, carrion insects have evolved to detect olfactory stimuli from the corpse, even at the early stages of decomposition. Those semiochemicals emitted from decomposing bodies are classified as 'apneumones'. These are chemicals which are emitted by non-living material, such as a corpse, and evoke a behavioural reaction from the receiver (see Fig. 11.2). The word apneumone was derived from the Greek word *ā-pneum*, meaning breathless or lifeless. This category was originally described by Nordlund and Lewis (1976) and has since become increasingly important when discussing carrion insects.

It is likely that chemicals emitted from a decomposing body may provide information about its location and suitability as a host and may even provide signals to their predators. It is well documented that insect succession occurs on a decomposing body and different species of insect are attracted at different stages of decomposition. While the production of volatile chemicals is caused by the decomposition of bodies due to bacterial and enzymatic activity, many of the volatiles released are as a result of the action of carrion insects on the corpse. While this is still in the early stages of investigation, other areas of research have revealed important findings that could be relevant to forensic entomology. For example, there are many examples in plant-insect interactions where mechanical damage caused by insect colonisation or feeding can alter the semiochemical profile of the host plant (Dewhirst and Pickett 2009). When plants are attacked by aphids (Hemiptera: Aphididae) the volatile profile of the plant is altered, and this profile can even be specific to a particular aphid species (Du et al. 1998). In such cases, plants release herbivore-induced signals that can alert and attract predators (e.g. parasitoids) to the plant, which indirectly protect plants against herbivory. An example of this can be seen when the pea aphid, *Aphis pisum*, attacks the broad bean, *Vicia faba*. Levels of attractive volatile chemicals (including 6-methyl-5-hepten-2-one, linalool, (*E*)-*β*-farnesene and *E*)-ocimene) increase following aphid feeding (Du et al. 1998). Similarly, lima bean plants infested with spider-mites *Tetranychus urticae* (Prostigmate: Tetranychidae), release an odour which attracts the spider-mites' natural predator, the predatory mite *Phytoseiulus persimilis* (Mesostigmata: Phytoseiidae) (Sabelis and van de Baan 1983; Dicke and Sabelis 1988). Similar instances of predation or parasitism are also witnessed on decomposing bodies. For example, the parasitoid wasp, *Alysia manducator* (Hymenoptera: Braconidae), is a common parasitoid of the blowfly, *Calliphora vicina* (Diptera: Calliphoridae), and will lay eggs on the fly larvae during specific stages of their development (Reznik et al 1992). *Nasonia vitripennis* (Hymenoptera: Pteromalidae), is another common parasitic wasp which comes to a decomposing body to lay eggs on larvae of the Muscidae (Order: Diptera). Their evidential value as indicators of time since death has been explored by Grassberger and Frank (2003) with some degree of success. Others, such as the yellow jacket wasp (Hymenoptera: Vespidae), the beetle *Necrodes littoralis* (Coleoptera: Silphidae) are predators of the adults blowflies and many Dipteran larvae, respectively (LeBlanc 2008). However, in these instances, the chemicals being released, attracting the predators and parasites, or their sources have not yet been identified. Whether it is volatiles from the decomposing body or those produced by the insects themselves which are used by the insects is still unknown.

11.3 Chemical Ecology Research

The identification of semiochemicals and their purpose can be achieved by examining the odour source and understanding their role in the biology of both the insect and the host. Various chemical ecology techniques can facilitate this. Volatile semiochemicals can be isolated using air entrainment (also referred to as headspace collection). This method allows the collection of volatile compounds produced by an odour source (e.g. decomposing body) onto a filter, commonly comprising a porous polymer, such as Porapak or Tenax. During this process contamination is kept to a minimum by, isolating the target source, with only the volatile chemicals produced by the source being collected (Agelopoulos et al. 1999). Subsequent analytical chemistry techniques, such as gas chromatography (GC) and GC–mass spectrometry (GC-MS) allow the chemicals to be identified accurately.

Headspace analysis is an established technique already used in forensic science for the collection of volatiles in blood and organ specimens (Statheropoulos et al. 2005; Hoffman et al. 2009). However, it has also been used successfully to investigate volatile odours from dead bodies. For example, Statheropoulos et al. (2005) used air entrainment to collect volatiles from the bodies of two males and identified and quantified over 80 different chemicals. The most prominent compounds were dimethyl disulfide, toluene, hexane, benzene 1,2,4-trimethyl, 2-propanone and 3-pentanone and they found marked differences in concentration between the two bodies. The authors suggest that these differences could reflect different rates of decomposition between the bodies and thus may provide valuable information about time of death (Statheropoulos et al. 2005). Later Statheropoulos et al. (2007) collected air entrainment samples of a body during the early stages of decomposition (4 days since death) during a period of 24 h. This time, over 30 volatile chemicals were identified. Eleven of these compounds were recovered throughout each sample, forming a “common core”. The “common core” was made-up of the following: ethanol, 2-propanone, dimethyl disulfide, methyl benzene, octane, *o*-xylene, *m*-xylene, *p*-xylene, 2-butanone, methyl ethyl disulfide and dimethyl trisulfide (Statheropoulos et al. 2007). More recently, a study was conducted to identify volatile compounds from 14 separate tissue samples (Hoffman et al. 2009). In total 33 compounds were identified and could be grouped into seven chemical classes such as alcohols, acid esters, aldehydes, halogens, aromatic hydrocarbons, ketones and sulfides. There were common compounds identified in all of these studies; however, it was found that, at this stage, no unique set of compounds could be used to create a “chemical signature” of the decomposing tissues.

Similarly, volatiles have been collected successfully from dead pigs (*sus scrofa*). In this case the pigs were contained inside a metal container during the time of sampling and the volatiles were extracted through Porapak and Tenax filters (LeBlanc 2008). These samples were taken daily in order to encompass the different stages of decomposition. A large number compounds were recovered, however, in this research the aim was to locate specific compounds which are detected carrion insects, specifically the blowfly *Calliphora vomitoria* (Diptera: Calliphoridae).

Although air entrainment followed by GC and GC-MS analysis can provide information regarding the general profile from an odour source (often hundreds of volatile chemicals), this alone does not indicate which volatiles are detected by the carrion insects antennae (i.e. those that are electrophysiologically-active). Therefore, additional techniques can be used to discriminate between electrophysiologically active and inactive compounds in the complex extract (Wadhams 1990). By combining different analysis methods, the identification of volatile compounds can be refined to those which have a behavioural impact on insects associated with decomposing bodies and thus, those that are characteristic of a particular stage of decomposition.

Electroantennogram (EAG) recordings were originally utilised by Schneider in 1957. Using microelectrodes, he found that it was possible to record depolarisation of the affected sensillum on the antenna stimulated by a volatile compound introduced over the antennal preparation (Fig. 11.3). In the case of Dipterans such as Muscidae or Calliphoridae, the antennae are connected to microelectrodes that record the response of the olfactory receptor neurones in the antennae. An odour stimulus can be delivered through an air stream flowing continuously over the preparation and a response (depolarisation) can be immediately recorded if the compound elicits an electrophysiological response. This is an effective means of initially identifying semiochemicals because EAG responses are recorded without the influence of environmental or neurological factors which could affect behavioural responses (Cork et al. 1990). EAG can be combined with gas chromatography to give GC-EAG which allows the location of active chemicals within a complex extract. This technique allows the location of EAG-active chemicals within complex extracts by taking advantage of the high-resolution of the GC while simultaneously utilising the

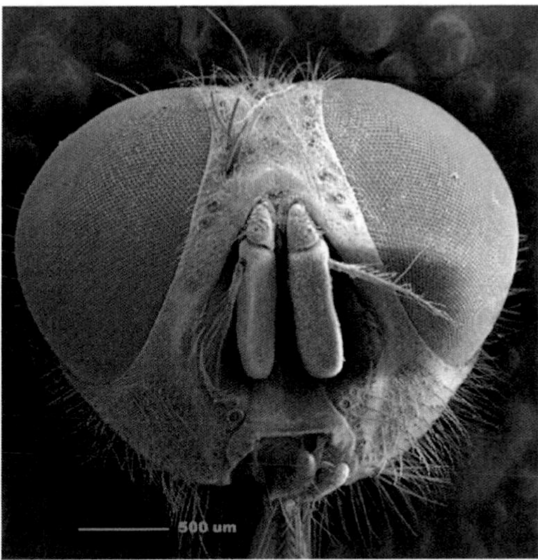

Provided by Images of Nature

Fig. 11.3 Scanning electron microscope (SEM) image of Muscidae

extreme sensitivity and selectivity of the antennal preparation of an insect. A volatile sample is injected into the GC and the sample is split in two – half of the sample travels to the flame ionisation detector (FID) and half is simultaneously passed over the insect preparation (Fig. 11.4). The compound is recorded and displayed within a chromatogram and the reaction from the fly, if any, is recorded as a depolarisation indicated along with the compound in the chromatogram (Fig. 11.5). GC-EAG was first reported in 1969 (Moorhouse et al. 1969) and was subsequently advanced to

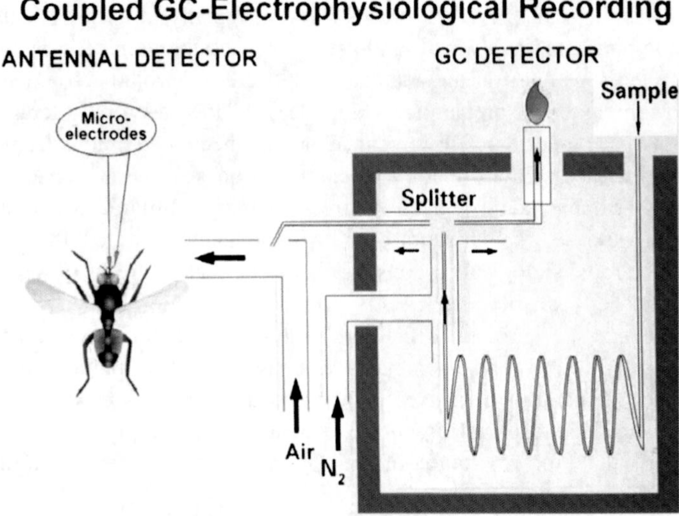

Fig. 11.4 System designed used for coupled gas chromatography and electroantennogram experiments

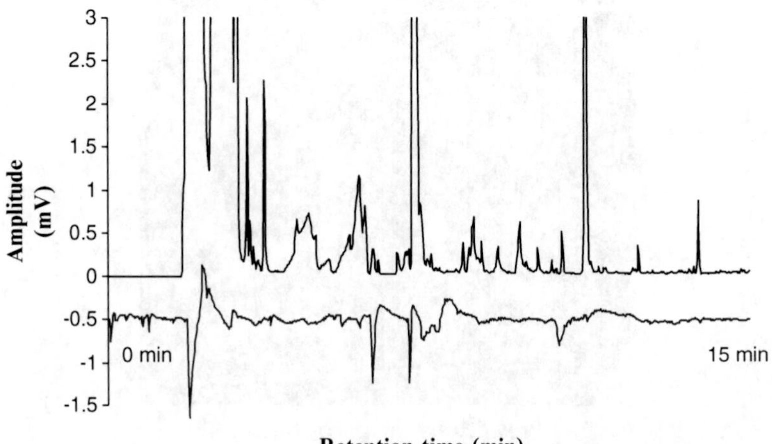

Fig. 11.5 Coupled GC-EAG trace. The upper trace shows the chromatogram of volatiles from the sample extract. The lower trace shows the corresponding EAG response of the insect preparation

possess the capabilities of also detecting responses of single olfactory cells (GC-SCR) (Wadhams et al. 1982). Although behavioural studies, have previously shown that *C. vomitoria* respond odours from bodies, to liver (Woolbridge et al. 2007) and to a compound, dimethyl trisulfide (Nilssen et al. 1996), no study had identified specific compounds associated with dead bodies which elicit a response from blowflies. However, recently the above chemical ecology methods have been used, for the first time in forensic entomology research, to identify semiochemicals from decomposing bodies that could be involved in the attraction of carrion insects (LeBlanc 2008). The identifications of chemicals from this study will be subsequently published.

While there are five recognisable stages in the physical decomposition of the carcass (Anderson and VanLaerhoven 1996), there is often no clearly defined beginning or end to a decomposition stage, especially during the later stages, making the decision of stage change somewhat subjective (Bornemissza 1957; Early and Goff 1986; Shoenly and Reed 1987; Archer 2004; Grassberger and Frank 2004). However, if the chemicals that are characteristic of specific stages of decomposition (i.e. "chemical fingerprinting") can be identified, this could aid confirmation of decomposition stage.

LeBlanc (2008) describes the daily collection of volatiles from decomposing pigs using an air entrainment method and the analysis of these compounds using GC-EAG with blowfly *C. vomitoria*. In these studies it was found that specific compounds, which included mainly sulphur compounds, within the volatile collections triggered an electrophysiological reaction from the blowfly. Most importantly differences were found between the different stages of decomposition. While the five stages of decomposition were determined through physical characteristics, the chemical composition and concentration of the volatiles sampled changed in a manner that closely followed these five stages. The Fresh and Dry stages exhibited the lowest concentrations of volatiles, however, the composition of volatile were different between these two stages. The EAG-active compounds, or semiochemicals, were at their highest concentration during the active decays stage. Visual observations supported these findings as the greatest number of Calliphoridae, adults and larvae, were recorded during this stage of decomposition.

Semiochemicals play a considerable role in mediating insect behaviour (Birkett et al. 2004; Pickett et al. 1998). Identifying these specific compounds and investigating the responses they elicit could provide a better understanding of the insects and, in some cases, allow manipulation of their behaviour. The period between death and the arrival of the first ovipositing blowflies is of great interest to the forensic entomologist, yet not fully understood.

Although some semiochemicals are likely to be attractants, others could have a negative effect on the insects behaviour. For example, Birkett et al. (2004) and Logan et al. (2008, 2009) found that at times when vertebrate hosts are not attractive to pests, there are elevated levels of certain semiochemicals which may act as active repellents or more passively by 'masking' attractants, thereby reducing the sensitivity of flies to these attractants. This could be relevant to decomposing bodies. Semiochemicals could be present at certain stages and may explain why certain carrion insects, such as *Calliphora* spp, are prevalent during the early stages of decomposition while others only appear later during decomposition. Competition and availability of food play an important role in insect succession on a corpse and therefore, repellents or 'masking' compounds could be used to prevent attraction of certain insects at particular times. For example, late colonizers could be repelled by early decomposition volatiles or by those produced by the immature stages (larvae) of the early colonizers to help avoid competition. However, the alternative is that the flies are simply not attracted to the odours present on a body at a particular time.

11.3.1 Decomposing Body Odour Mimicry

Olfactory stimuli associated with decomposing bodies is even exploited by plants. Certain plants of the Araceae family use mimicry and deception by releasing compounds similar to those associated with decomposing bodies to lure carrion insects such as blowflies and carrion beetles for pollination (Kite and Hetterschieid1997; Stensmyr et al. 2002). Semiochemicals associated with decomposing bodies are produced by the Mediterranean flower dead-horse arum, *Helicodiceros muscivorus* (Araceae: Aroideae) to trick the carrion flies in favour of the plant. This flower lures blowflies to act as pollinators by emitting dimethyl sulfide, dimethyl disulfide, dimethyl trisulfide, and dimethyl trisulfide derivatives (confirmed through GC-EAG experiments), which are also found in decomposing meat. The flies are thus enticed and trapped into a floral chamber that surrounds the female florets (Stensmyr et al. 2002). Other factors such as the appearance of the plant – the flower is said to resemble the anal section of a dead mammal – and the pseudo-thermogenic properties of the plant are also key factors in the fly's attraction to the plant (Stensmyr et al. 2003). Circadian activity may also play a role in fly attraction as the odours are only emitted between sunrise and noon. GC-EAG tests also showed that female *C. vicina*, *L. caesar*, and other Calliphoridae species were attracted to these compounds and were not able to discriminate between decomposing meat and the flower through olfaction alone (Stensmyr et al. 2002).

11.4 Future Prospects

11.4.1 Time of Death

To date, entomology still remains the most reliable method of determining the post mortem interval, however, it is anticipated that in the future volatiles associated with a decomposing body, a "chemical fingerprint", could be used to determine time of death and provide the pathologist with important forensic details (Vass et al. 2002; Statheropoulos et al. 2007). As described earlier, decomposition-related volatiles are greatly influenced by a variety of factors such as enzymatic and bacterial activity, temperature, humidity, body size, soil composition, the presence of clothing, as well as stomach content (Vass et al. 2002; Dent et al. 2004). Further investigations are required to account for such variations. Although insect development can also be affected by factors such as s temperature, humidity, and geographical location, these external influences have been studied in great detail in certain species and are thus highly predictable. Despite this, an accurate estimation of the postmortem interval still remains a difficult task (Amendt et al. 2004).

Although there are many parallels, in terms of chemical ecology-related mechanisms, between carrion flies that interact with decomposing bodies, and other insects that interact with plants, the production and release of volatile chemicals are not the same. Nevertheless, it is certain that decomposing bodies release different volatiles at different stages of decomposition. LeBlanc (2008) has found that the change in composition and concentration of volatiles released change in sync with the physical changes noted on the pig often termed Fresh, Bloated, Active Decay, Advanced Decay, and Dry stages of decomposition; therefore, giving information on the physical state of the body. Portable methods of collecting volatiles from a body have already been developed (LeBlanc 2008; Logan et al. 2009). This means that volatile samples from a decomposing body could be collected from the crime scene and analysed in order to gain added information about the stage of decay and the behaviour of the insects on the carcass. While estimating a more accurate PMI is the ultimate goal, it may also be possible to determine volatile compounds which insects have evolved to avoid, providing useful repellents (Pickett et al 1998).

11.4.2 Body Recovery

The methods of detection and analysis mentioned in this chapter have a wide application. Efforts are already being made to use the identified volatiles which are associated with decomposition to make portable detection systems which would locate decomposing bodies (Smedts 2004; Hoffman et al. 2009). It is hoped that this technique would compliment canine lead victim recovery searches and that the volatiles identified could be used to help train the canines (Hoffman et al. 2009).

11.4.3 Pest-Control

Another function of semiochemical studies could have great agricultural applications. Already baited traps are being tested and used to lure myiasis causing blowfly, *Lucilia sericata*, away from livestock in order to reduce the amount of sheep strikes (Ashworth and Wall 1994; Smith and Wall 1998). These often include coloured adhesive boards with a bait such as sodium sulphate and liver mixture or the chemical attractant Swormlure-4, developed to more efficiently trap the New World screwworm fly, *Cochliomyia hominivorax* (Diptera: Calliphoridae). (Freney 1937;

Mackley and Brown 1984; Hall 1995; Smith and Wall 1998; Fisher et al. 1998; Hall et al. 2003; Woolbridge et al. 2007). A similar technique could be used, and possibly improved, for use in slaughterhouses and butcher shops where meat must remain at a high grade and therefore free of any insect larvae.

As research continues in the area of decomposition volatiles, new groundbreaking findings will be made. However, insects remain the most accurate method of determining the post-mortem interval and, therefore, it is important that the two, in conjunction, are studied further in order to determine their specific link to decomposition and time since death.

References

Agelopoulos NG, Pickett JA (1998) Headspace analysis in chemical ecology: Effects of different sampling methods on ratios of volatile compounds present in headspace samples. J Chem Ecol 24(7):1161–1172

Agelopoulos NG, Hooper AM, Maniar SP, Pickett JA, Wadhams LJ (1999) A novel approach for isolation of volatile chemicals released by individual leaves of a plant *in situ*. J Chem Ecol 25(6):1411–1425

Amendt J, Krettek R, Zehner R (2004) Forensic entomology. Naturwissenschaften 91(2):51–65

Anderson GS (2001) Insects succession on carrion and its relationship to determining time of death. In: Byrd JH, Castner JL (eds) Forensic entomology: the utility of arthropods in legal investigations. CRC, LLC, USA, pp 143–175

Anderson GS, VanLaerhoven SL (1996) Initial studies on insect succession on carrion in southwestern British Columbia. J Forensic Sci 41(4):617–625

Archer MS (2004) Rainfall and temperature effects on the decomposition rate of exposed neonatal remains. Sci Justice 44(1):35–41

Ashworth JR, Wall R (1994) Responses of the sheep blowflies, *Lucilia sericata* and *L. cuprina* (Diptera: Calliphoridae), to odour and the development of semiochemical baits. Med Vet Entomol 8:303–309

Benecke M (2001) A brief history of forensic entomology. Forensic Sci Int 120(1–2):2–14

Birkett MA, Angelopoulos N, Jensen K-MV, Jespersen JB, Pickett JA, Prijs HJ, Thomas G, Trapman JJ, Wadhams LJ, Woodcock CM (2004) The role of volatile semiochemicals in mediating host location and selection by nuisance and disease-transmitting cattle flies. Med Vet Entomol 18:313–322

Blackwell A, Mordue AJ, Young MR, Mordue W (1992) Bivoltinism, survival rates and reproductive characteristics of the Scottish biting midge, *Culicoides impunctatus* (Diptera: Ceratopogonidae) in Scotland. Bull Entomol Res 82:299–306

Blight MM (1990) Techniques for isolation and characterization of volatile semiochemicals of phytophagous insects. In: McCaffery AR, Wilson ID (eds) Chromatography and isolation of insect hormones and pheromones. Plenum, New York, London, pp 281–288

Bornemissza GF (1957) An analysis of arthropod succession in carrion and the effects of its decomposition on the soil fauna. Aus J Zool 5:1–12

Borror DJ, Triplehorn CA, Johnson NF (1989) An introduction to the study of insects, 6 edn. Saunders College Publishing, USA

Bruce TJA, Wadhams LJ, Woodcock CM (2005) Insect host location: a volatile situation. Trend Plant Sci 10(6):269–274

Cadré RT, Millar JG (eds) (2004) Advances in insect chemical ecology. Cambridge University Press, UK, p ix

Castner JL (2001) General biology and arthropod biology. In: Byrd JH, Castner JL (eds) Forensic entomology: the utility of arthropods in legal investigations. CRC, LLC, USA, pp 17–42

Clark MA, Worrell MB, Pless JE (1997) Post mortem changes in soft tissues. In: Haglund WD, Sorg MH (eds) Forensic taphonomy: the postmortem fate of human remains. CRC, London, pp 151–164

Cork A, Beevor PS, Gough AJE, Hall DR (1990) Gas chromatography linked to electroantennography: A versatile technique for identifying insect semiochemicals. In: McCaffery AR, Wilson ID (eds) Chromatography and isolation of insect hormones and pheromones. Plenum, New York, London, pp 271–279

Cragg JB, Cole P (1956) Laboratory studies on the chemosensory reaction of blowflies. Ann Appl Biol 44:478–491

Dent BB, Forbes SL, Stuart BH (2004) Review of human decomposition processes in soil. Environ Geol 45(4):576–585

Dewhirst SY, Pickett JA (2009) Production of semiochemical and allelobiotic agents as a consequence of aphid feeding. Chemoecology In Press

Dicke M, Sabelis MW (1988) How plants obtain predatory mites as bodyguards. Neth J Zool 38(2–4):148–165

Dix J, Graham M (2000) Time of death, decomposition and identification: an atlas. Causes of death atlas series. CRC, London

Du YJ, Poppy GM, Powell W, Pickett JA, Wadhams LJ, Woodcock CM (1998) Identification of semiochemicals released during aphid feeding that attract parasitoid Aphidius ervi. J Chem Ecol 24:1355–1368

Early M, Goff ML (1986) Arthropod succession patterns in exposed carrion on the island of O'Ahu, Hawaiian Islands, USA. J Med Entomol 23:520–531

Fisher P, Wall R, Ashworth JR (1998) Attraction of the sheep blowfly, *Lucilia sericata* (Diptera: Calliphoridae) to carrion bait in the field. Bull Entomol Res 88:611–616

Freney MR (1937) Studies on the chemotrophic behaviour of sheep blowflies. Council for Scientific and Industrial Research, Australia. Pamphlet no. 74

Gikonyo NK, Hassanali A, Njagi PGN, Saini RK (2003) Responses of *Glossina morsitans morsitans* to blends of electroantennographically active compounds in the odors of its preferred (buffalo and ox) and non preferred (waterbuck) hosts. J Chem Ecol 29(10):2331–2345

Grassberger M, Frank C (2003) Temperature-related development of the parasitoid wasp *Nasonia vitripennis* as forensic indicator. Med Vet Entomol 17:257–262

Grassberger M, Frank C (2004) Initial study of arthropod succession on pig carrion in a Central European urban habitat. Journal of Medical Entomology 41(3):511–523

Greenberg B, Kunich JC (2005) Entomology and the law: flies as forensic indicators. Cambridge University Press, New York, USA

Hall MJR (1995) Trapping flies that cause myiasis: their response to host-stimuli. Annals of Tropical Medicine and Parasitology 89(4):333–357

Hall MJR, Hutchinson RA, Farkas R, Adams ZJO, Wyatt NP (2003) A comparison of Lucitraps® and sticky targets for sampling the blowfly *Lucilia sericata.* Medical and Veterinary Entomology 17:280–287

Hansson BS (2002) A bug's smell – research into insect olfaction. Trends Neurosci 25:270–274

Hoffman EA, Curran AM, Dulgerian N, Stockham RA, Eckenrode BA (2009) Characterization of the volatile organic compounds present in the headspace of decomposing human remains. Forensic Sci Int 186:6–13

Howse PE, Stevens IDR, Jones OT (1998) Insect pheromones and their use in pest management. Chapman and Hall, London

Ignell R, Hansson B (2005) Insect olfactory neuroenthology – an electrophysiological perspective. In: Christensen TA (ed) Methods in insect sensory neuroscience. CRC, Florida, pp 319–348

Kite GC, Hetterschieid WLA (1997) Inflorescence odours of *Amorphophallus* and *Pseudodracontium* (Araceae). Phytochemistry 46(1):71–75

LeBlanc HN (2008) Olfactory stimuli associated with the different stages of vertebrate decomposition and their role in the attraction of the blowfly *Calliphora vomitoria* (Diptera: Calliphoridae) to carcasses. The University of Derby for the Degree of Doctor of Philosophy

Logan JG, Birkett MA, Clark SJ, Powers S, Seal NJ, Wadhams LJ, Mordue AJ, Pickett JA (2008) Identification of human-derived volatile chemicals that interfere with attraction of Aedes aegypti mosquitoes. J Chem Ecol 34:308–322

Logan JG, Seal NJ, Cook JI, Stanczyk NM, Birkett MA, Clark SJ, Gezan SA, Wadhams LJ, Pickett JA, Mordue J (2009) Identification of human-derived volatile chemicals that interfere with attraction of the Scottish Biting Midge and their potential use as repellents. J Med Entomol 46:208–219

Mackley JW, Brown HE (1984) Swormlure-4: A new formulation of Swomlure-2 mixture as an attractant for adult screwworms, *Cochliomyia hominivorax* (Diptera: Calliphoridae). Journal of Economic Entomology 77:1264–1268

Mciver SB (1982) Sensilla of mosquitos (Diptera: Culicidae). J Med Entomol 19:489–535

Nilssen AC, Tømmerås BA, Schmid R, Evensen SB (1996) Dimethyl trisulphide is a strong attractant for some Calliphorids and a Muscid but not for the reindeer oestrids *Hypoderma tarandi* and *Cephenemyia trompe*. Entomologia Experimentalis et Applicata 79(2):211–218

Nordlund DA, Lewis WJ (1976) Terminology of chemical releasing stimuli in intraspecific and interspecific interactions. J Chem Ecol 2(2):211–220

Payne JA (1965) A summer carrion study of the baby pig *Sus scrofa* (Linnaeus). Ecology 46(5):592–602

Pickett JA, Wadhams LJ, Woodcock CM (1998) Insect supersense: mate and host location by insects as model systems for exploiting olfactory interactions. Reprinted from – *The Biologist*. Portland Press Ltd, London

Reznik SY, Chernoguz DG, Zinovjeva KB (1992) Host searching, oviposition preferences and optimal synchronization in *Alysia manducator* (Hymenoptera, Braconidae). A parasitoid of the blowfly, *Calliphora vicina*. Oikos 65(1):81–88

Sabelis MW, van de Baan HE (1983) Location of distant spider mite colonies by phytoseiid predators: demonstration of specific kairomones emitted by *Tetranychus urticae* and *Panonychus ulmi*. Entomologia Experimentalis et Applicata 33:303–314

Selby MJ (2003) Chemical ecology of the carrot fly, *Psila rosae* (F.): laboratory and field studies. The University of Nottingham for the Degree of Doctor of Philosophy

Shields VDC, Hildebrand JG (2001) Recent advances in insect olfaction, specifically regarding the morphology and sensory physiology of antennal sensillae of the female Sphinx Moth *Manduca sexta*. Microsc Res Tech 55:307–329

Shoenly K, Reed W (1987) Dynamics of heterotrophic succession on carrion arthropod assemblages: discrete seres or a continuum of changes. Oecologia 73:192–202

Smedts BR (2004) Detection of buried cadavers in soil using analysis of volatile metabolites: First results. The Royal Society of Chemistry. The RSC Conference: Forensic Analysis 2004, June 20–22

Smith KGV (1986) A manual of forensic entomology. Trustees of the British Museum, London

Smith KE, Wall R (1998) Suppression of the blowfly *Lucilia sericata* using odour-baited triflumuron-impregnated targets. Med Vet Entomol 12:430–437

Statheropoulos M, Spiliopoulou C, Agapiou A (2005) Study of volatile organic compounds evolved from the decaying human body. Forensic Sci Int 153(2):147–155

Statheropoulos M, Agapiou A, Spiliopoulou C, Pallis GC, Sianos E (2007) Environmental aspects of VOCs evolved in the early stages of human decomposition. Sci Total Environ 385:221–227

Stensmyr MC, Urru I, Collu I, Celander M, Hansson BS, Angioy A-M (2002) Rotting smell of dead-horse arum florets. Nature 420:625–626

Stensmyr MC, Gibernau M, Ito K (2003) Thermogenesis and respiration of inflorescences of the dead horse arum *Helicodiceros muscivorus*, a pseudo-thermoregulatory aroid associated with fly pollination. Funct Ecol 7:886–894

Sutcliffe JF, Steer DJ, Beardsall D (1991) Studies of the host location behaviour in the blackfly *Simulium arcticum* (IIS-10.11) (Diptera: Simuliidae): aspects of close range trap orientation. Bulletin of Entomological Research 85:415-424

Takken W, Knols BGJ (1999) Odor-mediated behavior of Afrotropical malaria mosquitoes. Annu Rev Entomol 44:131–157

Vass AA (2001) Beyond the grave – understanding human decomposition. Microbiology Today. 28(November):190–192

Vass AA, Bass WM, Wolt JD, Foss JE, Ammons JT (1992) Time since death determinations of human cadavers using soil solutions. J Forensic Sci 37:1236–1253

Vass AA, Barshick S-A, Sega G, Caton J, Skeet JT, Love JC, Synstelien BA (2002) Decompositional chemistry of human remains: a new methodology for determining the post mortem interval. J Forensic Sci 47(3):542–553

Vickers NJ (2000) Mechanisms of animal navigation in odor plumes. Biol Bull 198:203–212

Wadhams LJ (1990) The use of coupled gas chromatography: electrophysiological techniques in the identification of insect pheromones. In: McCaffery AR, Wilson ID (eds) Chromatography and isolation of insect hormones and pheromones. Plenum, New York, London, pp 289–298

Wadhams LJ, Angst ME, Blight MM (1982) Responses of the olfactory receptors of *Scolytus scolytus* (F.) (Coleoptera, Scolytidae) to the stereoisomers of 4-methyl-3-heptanol. J Chem Ecol 8:477–492

Williams DJ, Ansford AJ, Priday DS, Forrest AS (2001) Forensic pathology – colour guide. Churchill Livingstone, Harcourt, London

Woolbridge J, Scrase R, Wall R (2007) Flight activity of the blowfly, *Calliphora vomitoria* and *Lucilia sericata*, in the dark. Forensic Sci Int 172:94–97

Zhou JJ, Huang WS, Zhang GA, Pickett JA, Field LM (2004) "Plus-C" odorant-binding protein genes in two *Drosophila* species and the malaria mosquito *Anopheles gambiae*. Gene 327:117–129

Zumwalt RE, Bost RO, Sunshine I (1982) Evaluation of ethanol concentrations in decomposed bodies. J Forensic Sci 27:549–554

Chapter 12
Decomposition and Invertebrate Colonization of Cadavers in Coastal Marine Environments

Gail S. Anderson

12.1 Introduction

The decomposition of a body on land and the interaction between the body and its insect fauna have been well studied, and predictable insect development and colonization patterns have been described from many countries, geographical regions, habitats and seasons (reviewed in Anderson 2009). However, despite the oceans' vast area, little is known about human or animal decomposition and the associated faunal dynamics in the marine environment. Knowledge of the effects of body submergence is important as many homicide victims are disposed of in the ocean in an effort to get rid of the body and a much greater number of people are lost in the marine environment to drowning, boating or airplane accidents. In all these cases, it is important to the investigators and to family members to understand what has happened to the decedent from the time they were last seen alive, to the time of discovery. This has been particularly highlighted recently when six individual feet, shod in running shoes, washed ashore along the Southern British Columbia, Canada coastline, with a seventh found a few kilometres south in the US' San Juan Islands.

Forensic entomology can be of value in many parts of a death investigation, but its' primary purpose is to estimate the time elapsed since death. This is important in any death investigation but is of paramount value in a homicide, supporting or refuting an alibi, and allowing the investigators to understand the victim's timeline prior to death. However, estimating elapsed time since death is difficult (Chorneyko and Rao 1996; Teather 1994; Wentworth et al. 1993; Lawlor 1992; Mant 1960) or even impossible after a body has been submerged for a period of time (Wentworth et al. 1993; Picton 1971). In most cases when a body is recovered from the marine environment, the elapsed time since death is based on when the victim was last seen alive (Wentworth et al. 1993). This may be perfectly valid when the victim is, perhaps, a person who was last seen alive on a

G.S. Anderson
School of Criminology, Simon Fraser University, Burnaby, British Columbia, Canada

J. Amendt et al. (eds.), *Current Concepts in Forensic Entomology*,
DOI 10.1007/978-1-4020-9684-6_12,

fishing vessel and was expected back shortly. It is much less likely to be valid when a person has been murdered, as witness statements as to last time seen alive may not be truthful (Anderson 2008).

There have been very few studies on decomposition of either human or animal remains in aquatic environments (Sorg et al. 1997), and those that have been conducted are mostly from freshwater (O'Brien and Kuehner 2007; Petrik et al. 2004; Hobischak and Anderson, 1999, 2002; Hobischak 1998;O'Brien 1997; O'Brian 1994; Kelly 1990; Dix 1987; Tomita 1976; Payne and King 1972). Most of our knowledge of the marine taphonomy of a body is anecdotal, based on case studies of recovered bodies and known drownings (Dumser and Türkay 2008; Ebbesmeyer and Haglund 2002; Haglund and Sorg 2002; Kahana et al. 1999; Boyle et al. 1997; Ebbesmeyer and Haglund 1994; Haglund 1993; Davis 1992; Giertson and Morild 1989; Donoghue and Minnigerode 1977). The most valuable reviews are from Sorg and colleagues (1997) and Teather (1994), a Royal Canadian Mounted Police (RCMP) Officer who was an investigative police diver and published the book 'Encyclopedia of Underwater Investigations'. Some research has been conducted in the more accessible intertidal regions (Davis and Goff 2000; Lord and Burger 1994) but these primarily describe colonization by terrestrial insects until the remains were submerged, when bacterial decomposition became prominent (Davis and Goff 2000). Large carcass falls, such as those of whales, have been observed since the 1850s. These do provide interesting information on the ecology related to large nutrient injections into normally nutritionally poor environments (Baco and Smith 2003). However, these are extremely large animals and usually are found at great depths (~4,000 m or more) (Smith and Baco 2003; Witte 1999) so do not relate well to a recovered human body.

Decomposition in the ocean is impacted by a large number of interacting variables including water temperature and salinity, tides, currents, depth in the water column, substrate type, season, body covering, water chemistry and the species and numbers of colonizing animals (Anderson and Hobischak 2002, 2004; Sorg et al. 1997; Teather 1994; Keh 1985; Polson and Gee 1973; Mant 1960). A body, whether human or animal, provides a sudden input of nutrients. Sorg and colleagues describe the four ecological roles of such a carcass on the seafloor. They state that, depending on depth, substrate type and other factors, the remains may become food for many vertebrate and invertebrate scavengers, may provide shelter for other species, may attract secondary scavengers and the bones may eventually provide a source of minerals and act as a substrate for bacterial grazers (Sorg et al. 1997).

The lack of research from marine environs is hardly surprising as the oceans cover such vast territory and are mostly inaccessible. Marine decompositional research requires boats, divers, or deep sea equipment, often making it prohibitively expensive. In an effort to fill this gap, this author has been conducting a number of studies in the Pacific Ocean, in the coastal waters of Southern British Columbia. The first set of experiments was conducted in shallow, well oxygenated waters in Howe Sound. The second set of experiments is being conducted in the deeper and much more anoxic waters of the Saanich Inlet. Pig carcasses were used as models for humans as pigs have been shown to decompose in a manner very similar to that of humans (Catts and Goff 1992).

12.2 Decomposition and Invertebrate Colonization of Carrion in a Shallow, Coastal Marine Environment

12.2.1 Material and Methods

Six freshly killed pig (*Sus scrofa* L.) carcasses were placed at two depths in the Howe Sound, close to a small private island. Howe Sound is a fjord northwest of Vancouver, British Columbia, running from West Vancouver north up to Squamish, along the Sunshine Coast, with its' mouth at the Strait of Georgia. Some of this work is published in Anderson (2009) and Anderson and Hobischak (2002, 2004). The pigs weighed 20–25 kg each and were euthanized using a pin-gun. After death, the carcasses were taken immediately by boat to the research sites. Each carcass was taken by divers to a pre-established site at a depth of either 7.6 or 15.2 m. Each site was separated by at least 150 m and each carcass was attached to a concrete weight by a 2 m long nylon rope. This allowed the carcass to float or sink naturally depending on decompositional stage, but still maintained the carcass in the correct area. Divers examined the remains at varying times from submergence until only a few bones remained. Video and still photographs were taken at each examination time and samples were collected. Examination times varied with diver and boat availability. The first set of experiments was begun in May and terminated in October (spring experiments) and a second set of experiments was begun in October and terminated the following May (fall experiments).

12.2.2 Decomposition

Decomposition seemed much slower than on land in similar geographic areas (Anderson and VanLaerhoven 1996). On land, carcasses progress through fresh, bloat, active decomposition, advanced decomposition and remains stages (Anderson and VanLaerhoven 1996, adapted from Payne 1965). In these experiments, the carcasses did go through most of these stages, although they were modified from those seen on land (Anderson and Hobischak 2004). When first placed, most of the carcasses floated with the head towards the ground for a few hours, then sank after 18–28 h at both depths (Table 12.1). The remains appeared to be in the fresh stage from submergence until 3 days after submergence, when bloat was first observed. Bloat lasted from Day 3 to Day 11 in most carcasses, despite depth. In some carcasses, the body remained suspended in the water column for several weeks, apparently being held up by gases maintained in pockets in areas of the body rather than true bloat (Anderson and Hobischak 2002, 2004). In the fall experiments, the first examination was not possible until Day 19, and at that time, most of the pigs were still floating, except one or two that had become wedged between rocks (Table 12.2).

By Day 11 in spring, the flesh was decomposing, with exposed bone in places at both depths. Significant scavenging had occurred and the skin was sloughing in places, and hair was shedding (Anderson and Hobischak 2002).

Table 12.1 Description of pig carcasses placed out in May (spring) in Howe Sound at 7.6 and 15.2 m. See Table 12.3 for details of fauna collected and observed on remains (Anderson and Hobischak 2004; 2002)

Elapsed time since submergence	7.6 m	15.2 m
13.5 h	No fauna present, floating with snout towards seabed. Coagulated blood still at wounds.	Two of three carcasses floating, snout down. Two carcasses one floating and one on seabed, have fauna present, including whelks in nostrils and on skin and sea stars on and actively moving towards remains. Crab, shrimp, copepods, annelids, and sea urchins also collected.
19.5 h	All three carcasses have sunk to seabed. Many small amphipods "sea lice" seen all over carcasses, several species present. Carcasses look fresh, slight stomach distention in one, lividity visible in some. Larval herring present, feeding on amphipods. All coagulated blood gone.	Two of three carcasses have sunk, one still floating. Some small amphipods present but not as many as at 7.6 m. Coon striped shrimp seen but scared away by lights. More whelks getting into wound. Seas stars present, including large sunflower sun star enveloping rump of one. Carcasses same as at 7.6 m, look fresh, lividity visible in some. Larval herring present, feeding on amphipods. All coagulated blood gone. Silt accumulating over body.
28.5 h	No fauna seen on carcasses. Strong current present. Silt on carcasses. No outward signs of decomposition.	Much less current. Large sunflower sea star still present on rump as well as vermillion sea stars . One carcass still floating, only one shrimp present on it, but red marks on hocks, may be feeding damage.
3 days	Bloating starting in some carcasses, a few amphipods on one carcass, a mottled sea star and 3 Oregon tritons present. One carcass has no fauna, but ears appear frayed from feeding.	Bloating starting in some carcasses. Large sea star gone, but large circular mark left at site of sea star, appears partially 'digested'. Some vermillion sea stars and Oregon tritons present, 2 red rock crabs tearing at flesh.
11 days	Decay apparent, exposed areas of flesh, some exposed bone, skin sloughing, hair falling off, hind ends, thighs, tail, and ears show signs of feeding. Silt and algae on carcasses. Bloat apparent, adipocere seems to be forming on snout of one carcass floating with snout in contact with sediment. Fauna only on carcass on seabed, 1 leather star and some sand dabs near body.	Decay apparent, exposed areas of flesh, some exposed bone on legs, bloat still present. Evidence of scavenging on ears, abdomen and hind end. One carcass less bloated, lying on seabed and adipocere appears to be forming on snout. Many large sunflower sea stars on two carcasses, completely covering one. Hermit crabs and sand dabs present and feeding. Two carcasses covered in algae and silt.

30 days	One carcass lost, rope fouled on concrete blocks, and cut by wave action. Both remaining carcasses floating. Feet gone in one and one foot remaining in other. Skin and hair remaining in some areas, but parts of skull and much of spine and legs skeletonized. Skin of ears remaining, Abdomen and torso intact. No fauna observed.	Part of intestine 'balloon-like' and holding up two carcasses, although most of the carcasses are skeletonized. Rib cage appears empty of organs in one, with abdomen intact in two. Skull skeletonized in all three. Limbs gone. Muscle falling off, and bones on sediment in carcasses still afloat. Only fauna are sunflower and vermillion sea stars.
40 days	Only one carcass observed. Still floating. No skin present, spinal column and part of skull fully exposed. Adipocere tissue present. Bones fallen to seabed below carcass. Ribs skeletonized. Gut still intact and floating. Seal nearby and schools of fish.	Only one carcass observed. Floating, snout down. Head heavily scavenged, with bones on face exposed. Brain still present. Internal organs exposed. Adipocere tissue present. No fauna on carcass.
47 days	Both carcasses still floating, although one not as high as before. Vertebrae exposed. Adipocere. Barnacles appear to be growing in one eye socket. Very little fauna.	Two floating, even though one is almost completely skeletonized. One on ground also mostly skeletonized. Sunflower sea star on bones of one. Small halibut under body of one. No other fauna.
54 days	Remaining tissue being eroded by surge of tide against concrete blocks. Body showing extreme adipocere formation. Mandible still articulated. Stomach in one is still intact and hanging out as a bag. Other carcass is now touching seabed at snout, but hind end is still bloated and floating, with intestines present near rectal area. Only fauna present are mussels and some small fish.	Skin remaining on bones and some bones now disarticulated and on ground. Two carcasses now on seabed. Bag of skin and bones is all that is left of one, still floating. Mandibles disarticulated, most of remains are skeletonized. Large hole seen in chest area of one carcass. Stomach present in one and hanging out like a bag, although now supine on ground. Only animals are sea stars and small fish.
116 days	Two remaining carcasses are reduced to just a few bones. Some bones missing. White moss algae all over one set of bones. Bones appear brown, no cartilage present.	One carcass gone, no bones recovered. Remaining carcasses are only bones, some teeth and bones missing. No fauna. Leaf kelp present.
140 days	Fifty-nine bones recovered from one carcass and 12 from other. Many small animals recovered on bones.	Twenty-four bones recovered from one carcass and 57 from second. Most bones black in colour. Many small animals recovered on bones.

Table 12.2 Description of pig carcasses placed out in October (fall) in Howe Sound at 7.6 and 15.2 m. See Table 12.4 for details of fauna collected and observed on remains (Anderson and Hobischak 2004; 2002)

Elapsed Time Since Submergence	7.6 m	15.2 m
19 days	Head of one carcass wedged between two rocks. Algae covering one carcass more than others. Not much silt. Carcasses mostly intact. Scavenging damage on surface of body, such as ears, feet, along spine. Only one leg intact in one carcass. Skin sloughing. Very few fauna present.	One carcass on sediment, other two floating or partially floating. Heavily scavenged, internal organs exposed and hanging out of body cavity, algae and silt on all carcasses. Partial skeletonization. Two legs missing in one carcass. Very few fauna present.
26 days	One carcass very heavily scavenged. Tissue hanging off, extremities missing, skin sloughing. Adipocere tissue. Algae present. Very few fauna present.	Heavily scavenged, adipocere tissue and skeletonization. Remains of one on ground. Tissue present on ground nearby. Very few fauna present.
33 days	Two carcasses still floating, one on seabed. One leg remaining on one carcass, rest of limbs gone or skeletonized. Adipocere on nose of one carcass. Surface scavenging on head, ears, partial skeletonization, spine exposed. Head and hind quarters still intact in one. Algae present. No fauna seen.	All three floating again. Adipocere formation, tissue lying on ground below carcasses. Tissue on ground attracting sea stars. Algae and silt on carcasses. Skin, tissue, hair and adipocere tissue sloughing. Partial skeletonization. Feet absent in one carcass, three intact on another.
35 days	Only one shallow carcass recovered for close examination then replaced after 1 h. Carcass still bloated and floating. Skin and hair sloughing, surface scavenging on neck, lower abdomen and hind quarters, three legs gone, most of abdomen covered in adipocere, face still intact, silt collecting in shedding hair. Deep hole in one flank, and eroded area on other flank. Fauna included annelids, copepods and some arthropods.	No carcasses examined.
48 days	No carcasses examined.	Only one deeper carcass recovered for close examination then replaced after 1 h. Carcass still bloated. Skull partly skeletonized, with spinal column and muscle tissue exposed. Remaining skin looks bleached and speckled. Hair mostly gone. Three legs gone.
225 days	Only bones remaining, many animals, mostly mollusks associated with the bones. Mussel and kelp bed forming in area on one carcass. Four bones recovered from one carcass, 57 from another and nothing recovered from the last. Bones blackish in colour. Most bones from torso rather than limbs.	Only bones remaining, many animals, mostly mollusks associated with the bones. Twenty-eight bones recovered from one, with some mollusks attached. Most bones associated with torso not limbs. Some bones have spongy appearance. Twenty-seven bones recovered from a second, and none from the last, although all rope apparatus tethering carcass was intact, so bones removed.

Bloat was still apparent in some carcasses at this time. Also, at this time, two carcasses, floating but with the nose in contact with the sediment, appeared to be forming adipocere on the tissue in contact with the seabed. In the spring experiments, extensive adipocere formation was seen by Day 40, and by Day 33 in the fall experiments (Tables 12.1 and 12.2).

In the spring experiment, by Day 30, half the carcasses were still floating, held up still by gases in the intestines. Those that had sunk to the seabed were much more scavenged and skeletonized than those that floated, and the degree of skeletonization also appeared to be a function of whether the carcass fell onto sand or rock. One carcass was lost at the shallower depth by the rope rubbing against the concrete weight and breaking. In the fall experiment, by Day 33, five of the six carcasses were still floating. The limbs had been scavenged, and most carcasses were missing some or all the legs, although the torso was mostly intact and skin and hair remained in places (Anderson and Hobischak 2002). There was no clear distinction seen between active and advanced decay.

In spring, the remaining shallow carcasses and two of the deeper carcasses were still suspended in the water column by Day 47, although all were partially skeletonized. In the fall experiments, two different carcasses (one from 7.6 m and one from 15.2 m) were each briefly brought to the surface for closer examination on Day 35 and Day 48, and then immediately replaced. On both occasions, the remains were still bloated and adipocere tissue was present. At Day 35, only surface scavenging was noted on the abdomen, hind end and neck. At Day 48, the carcass was partially skeletonized and both muscle tissue and the spinal column were visible (Anderson and Hobischak 2002). In one case, part of the remains was externally coated in adipocere tissue, which protects the inner areas (Hobischak and Anderson 2002; Hobischak 1998).

During active decay, parts of the carcasses were skeletonized but in many cases, the carcasses continued to exhibit bloat and remained off the sea bed. Organs were still present in some carcasses, although ribs, skull and spine became exposed. There was no clear delineation between active and advanced decay as there is on land, and by 40 days post submergence in the spring experiments, most carcasses were entering the remains stage, although tissue and organs were still present in some carcasses (Anderson and Hobischak 2004, 2002). The bones were recovered on Day 140 after the spring experiment.

Only six examination days were possible in the fall due primarily to a lack of divers and vessels. No examination was possible until 19 days after submergence. Therefore, much of the decomposition and faunal colonization was missed. By 19 days post submergence, the carcasses had passed the bloat stage, although some continued to float, as in spring (Table 12.2). Those on the ground had been heavily scavenged although tissue and organs did remain and skin was sloughing. By 26 days post submergence, several of the carcasses showed extensive adipocere formation, and some had been scavenged heavily although few animals were present. At Day 33, all but one carcass was still floating. In the fall experiments, the remains were not examined between Day 48 and Day 225, when the bones were recovered and at Day 48, the remains were still bloated and the torso was intact.

12.2.3 Faunal Colonization

In terrestrial environments, wounds are the most attractive to arthropods, followed by the natural orifices. The only wounds on the experimental carcasses were the head wound inflicted by the pin-gun, as well as various cuts and sores that had been present in life. Only the spring carcasses were observed for the first few weeks after submergence. Very shortly after submergence, a periwinkle, the wrinkled amphissa (*Amphissa columbiana* Dall) were attracted to the head wounds at 15.2 m (Anderson and Hobischak 2002). However, after that the wounds and orifices did not appear to be any more attractive than undamaged areas. Periwinkles are grazers and usually feed on algal films although they have been found to scavenge corpses of humans, marine mammals and fish (Sorg et al. 1997). It has been suggested that they are not feeding directly on the carrion, but rather on the algal films developing on the carcass (Sorg et al. 1997). However, such early arrival suggests they are directly attracted to the carcass.

The carcasses at 15.2 m were more rapidly attractive to fauna than those at 7.6 m, but within 19 h, both sets of carcasses were attracting a variety of species (Table 12.3). The remains became covered in silt within hours of death and much of the faunal activity appeared to be related to this, presumably feeding on microbes, and micro flora and fauna within the silt. Many of the larger invertebrates caused abrasions or deep openings into the remains. In the spring experiments, within 24 h of submergence, some feeding damage was noted and by 3 days post submergence, the ears of most of the carcasses had a 'frayed' appearance, suggesting feeding activity also. Table 12.3 lists the species present on the carcasses from the spring experiments over time.

In the first 24 h, a green sea urchin (*Strongylocentrotus droebachiensis* (Müller)) and red or vermillion sea stars (*Mediaster aequalis* Stimson) were seen on some of the deeper carcasses, with more sea stars actively moving towards the deeper carcasses. A large sunflower sea star, *Pycnopodia helianthoides* (Brandt) was attracted shortly after death to the deeper carcasses only and was rapidly joined by several others that completely enveloped several carcasses. Several of these sunflower sea stars were seen moving rapidly towards the carcasses, actively attracted. On only one occasion was this sea star collected from remains at the shallower location and this at the remains stage. Some larger sea stars are known to feed on carrion as well as invertebrates (Sorg et al. 1997). After the sunflower sea stars had left, there appeared to be large grazed or damaged areas left behind. It may have been caused by the sun star when grazing on epifauna on the surface of the pig (Sloan and Robinson 1983). These sunflower sea stars were attracted from immediately after submergence until only bones remained. At one point, when only bones and skin remained, the sea star was found inside the bag of skin.

Other fauna noted on the remains within 24 h of submergence included the Western Lean Nassa (*Nassa mendicus* (Gould)), marine Oligochaetes (family Enchytraeidae), copepods, proboscis worms (*Glycera* sp.), bloodworms (*Euzonus* sp.), red rock crabs (*Cancer productus* L.), blue mud shrimp (*Upogebia pugettensis* Dana), coon striped shrimp (*Pandalus danae* Stimson), Oregon triton (*Fusitriton*

Table 12.3 Species collected in the spring experiment. S = 7.6 m, D = 15.2 m. At 140 days post submergence, the remains were recovered and examined very carefully resulting in a more detailed collection

Species	13.5H	19.5H	28.5H	3 D	11D	30 D	40 D	47 D	54 D	116 D	140D
Phylum Arthropoda											
Alaskan pink shrimp *Pandalus eous* Makarov		D									
Hermit crab *Pagurus beringanus* (Benedict)				D							
Pacific red hermit *Elassochirus gilli* (Benedict)											D
Barnacles (*Balanus* sp.)											S
Small amphipods		S, D		S							
Red rock crab *Cancer productus* L.	D			D							S, D
Pacific lyre Crab *Hyas lyratus* Dana											D
Blue mud shrimp *Upogebia pugettenis* Dana	D										S
Large amphipods											
Coon striped shrimp *Pandalus danae* Stimson		D	D								
Sitka shrimp *Heptacarpus sitchensis* Brandt											S
Ostracod											S
Diastylis rathkei Krøyer											S
Copepods	D										S, D
Phylum Annelida											
Marine oligochaetes Family Enchytraeidae	D										S, D
Proboscis worms *Glycera* sp.	D										D
Bloodworms *Euzonus* sp.	D										S, D
Pile worms *Nereis vexillosa* Grube											S
Marine worms *Nereis* sp.											S
Scaleworm *Arctonoe* sp.											S, D
Polychaetes *Ammotrypane aulogaster* Rathke											S, D

(continued)

Table 12.3 (continued)

Species	13.5H	19.5H	28.5H	3 D	11D	30 D	40 D	47 D	54 D	116 D	140D
Phylum Mollusca											
Oregon triton *Fusititron oregonensis* (Redfield)		D		S, D							
Aleutian macoma *Macoma lama* Bartsch											S
Sitka periwinkle *Littorina sitkana* Philippi											S
Periwinkle *Littorina scutulata* Gould											S
Wrinkled amphissa *Amphissa columbiana* Dall	D	D									S
Pandora sp.											
Western lean nassa *Nassarius mendicus* (Gould)	D	D									S
Giant western nassa *Nassarius fossatus* (Gould)											D
Mussels								S	S		
Phylum Echinodermata											
Leather star *Dermasterias imbricata* (Grube)					S						
Mottled sea star *Pisaster brevispinus* (Stimson)				S							
Sunflower sea star *Pycnopodia helianthodes* (Brandt)		D	D		D	D		D	D		
Red sea star *Mediaster aeqalis* Stimson	D	D	D	D		D					
Green sea urchin *Strongylocentrotus droebachiensis* (Müller)	D										
Phylum Chordata											
Gobies Family: Gobiidae									D		
Sculpins Family: Cottidae									D		
Ling cod *Ophiodon elongatus Girard*										D	
Pacific halibut *Hippoglossus stenolepis (Schmidt)*								D			
Sand dabs *Citharichthys* sp.					S, D						
Larval herring Family: Clupeidae		S, D									
Phylum Nematoda											S, D

oregensis (Redfield)), Alaskan pink shrimp (*Pandalus eous* Makarov) as well as larval herring (*Clupea* sp.) and two species of small amphipods, which were not caught for identification (Anderson and Hobischak 2002). The majority of the fauna were seen on the deeper carcasses (Tables 12.3 and 12.4). There was a stronger current at the shallower sites which may have accounted for the lower numbers of animals seen. Some of the fauna were clearly predating on other fauna, while some appeared to be feeding on the carcass. In most cases, only one or two members of each species were found on the remains, and no species was found in large numbers. On several examination days, some carcasses at 7.6 m had no visible fauna present, although feeding damage was evident.

When the remains sank, they fell randomly on either a sandy silt seabed or on rocks. This appeared to influence the fauna, with sea stars (*Henricia* sp), coon striped shrimp (*Pandalus danae*), sand dabs (*Citharichthys* sp.) and gobies found almost exclusively on carcasses on sand, while sunflower sea stars, pile worms (*Nereis vexillosa* Grube), sea strawberries (*Gersemia rubiformis* (Erhrenberg)), Oregon tritons, smooth cockles (*Clinocardium blandum* (Gould)), leafy hornmouth (*Ceratostoma foliata* (Gmelin)), chalky macoma (*Macoma calcarea* (Gmelin)), mussels and oysters were more common when the carcass rested on a rocky substrate (Anderson and Hobischak 2004).

When the bones were recovered at the end of the spring experiment, a more exhaustive examination of the remains could be made and marine oligochaetes (Family Enchytraeidae), bloodworms (*Euzonus* sp.), juvenile red rock crab, blue mud shrimp, pile worms (*Nereis vexillosa*), *Nereis* sp., *Pandora* sp., scaleworm, Aleutian Macoma (*Macoma lama* Bartsch), Sitka periwinkle (*Littorina sitkana* Philippi), *Diastylis rathkei* Krøyer, *Ammotrypane aulogaster* Rathke, copepods, nematodes, ostracoda, wrinkled amphissa, Sitka shrimp (*Heptacarpus sitchensis* Brandt)) were found on the shallow remains (Table 12.3). On the remains at 15.2 m a Red rock crab, Pacific lyre crab (*Hyas lyratus* Dana), Pacific red hermit (*Elassochirus gilli* (Benedict)), proboscis worms, copepods, bloodworms, marine oligochaetes (Family Enchrytraeidae), scaleworms, giant Western Nassa (*Nassarius fossatus* (Gould)), *Ammotrypane aulogaster*, juvenile red rock crab and nematodes were recovered (Table 12.3) (Anderson and Hobischak 2002). However, at this time, only bones remained and the remains had been skeletonized for some time, so many of these species may have only been incidentally associated with the remains.

Very low numbers of animals were seen on the fall carcasses, although again, the deeper remains were more attractive (Table 12.4). When the bones were recovered at the end of the fall experiment, both shallow and deeper remains were found to have shrimp, small amphipods, marine oligochaetes, sunflower sea stars, and a large number of mollusks, although by far the greater number of mollusks were found on the deeper remains (Table 12.4). A mussel bed and a kelp bed had developed near the remains of one of the shallow carcasses. Many more mollusks were found on the fall carcasses than on the spring carcasses. A large sunflower sea star was seen on the bones at 225 days on a shallow pig, the first time this species was recorded on a carcass at 7.6 m. This species appeared to be much more attracted to

Table 12.4 Species collected in the Fall experiment. S = 7.6 m, D = 15.2 m. At 225 days post submergence, the remains were recovered and examined very carefully resulting in a more detailed collection. At 35 D, only one carcass, from 7.6 m, was brought to the surface for examination and at Day 48 a carcass from 15.2 m was brought up for examination then returned

Species	19 D	26 D	33 D	35 D	48 D	225 D
Phylum Arthropoda						
Shrimp *Pandalus* sp.						S, D
Barnacles (*Balanus* sp.)						S
Small amphipods				S		S, D
Diastylis rathkei Krøyer				S		
Copepods				S		
Phylum Annelida						
Marine oligochaetes Family Enchytraeidae				S	D	S, D
Proboscis worms *Glycera* sp.						
Bloodworms *Euzonus* sp.				S	D	
Pile worms *Nereis vexillosa* Grube						S
Marine worms *Nereis* sp.				S		
Sandworms*Nephtys* sp.						S
Polychaetes *Ammotrypane aulogaster* Rathke				S		
Phylum Mollusca						
Three rib chiton *Lepidozona trifida* (Carpenter)						S,D
Copper's chiton *Lepidozona cooperi* (Dall)						D
Wrinkled amphissa*Amphissa columbiana*Dall	S					
Smooth cockle *Clinocardium blandum* (Gould)						S, D
Wrinkled slipper-shell *Crepidula lingulata* Gould						D
Pacific blue mussel *Mytilus edulis* L.						D
Smooth pink scallop *Chlamys rubida* Hinds						D
Thin shell littleneck *Protothaca tenerrima* (Carpenter)						D
Red turban *Astraea gibberosa* (Dillwyn)						D
Hooded puncturella *Cranopsis cucullata* (Gould)						D
Tucked margarite *Margarites succinctus* Carpenter						D
Smooth margarite *Margarites helicinus* (Phipps)						S
Chalky macoma *Macoma calcarea* (Gmelin)						S, D
Macoma *Macoma* sp.						S, D
Hind's mopalia *Mopalia hindsii* (Reeve)						D
Oyster *Ostrea* sp.						D
Edible flat oysters *Ostrea edulis* L.						S
Olympia oyster *Ostrea conchaphila* Carpenter						S

(continued)

Table 12.4 (continued)

Species	19 D	26 D	33 D	35 D	48 D	225 D
Sitka periwinkle *Littorina sitkana* Philippi						D
Slender bittium *Bittium attenuatum* Carpenter						D
Abalone piddock *Penitella conradi* Valenciennes						D
Variegated chink-shell *Lacuna variegata* Carpenter						S
Leafy hornmouth *Ceratostoma foliata* (Gmelin)						S
Tusk shells Family Dentaliidae						D
Mussels						S, D
Phylum Echinodermata						
Sea star *Henricia* sp.	D	D				
Feather star *Florometra serratissima* (Clark)	D	D				
Sunflower sea star *Pycnopodia helianthodes* (Brandt)			D			S, D
Brittle star *Ophiopsilla* sp.						D
California sea cucumber *Parastichopus californicus* (Stimson)						S
Phylum Chordata						
Gobies Family: Gobiidae						D
Phylum Nematoda						S
Phylum Cnidaria						
Sea strawberry *Gersemia rubiformis* (Erhrenberg)						S, D

carcasses placed out in spring rather than those in fall, actively moving rapidly towards the carcasses, and preferred the carcasses at 15.2 m, obviously preferring the slightly deeper waters. Although a larger diversity of fauna was collected when the skeletal remains were recovered, many species were only represented by a single specimen, or two or three specimens.

A much greater diversity of species were recovered on the carcasses placed out in spring rather than those in the fall, although a large number of species of mollusks were recovered 225 days after submergence in the fall experiment, but many were probably incidental, merely settling on a substrate (Tables 12.3 and 12.4). At no time were very large numbers of animals found on the remains in either season. As well, due to inclement weather and difficulty in getting vessels, the fall experiments were sampled much less often than in spring.

In many cases, only one or two specimens of a species were present at any time. As can be seen from Tables 12.3 and 12.4, on many occasions, no fauna were observed on the remains at all. The large diversity of specimens collected when the bones were recovered suggests that many of the invertebrates were missed by divers, or dispersed due to disturbance.

12.2.4 Discussion

Pig carcasses are commonly used in forensic entomology studies as they have been found to decompose in a very similar manner to that of humans on land (Catts and Goff 1992). Our work has shown that pigs also decompose in the aquatic environment in a similar manner to submerged humans (Petrik et al. 2004; Hobischak and Anderson 1999; Hobischak 1998).

In these experiments, the carcasses decomposed much more slowly than was expected. Public safety divers frequently report complete or partial skeletonization of a body within hours of death in certain situations (Teather 1994), so the slow decomposition and lack of anthropophagy were unexpected. Surprisingly, although a number of different species did colonize and feed on the remains, the actual numbers of animals present on the remains were extremely low at all examination times, at both depths and in both seasons. Scavenging and feeding damage was seen on the carcasses but most tissue loss seemed to be due to decomposition, current action, tidal action, and movement of the carcasses against the substrate. The lack of numbers of animals may have been due to disturbance by divers, as scavenging did occur, even when carcasses seemed to be relatively free of invertebrates. However, the much greater numbers of animals seen in the next set of experiments suggests that disturbance was not the only reason for such low numbers. It is quite possible that many species were missed, either due to disturbance, or lack of visibility as many more species were collected when the bones were recovered at the end of each experiment, and on the two occasions when remains were brought to the surface briefly.

Adipocere formation was visible to the naked eye by Days 33 and 40. Adipocere is the result of the saponification or hydrolysis and hydrogenation of the adipose or fatty tissue and occurs most commonly in moist anaerobic situations (Lo 2007). It has been reported that the timing of adipocere formation can be used to estimate elapsed time since death (Mellen et al. 1993) particularly when temperature is known (Kahana et al. 1999). It was generally believed that adipocere took several months to form unless in very warm conditions (Simonsen 1977). However, in a study of 15 bodies recovered over a period of months from a ship wreck, adipocere formation was seen in 38 days after submergence, despite cold waters (Kahana et al. 1999), and in a study of aircraft accidents in the ocean, adipocere formation was seen after 34 days in the Mediterranean Sea at a depth of 540–580 m (Dumser and Türkay 2008). A recent review of the literature has shown that the timing of onset of adipocere formation is very variable. (Lo 2007; O'Brian 1994).

There was no evidence of any clear successional patterns at either depth or season and most animals seemed to be opportunistic feeders, feeding on many kinds of food, including carrion if it was present and on its' microfauna and flora. No species appeared to be carrion dependent, as is seen with many carrion insects on land. Most of the species were not time dependent and were found on the carcasses at any time from fresh to the remains stages.

In both seasons and depths, the carcasses sank at first, then refloated with bloat, but surprisingly, remained floating for a considerable period of time, due to gases in the tissue and, when almost skeletonized, in the remains of the organs, which appeared to act like balloons. In contrast, pig carcasses in both still and running freshwater remained bloated from 9 to 35 days in the same geographic area (Hobischak 1998) and 2–10 days in nearby terrestrial environments (Anderson and VanLaerhoven 1996). This prolonged flotation was also noted in brackish waters (Zimmerman and Wallace 2008).

A much greater diversity of fauna was seen at 15.2 m, and some species seemed only to be attracted to the deeper carcasses, although this was less apparent in the fall experiments. In general, the difference in depth did not seem to have as much impact as flotation of the remains. Those carcasses that remained floating for much of the experiment were much less scavenged than those that had contact with the sea bed. The seabed at this site was very variable with rocky areas close to sandy areas. For those carcasses that remained in contact with the seabed, or sank down, those that fell onto sand or silt were much more rapidly scavenged than those that fell to rock. This is presumably related to the greater abundance of fauna within the silt rather than on rocks and the accessibility of the carcass, with those floating only being accessible to fauna that can swim.

The obvious limitation with this study is that divers and boats were required to attend the scenes. This is limiting financially and was dependent on divers who donated their time to this project and the availability of RCMP, Canadian Coast Guard, Vancouver Aquarium Marine Research Centre and Canadian Amphibious Search Team vessels, whose time was donated. This meant that the carcasses could not be examined as frequently as is desirable and this was particularly true in experiments begun in October, when weather conditions were less favourable than in summer. It is also limiting with regards to human safety, as bad weather conditions prevented diving, and diver safety also limited the depths and situations that could be examined. As well, divers approaching and examining the remains may have disturbed the fauna on the carcasses. These problems were alleviated in the next set of studies which were performed with the Victoria Experimental Network Under the Sea (VENUS).

12.3 Decomposition and Invertebrate Colonization of Carrion in a Deep Coastal Marine Environment

The following studies are ongoing and are performed in collaboration with the Victoria Experimental Network Under Sea or VENUS (www.venus.uvic.ca) (Anderson 2008, 2009). VENUS is an underwater laboratory consisting of an array of oceanographic instruments, cameras and sensors connected to a node via fibre optical cables and to a remote station on shore (www.venus.uvic.ca). This allows researchers to access real-time data from the ocean bed and from their experiments, using an internet connection, from anywhere in the world. VENUS presently has nodes in the Saanich Inlet, a glacially carved 24 km long fjord , close to Vancouver Island in British Columbia (Herlinveaux 1962), as well as the Strait of Georgia and the Fraser Delta, with another node expected to be deployed in the Strait of Juan de Fuca, just south of the Canadian/US border. These present studies were conducted in Saanich Inlet, but plans for the future include all VENUS sites.

The Saanich Site includes the Node which provides and distributes power and communication, the VENUS Instrument Platform (VIP), and a remote still and video camera. The camera, an eight megapixel Olympus C8080, is housed in a copper jacket to prevent fouling and is placed on a camera platform, which supports the camera 2 m above the seabed. The camera is operated remotely and has an array of lights and lasers for scale. All images were taken by this author with the VENUS camera and images are shown here with permission from the VENUS project, University of Victoria. The VIP housed the Seabird Electronics Inc. CTD and the Falmouth Scientific, Inc. CTD, measuring conductivity, temperature and depth via pressure at 1 and 60 s intervals, respectively, as well as a Gas Tension Device, measuring dissolved gas pressure, an oxygen optode, measuring dissolved oxygen and a Sea-Tech Transmissometer measuring clarity of the water (www.venus.uvic.ca). The data from these instruments can be seen and downloaded from the VENUS website. The Saanich Site was at a depth of 94 m for the first two deployments and 99 m for the third. The seabed was made up of fine silt which was approximately 10–20 cm deep, over rock.

Saanich Inlet is a deep water fjord, with a maximum depth of 230 m, which is separated from the Strait of Georgia by a shallow sill which restricts the flow of water in and out of the Inlet (www.venus.uvic.ca). This means that, for much of the year, the inlet is oxygen depleted or hypoxic.

The first carcass was placed at the site on 5 August 2006. It weighed 26 kg and was killed by electricity, then taken to the site by boat. On the boat, it was refrigerated at 4°C until it could be deployed. Due to camera problems, the pig was kept under refrigeration for almost 44 h. The carcass was weighted at the head and rear end by lead weights and rope, and the weights roped together (Fig. 12.1). The weights were required to keep the carcass within view of the camera at all times, rather than to prevent it from floating away, as bloat and therefore refloat do not occur at such depths (Teather 1994).

The carcass was attached to an acoustic transponder then dropped into the ocean. It was then located using the transponder by a remote submarine vehicle, ROPOS (Remote Operated Platform for Oceanic Science). ROPOS retrieved the carcass and carried it to the camera platform, where it was laid on its side, under the tripod holding the camera. The camera platform was approximately 100 m from the VIP where the chemical sensors resided. The transponder was then recovered by ROPOS. Once ROPOS left the site, there was no further direct contact with the carcass. The carcass was observed remotely several times a day, until the carcass was no longer in view of the camera. At 94 m, the ocean is completely dark so lights had to be turned on in order to observe activity at the carcass. An array of lights and light types were available. Each observation session was recorded in its entirety by video tape and still images were taken at will. Salinity, clarity, temperature, and oxygen level were monitored at 1 and 60 s intervals (Anderson 2008, 2009). Days are measured in days since submergence. Once the carcass had been placed it was not disturbed again, except by the turning on of lights during observation times.

The second carcass was deployed at 0800 h, 16 September 2007. It weighed 24.7 kg and was killed by pin-gun on 15 September and kept on ice until deployed by ROPOS. The third carcass was deployed at 0835 h 29 September 2008 at a site close by, which was 4 m deeper, at 99 m. It weighed 23 kg and was killed by pin gun on 28 September and kept in a refrigerated room until deployed by ROPOS.

Further carcass deployments are planned at 65 m in the Saanich Inlet and 170 and 300 m in the Strait of Georgia.

12.3.1 First Deployment, 2006

The carcass sank immediately, and at no time showed any evidence of bloat or re-flotation (Fig. 12.1). A large herring ball (*Clupea* sp.) was present when the pig was deployed but showed no interest in the carcass. However, a number of arthropods showed immediate interest. Squat lobsters (*Munida quadrispina* Benedict, Family Galatheidae) arrived in large numbers and squads of *M. quadrispina* were seen marching across the sand towards the carcass. The *M. quadrispina* picked at all areas of the carcass, but particularly at the face and nose (Fig. 12.2). Three spot shrimp (*Pandalus platyceros* Brandt) and Dungeness crabs (*Cancer magister* Dana.) were also immediately attracted and picked at the carcass and at the silt that had settled on it. By Day 1, *M. quadrispina*, *C. magister* and *P. platyceros* were feeding at all the orifices and the carcass was moved and rocked

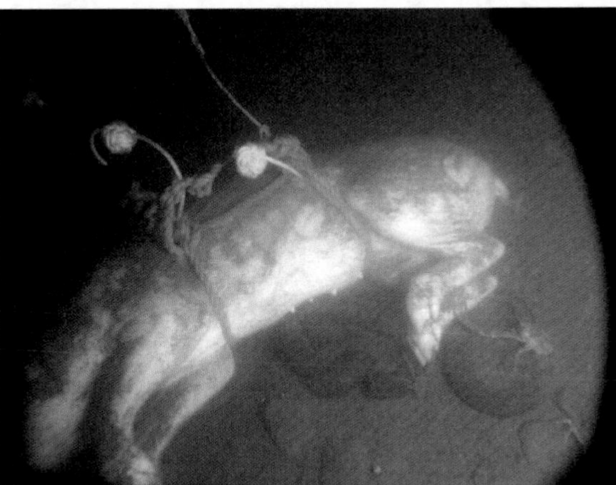

Fig. 12.1 Pig carcass placed at a depth of 94 m in August, in Saanich Inlet, British Columbia, Canada, 2006 (VENUS Project, University of Victoria)

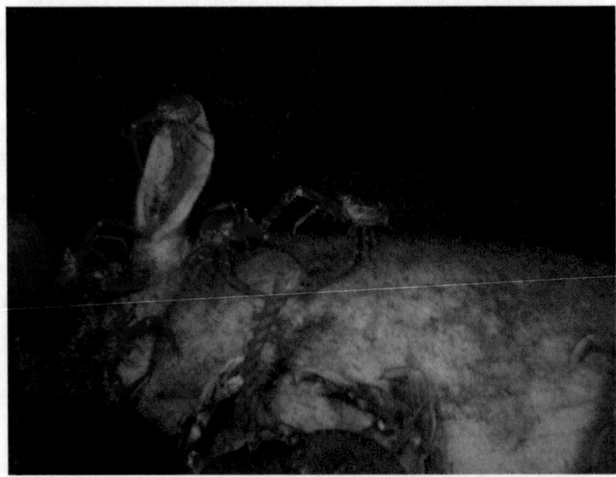

Fig. 12.2 Squat lobsters (*Munida quadrispina*) attracted to the face of a pig carcass at a depth of 94 m, 2006 (VENUS Project, University of Victoria)

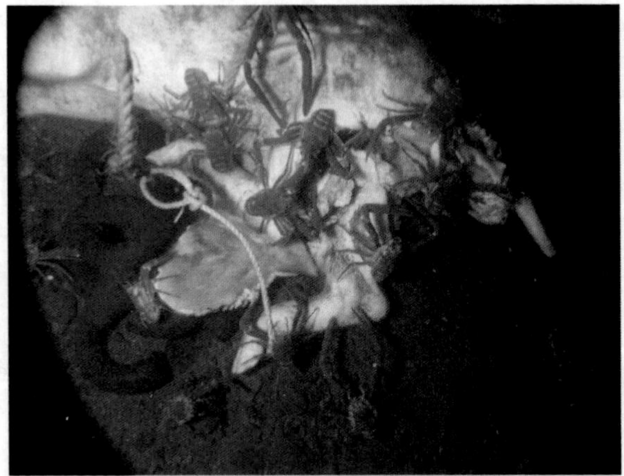

Fig. 12.3 Day 3 post submergence 2006. A probable sixgill shark (*Hexanchus griseus*) bite in the rump of the carcass became a major site of attraction (VENUS Project, University of Victoria)

on occasion by the crab activity. On Day 2, the pig had been moved 1.5 m and 180° from its original position, a large piece of flesh had been removed from the hind quarter and a flap of skin and tissue from the stomach had been laid open (Fig. 12.3). The culprit was not observed, but the shape, size and pattern of the bite mark indicate that it was probably caused by a sixgill shark (*Hexanchus griseus* Bonneterre) (Tunnicliffe 2006). From this moment on, the majority of feeding activity and invertebrate attraction was centred on this area (Anderson 2008). The main invertebrate species on the carcass were *M. quadrispina* which fed on the

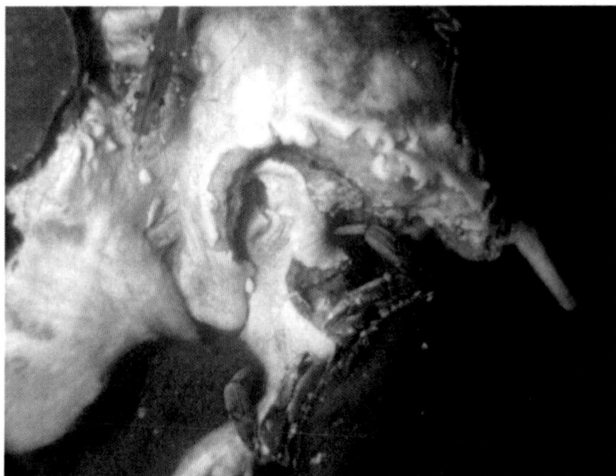

Fig. 12.4 Dungeness crabs (*Cancer magister.*), squat lobsters (*Munida quadrispina*) and three spot shrimp (*Pandalus platyceros*) feed at the bite area in the rump of a pig carcass on Day 4 post submergence 2006. Lasers indicate 10 cm between light spots (VENUS Project, University of Victoria)

carcass in large numbers. A separate study done at the same time as the 2006 VENUS experiment showed that the density of the scavenging organisms, in particular, *M. quadrispina* increased when the carcass was present, with an *M. quadrispina* density of up to 1.6 per dm^2 when the carcass was present but only up to a maximum of as 0.7 per dm^2 when it was not (Peters 2007).

The lights did not seem to have a large impact on either attracting or repelling invertebrates to the carcass, as observed by the camera, although it is likely that the lights did impact the fauna as divers in the earlier experiments did report attraction to their lights (MacFarlane 2001).

Over the subsequent few days, the major fauna on the carcass were consistently *M. quadrispina*, *P. platyceros* and *C. magister*, which were seen to be eating the tissue and hollowing out the inside of the carcass from the wounded area (Fig. 12.4). Large numbers of all three species were seen on the carcass and camera scans of the surrounding area showed these species being actively attracted, moving towards the remains. *Cancer magister* were capable of rocking the entire carcass, and removed large pieces of tissue and organs from the remains. *Munida quadrispina* still showed an interest in the head region although little damage was observed. They were seen feeding on the carcass but also sifting through the fine layer of silt on the carcass. However, the majority of the *M. quadrispina* were seen at the wound. *Pandalus platyceros* fed at the wound and also on the inside of the flap of skin removed from the stomach region (Fig. 12.5).

By Day 5, a large amount of tissue had been removed from the hind quarters and the abdominal cavity had been opened (Anderson 2008). The spinal column and the intestines were exposed and *C. magister*, *M. quadrispina* and *P. platyceros* continued to dominate the carcass (Fig. 12.6). The fauna on this carcass was distinctly different

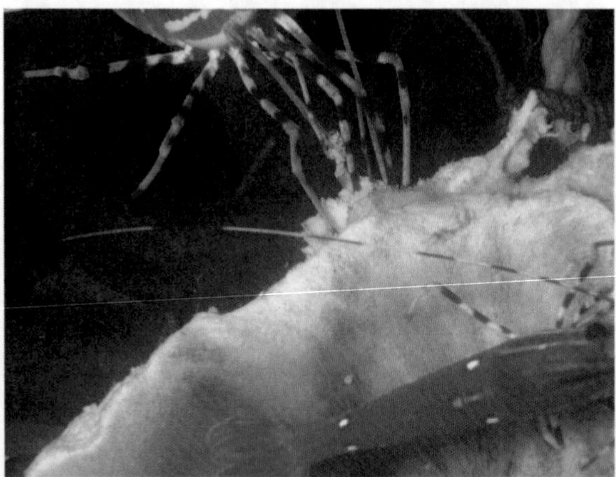

Fig. 12.5 Three spot shrimp (*Pandalus platyceros*) feeding on tissue 2006 (VENUS Project, University of Victoria)

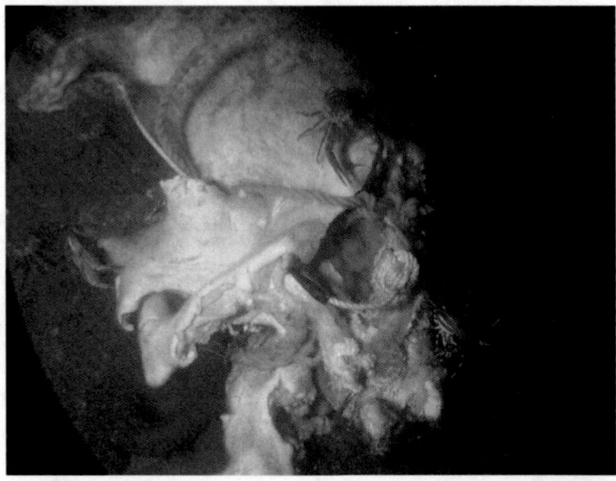

Fig. 12.6 Day 5, 2006. Spinal column and intestines exposed. Dungeness crab (*Cancer magister)*, three spot shrimp (*Pandalus platyceros)* and squat lobsters *Munida quadrispina* dominate (VENUS Project, University of Victoria)

from those seen on the shallower carcasses, with much lower diversity but much greater numbers of individual specimens present. In some cases, the *M. quadrispina*, *C. magister* and *P. platyceros* almost obscured the carcass. A small sunflower sea star (*Pycnopodia helianthoides*) was observed on Day 6, but larger specimens such as those observed on the shallower carcasses, were never observed. A ruby octopus was also briefly attracted on Day 6 (*Octopus rubescens* Berry). Tissue continued to be removed from the hind end, but although *M. quadrispina* were seen picking at the face and ears, very little damage was observed on the head.

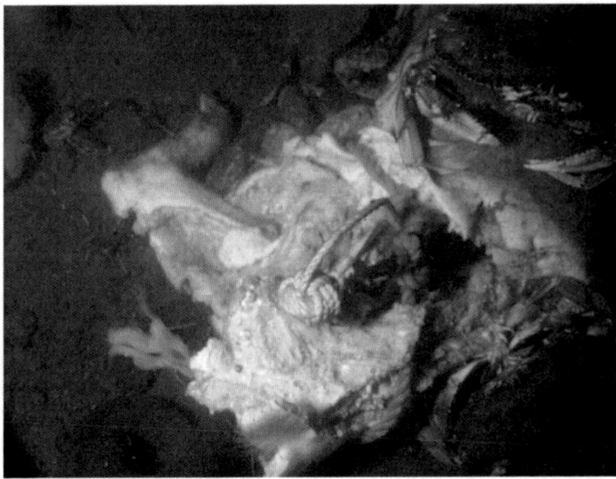

Fig. 12.7 Day 8, 2006. Much of the hind quarters have been removed (VENUS Project, University of Victoria)

By Day 8, the hind quarters of the carcass had almost been removed and the back legs were partially eaten (Fig. 12.7). As most of the hind quarters had been removed, this meant that the rope securing the carcass to the weights, keeping it in camera view, had mostly slipped off the carcass. As the rear set of weights were attached to the front set by rope, this meant that all the weights were compromised and no longer kept the carcass in one spot. Therefore, the carcass was moved by animal activity several times from this point on, probably by *C. magister* as these were seen to rock the carcass almost over at one point. By Day 10, the rear third of the carcass was mostly reduced to bones and cartilage with some soft tissue still adhering and the lowest rib was exposed. *Cancer magister* were seen reaching into the cavity and tearing out flesh. Any tissue dropped by *C. magister* was eaten rapidly by waiting *M. quadrispina*. In some cases, *C. magister* would eat the *M. quadrispina*.

Cancer magister, *M. quadrispina* and *P. platyceros* were present day and night. Despite the rapid removal of tissue from the rear end of the carcass, the rest of the carcass, including face, ears and front legs appeared to be completely intact (Fig. 12.8). By Day 11, *C. magister* had begun to feed at the abdominal region, further up the torso. By Day 13, the carcass had been pulled free of the weights and ropes. For the first time, small red amphipods (*Orchomenella obtusa* Sars) were seen in small numbers on the exposed tissue of the carcass. By Day 14, the rear half of the carcass was almost entirely skeletonized, with very little tissue holding on the rear left leg (Fig. 12.9). The carcass had been moved several times and was now partway past one of the tripod legs, making viewing the front half of the carcass difficult. *Cancer magister*, *M. quadrispina* and *P. platyceros* continued to dominate the fauna, although several fish including slender sole (*Lyopsetta exilis* (Jordan and Gilbert)), herring and dogfish, swam through the area, apparently showing little interest.

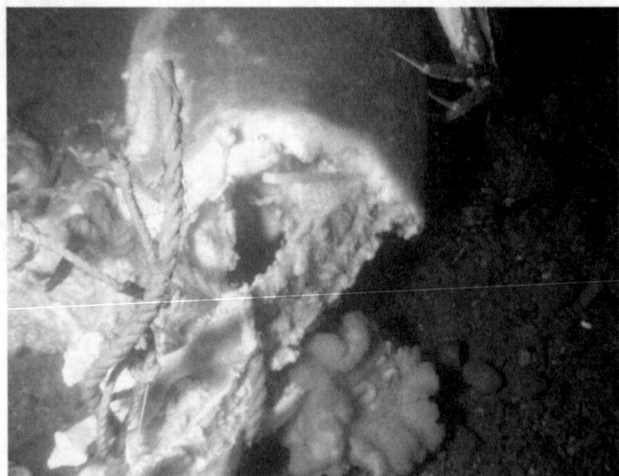

Fig. 12.8 Day 11 2006, the rear third of the carcass was mostly skeletonized but the upper torso, front legs and head, were still intact (VENUS Project, University of Victoria)

Fig. 12.9 Day 14, 2006. Rear half of carcass is skeletonized, and carcass is no longer restrained by weights so has been pulled away from camera. (VENUS Project, University of Victoria)

By early on Day 15, the carcass had been moved further away and the snout region showed some evidence of feeding, although the upper body remained mostly intact. A single bone was seen near the carcass, but the rest of the carcass remained articulated. Later that day, the carcass was moved a further 30 cm and was now lying outside the tripod area. The left rear leg had been disarticulated and was lying close to the carcass (Fig. 12.10). By Day 16, the carcass had been completely turned around and the only animals seen were large numbers of *M. quadrispina* which were feeding on both the main carcass and the disarticulated leg (Fig. 12.11). Some feeding damage could now be seen on the face, with the eye socket empty

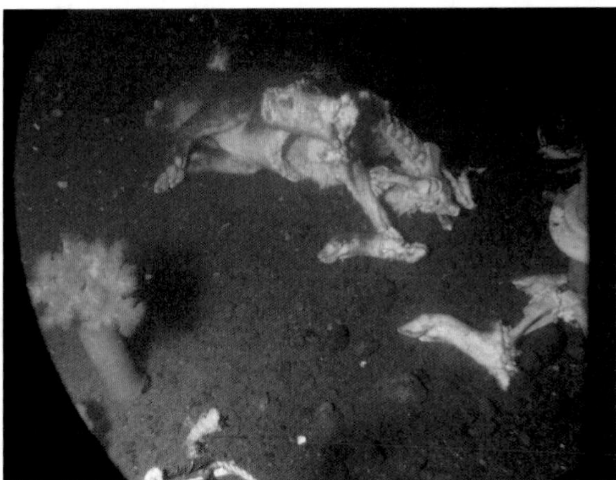

Fig. 12.10 Day 15, 2006. The carcass has been dragged completely out of the tripod area and one leg has been disarticulated (VENUS Project, University of Victoria)

Fig. 12.11 Day 16, 2006. Carcass has been turned 180°. Some feeding damage now seen on face. Disarticulated leg is still present. Only macro fauna are squat lobsters, *Munida quadrispina* (VENUS Project, University of Victoria)

and the upper part of the nasal bones exposed. In general, however, the upper body was intact, with skin and hair present, whereas the lower half of the remains was skeletonized. On Day 20, the carcass was again turned around (Fig. 12.12). Most of the rear part of the carcass was gone, with both legs and pelvis detached and only the lower part of the spinal column remaining. One femur had been picked clean and was still in camera range. By Day 22, the bulk of the carcass had been removed from camera range and by Day 23, only a single femur remained in range.

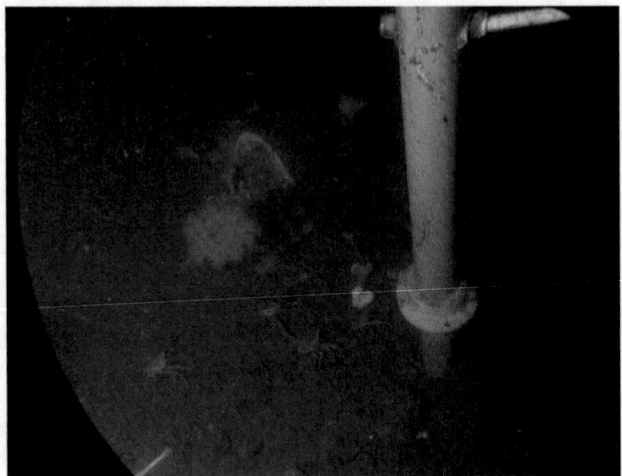

Fig. 12.12 Day 20, 2006. Carcass has again been turned 180°. Pelvis and remaining rear leg gone (VENUS Project, University of Victoria)

For the duration of the experiment, by far the most common animals on the carcass were *C. magister*, *P. platyceros* and *M. quadrispina*. At no point were any signs of classic decomposition observed and biomass removal appeared to be entirely due to animal activity. In November, a search was made for any remaining bones but none were recovered.

The levels of dissolved oxygen in the water were around 2.1 mL/L at the start of the study in early August and dropped slowly over the subsequent days to lows of 0.2 mL/L and means around 0.7 mL/L/ (Fig. 12.13) (Anderson 2008). Salinity remained stable throughout the experiment at 31 PSU (Practical Salinity Units), and temperature remained constant around 9.5–9.8°C.

12.3.2 Second Deployment, 2007

The second study was conducted the following year in an almost identical manner to the first; the only differences were that the study was begun in early September, almost 1 month later than the first study, the carcass was killed by pin gun so a small head wound was present, and the carcass was weighted differently, with three separate sets of weights placed on the carcass. These were not connected, so if one weight was lost, the others would remain. The carcass was deployed at the same site as before.

As soon as the carcass was placed on the seabed, large numbers of *M. quadrispina* were immediately attracted. These were already present on the sea floor, but arrived in much greater numbers so were clearly attracted by the presence of the carcass. Within a few hours, the carcass was literally covered with *M. quadrispina*. Shortly afterwards, *C. magister* and *P. platyceros* arrived at the carcass (Fig. 12.14).

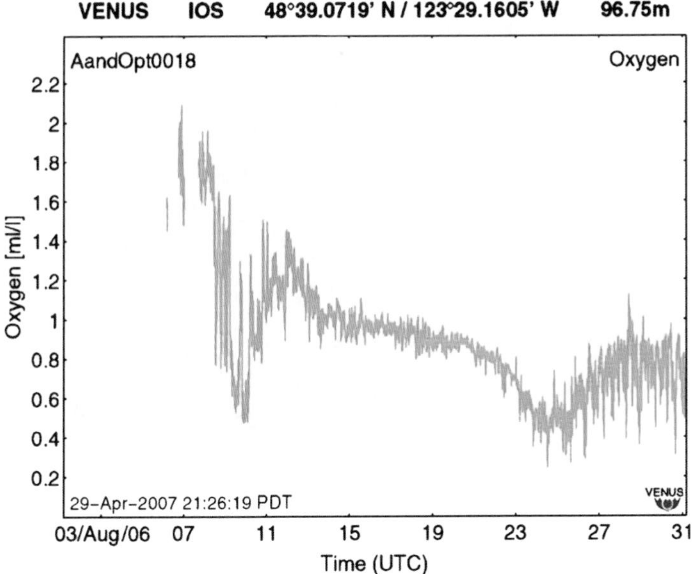

Fig. 12.13 Dissolved oxygen levels during the 2006 VENUS pig study (VENUS Project, University of Victoria)

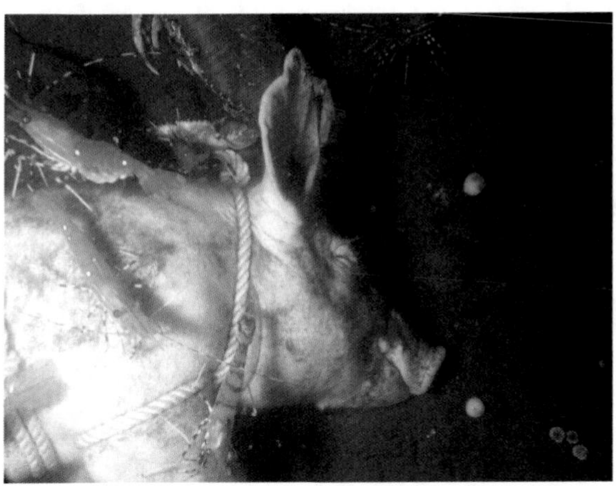

Fig. 12.14 Three spot shrimp (*Pandalus platyceros*) and Dungeness crabs (*Cancer magister*) attracted immediately to the 2007 carcass (VENUS Project, University of Victoria)

When first turned on, the camera lights seemed to repel the *C. magister* and *P. platyceros*, although only for a few seconds, as they rapidly returned, but they seemed to acclimatize to them after a few days and later observations did not show the lights particularly attracting or repelling the macro fauna although zooplankton, in particular arrow worms, (Chaetognath, *Sagitta elegans* Verrill) were greatly

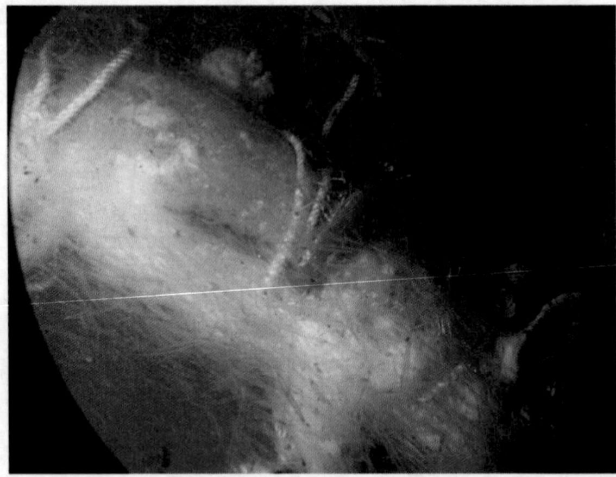

Fig. 12.15 2007. Zooplankton, in particular, arrow worms (*Sagitta elegans* Verrill), attracted by the camera lights (VENUS Project, University of Victoria)

attracted to the lights and after a few minutes of observation would almost obliterate the view (Fig. 12.15). The animals seemed to be attracted to all areas of the carcass and a small round grazed area was noted in the skin. By Day 1, 24 h after deployment, a rip had appeared in the skin in the central line of the abdomen, approximately 7 cm long. A crab was seen reaching deeply into the abdomen (Tunnicliffe 2006). The crabs, *C. magister*, frequently had barnacles on their carapace and the patterns of these barnacles allowed identification of some individual specimens. It was seen that the same specimens returned to the carcass repeatedly. On Day 1, the dominant fauna were *P. platyceros* and *C. magister*. Also seen, but in lesser numbers, were tanner crabs (*Chionectes tanneri* Rathbun).

By Day 2, tissue bulged from the abdominal rip, which had been extended, and a flap of skin had been pulled away. From this point on, the majority of animal activity was concentrated at this area (Fig. 12.16). The extruded tissue and skin were voraciously fed upon by *C. magister* and *P. platyceros* and rapidly removed (Fig. 12.17). Once the extruded tissue was removed, *C. magister* reached into the abdominal cavity and pulled out more tissue, organs and coils of gut. The snout area of the head had also been grazed, although this area was of much less interest than the abdominal area. *Munida quadrispina* feeding and picking at the head region did not appear to be able to break the skin and left only grazed areas. By Day 3, three small, round, ragged marks were seen in the side of the carcass above the abdominal area (Fig. 12.18). These are thought to be caused either by the rear legs of *C. magister* as they anchored themselves above the abdominal area, to feed inside the abdomen, or may have been caused by the chelicerae as they picked at the skin (Wallace 2008). These areas later became attractive themselves.

Over the following days, the abdominal rip was extended and large amounts of tissue were removed. *C. magister* and *P. platyceros* were the predominant scavengers,

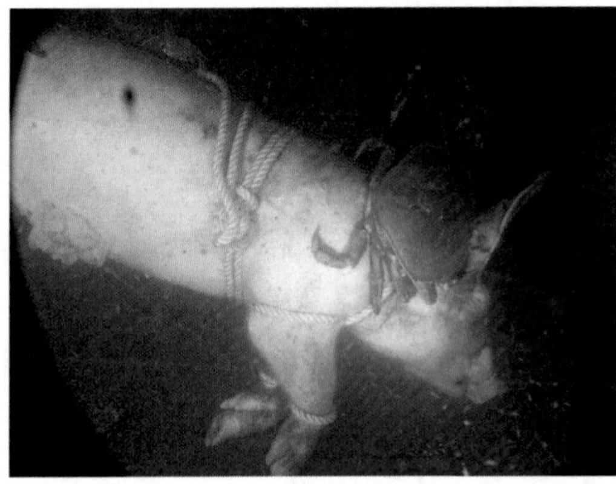

Fig. 12.16 Day 2, 2007. Tissue pulled out of abdominal rip. Note squat lobster pulling at tissue (VENUS Project, University of Victoria)

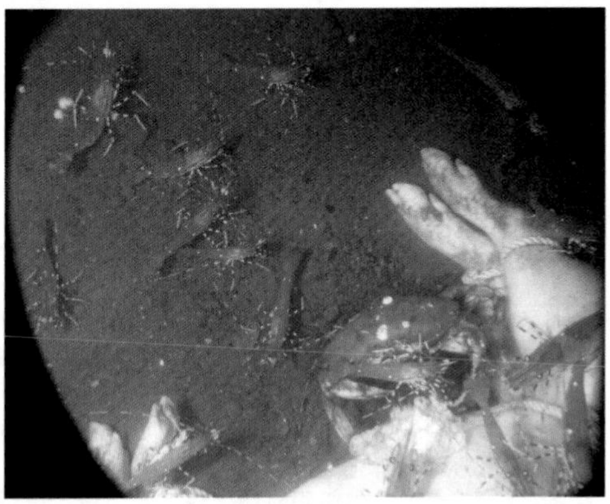

Fig. 12.17 Day 2, 2007. Crabs feeding voraciously at abdominal rip (VENUS Project, University of Victoria)

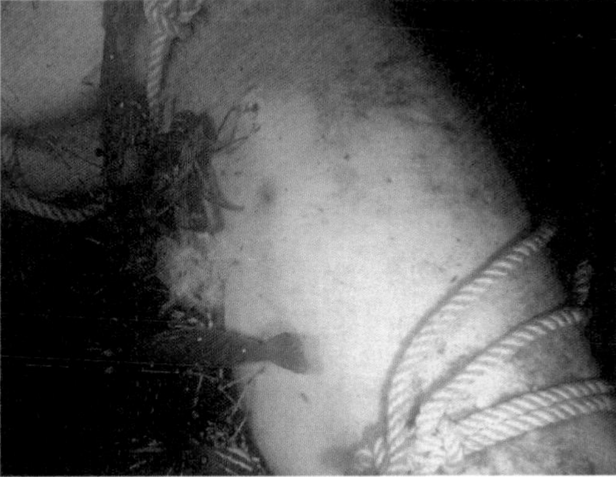

Fig. 12.18 Day 3, 2007. Tears in the skin above the abdominal area, presumably caused by crab rear claws as they maintain a purchase on the tissue while reaching into the abdomen or by crabs picking at skin (VENUS Project, University of Victoria)

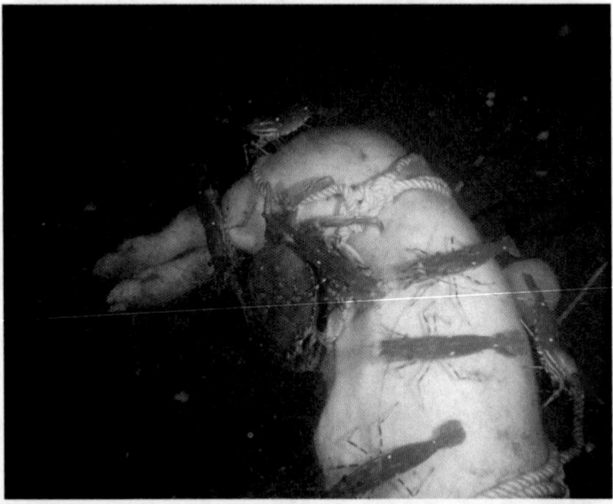

Fig. 12.19 Day 5, 2007. Crabs and shrimp have removed much of the internal tissue and organs and extended the abdominal rip to the sternum (VENUS Project, University of Victoria)

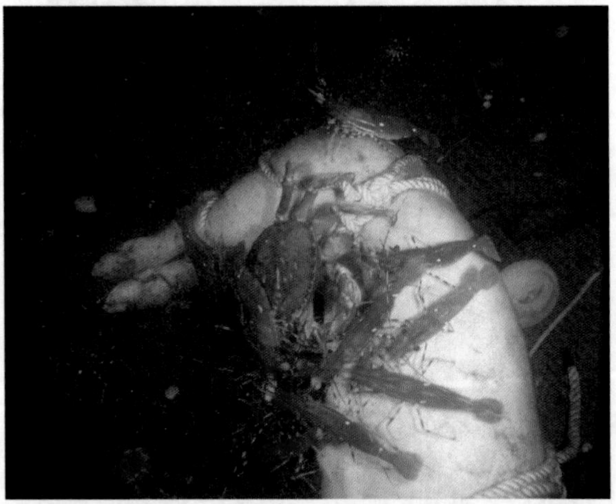

Fig. 12.20 Day 5, 2007. Dungeness crab (*Cancer magister)* and large numbers of three spot shrimp (*Pandalus platyceros*) feeding at abdomen (VENUS Project, University of Victoria)

with *M. quadrispina* also present and feeding. *C. magister* were seen to pursue, kill and eat the *M. quadrispina* on occasion. By Day 5, although little could be seen externally, clearly a large amount of tissue had been removed internally through the abdominal slit as the abdomen had an exaggerated concave appearance (Fig. 12.19). *Cancer magister* and *P. platyceros* crowded this area to feed, almost obscuring it (Fig. 12.20). Arrow worms continued to be present in large numbers. Crab chelicerae made cutting and snipping marks in the edge of the skin, and rear legs were used to anchor the crab firmly in the tissue as they rocked to pull organs out.

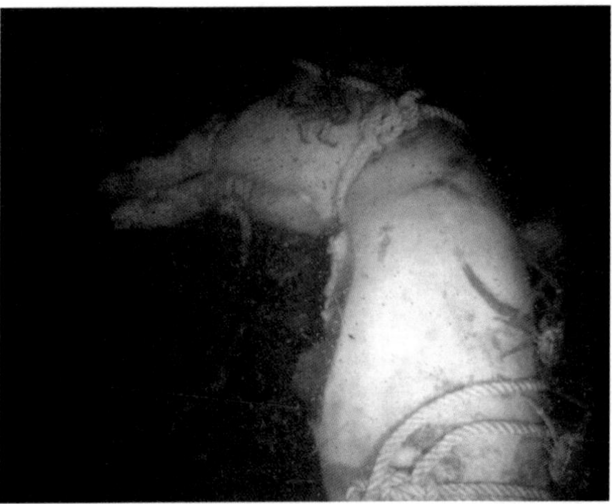

Fig. 12.21 Day 9, 2007. Abdominal damage extended down between back legs and organ tissue being removed (VENUS Project, University of Victoria)

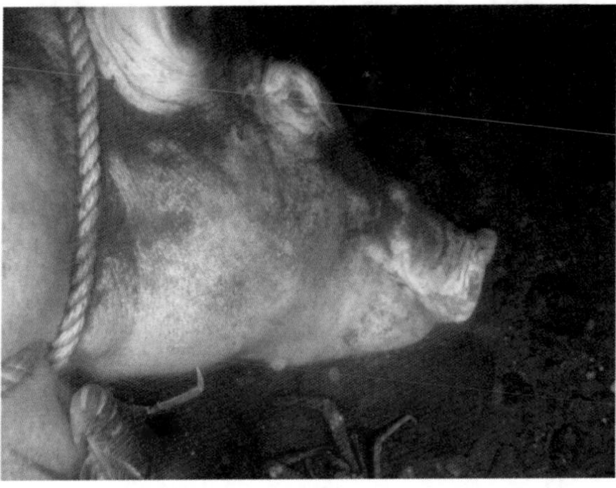

Fig. 12.22 Day 9, 2007. Head region still intact, with eyeball present, although eyelid has been removed (VENUS Project, University of Victoria)

By Day 9, the abdominal damage had been extended down between the back legs and organ tissue was still being removed (Fig. 12.21), although the head was still barely damaged and the eye still intact, although the eyelid itself had been removed (Fig. 12.22). *Pandalus platyceros* were seen picking at the ears. The abdominal opening had been extended to approximately 20–30 cm long and 10 cm wide. The predominant fauna continued to be *C. magister*, *M. quadrispina* and *P. platyceros*, with *Sagitta elegans* frequently attracted to the lights, appearing in large clouds. On Day 10, the dominant fauna were *M. quadrispina*, with these seen

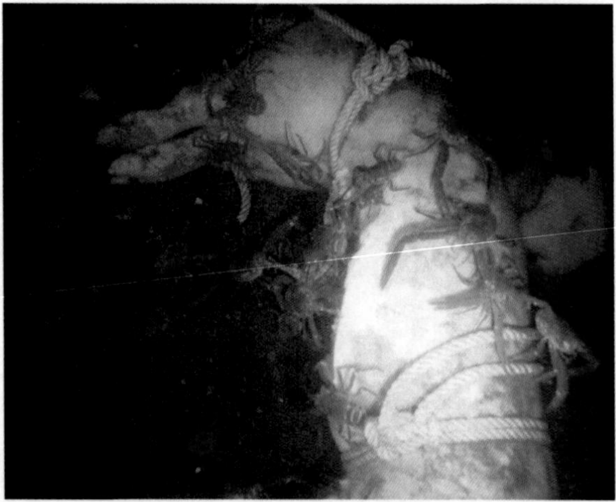

Fig. 12.23 Day 10, 2007. Dominant fauna are *Munida quadrispina* (VENUS Project, University of Victoria)

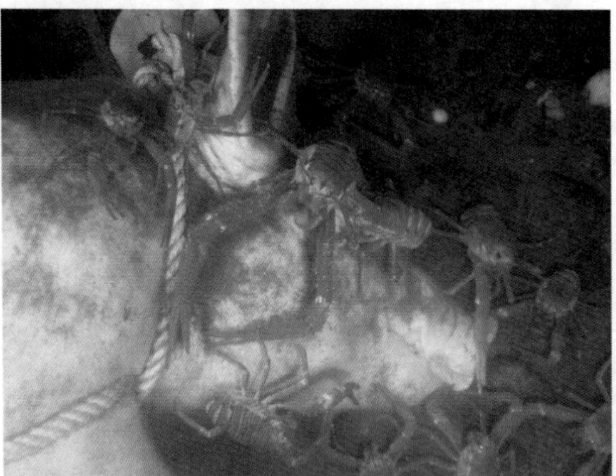

Fig. 12.24 *Munida quadrispina* on the head (VENUS Project, University of Victoria)

to be actively pulling tissue out of the abdominal cavity (Fig. 12.23). Large numbers of *M. quadrispina* were all over the carcass as well as the entire surrounding area. Some slender sole (*Lyopsetta exilis*) were seen to swim over the carcass but showed no interest in the carcass itself. Such fish were seen on and around the carcass throughout the study. Over the following days, as oxygen levels dropped, *M. quadrispina* were still seen in large numbers, but much fewer *C. magister* and *P. platyceros* were present, although these were still seen in lower numbers (Fig. 12.24). On Day 11, *C. magister* were seen removing pieces of tissue from between the back legs and two circular areas of feeding damage could be seen on the hock region of the rear legs (Fig. 12.25).

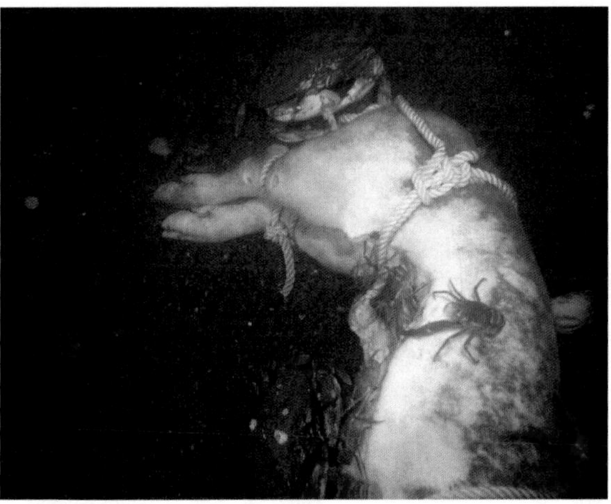

Fig. 12.25 Day 11, 2007. Dungeness crab *Cancer magister* removing tissue from between back legs and abdominal region. Grazing circles seen on hocks (VENUS Project, University of Victoria)

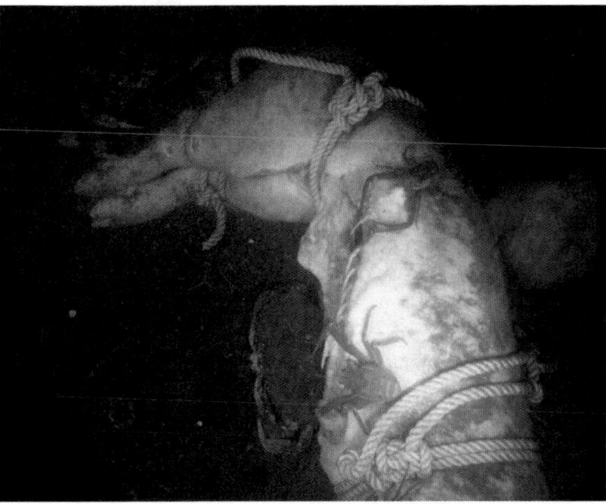

Fig. 12.26 Day 12, 2007. Rib ends now exposed. Laser lights indicate 10 cm distance between lights. Note remains of squat lobster (*Munida quadrispina*) eaten earlier by Dungeness crab (*Cancer magister*) (VENUS Project, University of Victoria)

By Day 12, the cartilaginous ends of the ribs were exposed (Fig. 12.26) and feeding continued all over the carcass including the head end. *Munida quadrispina* were seen going into the abdominal cavity and pulling out parts of organs. On Day 14, small gammarid shrimp (Family Lysianassidae), *Orchomenella obtusa* (Tunnicliffe 2006), were first seen on the exposed tissue of the back legs and ropes nearby (Fig. 12.27). These increased greatly in numbers and began to cover the carcass several mm deep by Day 17 and Day 18, concentrating on areas of exposed

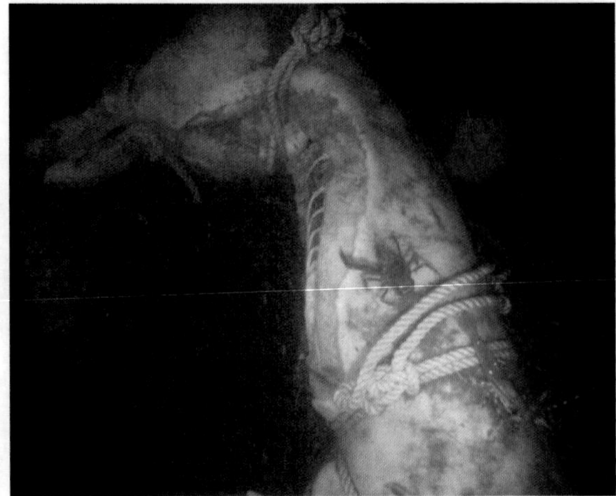

Fig. 12.27 Day 14, 2007. Small gammarid shrimp (*Orchomenella obtusa,* Family Lysianassidae) first seen on carcass (VENUS Project, University of Victoria)

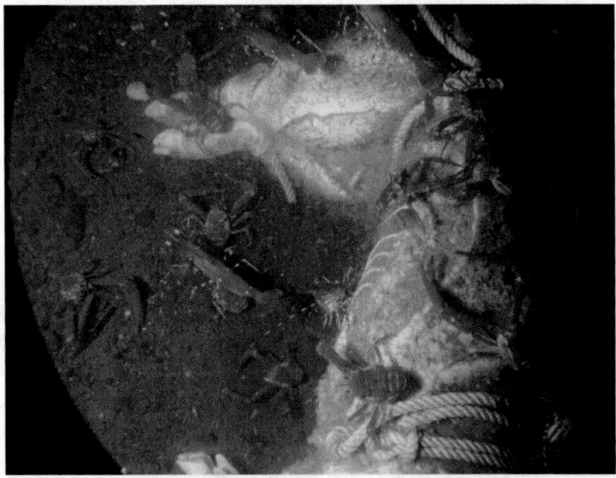

Fig. 12.28 Day 18, 2007. Large numbers of *Orchomenella obtusa* on remains (VENUS Project, University of Victoria)

muscle and tissue (Fig. 12.28). Such amphipods are often seen by public safety divers when recovering bodies and are colloquially referred to as "sea lice" (MacFarlane 2001; Teather 2000). Both *M. quadrispina* and *P. platyceros* were seen to feed on the small amphipods as well as on the tissue. *Cancer magister, M. quadrispina* and *P. platyceros* continued to feed and extended the opening at the abdomen to expose more ribs, although as oxygen levels were low at this time, and remained below 1.0mL/L, the dominant fauna remained *M. quadrispina.* The skin remaining on the torso appeared to be "rucked up" as if it were a shirt with little

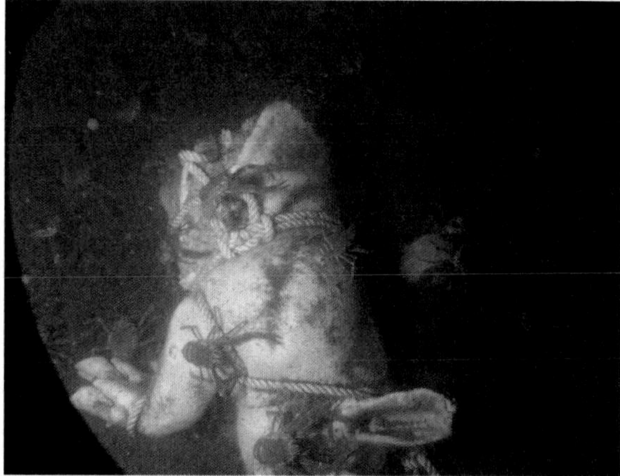

Fig. 12.29 Day 18, 2007. Lower part of pig has been completely removed. The skin appears loose below the rope line (VENUS Project, University of Victoria)

connection to the tissue underneath, suggesting that much of the tissue underneath had been removed. The skin itself remained intact, and hair could still be seen to be present.

Between the evening of Day 17 and the morning of Day 18, the lower half of the pig was removed completely, leaving just the upper part of the remains in sight of the camera (Fig. 12.29). As the camera tripod is static, it was not possible to determine whether the lower half had simply been removed a few metres out of camera view, had been taken a distance away or had been consumed. No clear mark was left as the remains were already extensively scavenged, but it seems likely that the culprit was as suspected before, the sixgill shark (*Hexanchus griseus*), which took not only part of the remains but also dragged the rope and chain from the lower body out of camera range. When the remains were recovered 5 months later, no sign of the lower body, or the lower body weights were recovered despite an extensive search of the surrounding area. As the remaining weights were still attached to the upper part of the remains and not interconnected to the other weights, the rest of the carcass remained in sight of the camera for the duration of the experiment, although the carcass had been dragged about 30 cm. The skin on the remaining part of the carcass appeared to be moved further up the remains, as if it were a loose shirt.

By Day 22, the skin was wrinkled over the entire area of thorax, neck and head as if no muscle or tissue lay beneath (Fig. 12.30) suggesting that all the tissue beneath had been removed by animal action. Large numbers of *O. obtusa* were present on the areas of skin that had been pulled back and large numbers of *M. quadrispina* were present, feeding on the remains as well as the small shrimp. *Cancer magister* and *P. platyceros* were present at times, although in lesser numbers. By the next day, the lack of tissue beneath the skin was confirmed as the majority of the skin had been pulled over the head like a sweater, revealing a clean,

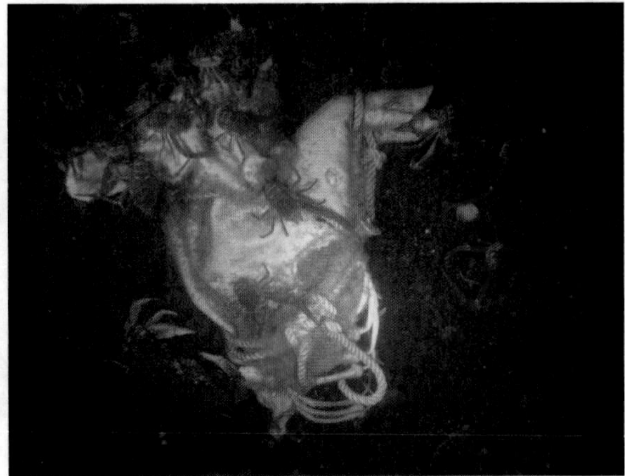

Fig. 12.30 Day 22, 2007. Tissue beneath skin has been removed and skin now appears to be lying directly on bones, with most of tissue removed (VENUS Project, University of Victoria)

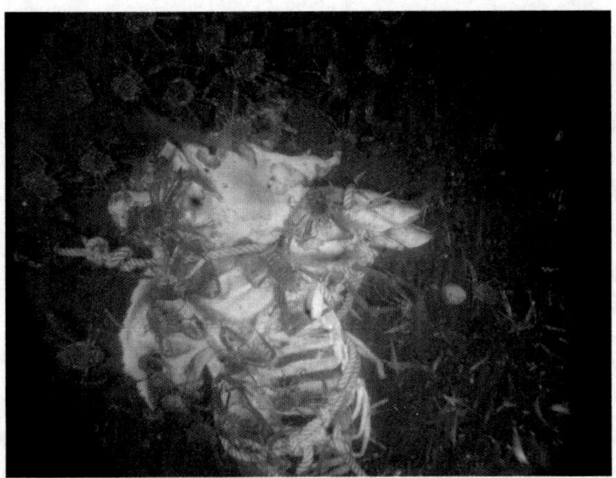

Fig. 12.31 Day 23, 2007. The skin has been pulled up above the head to show that the remains beneath are clean bones with almost no tissue attached. (VENUS Project, University of Victoria)

fully articulated skeleton with cartilage intact (Fig. 12.31). In the subsequent few days, the entire skin was pulled over the head of the carcass and torn apart by *M. quadrispina,* with pieces of skin being pulled in different directions by large numbers of these animals (Fig. 12.32). As the skin was pulled further, the skull was also shown to be completely skeletonized (Fig. 12.33). At this time, the only animals on the carcass were *M. quadrispina*. In the following days, the rest of the skin was consumed, with the last piece of recognizable tissue being the ears. Once only bones were left, *M. quadrispina* removed the cartilage from the end of the ribs and scapula, gradually disarticulating the bones. Although some *P. platyceros* were still

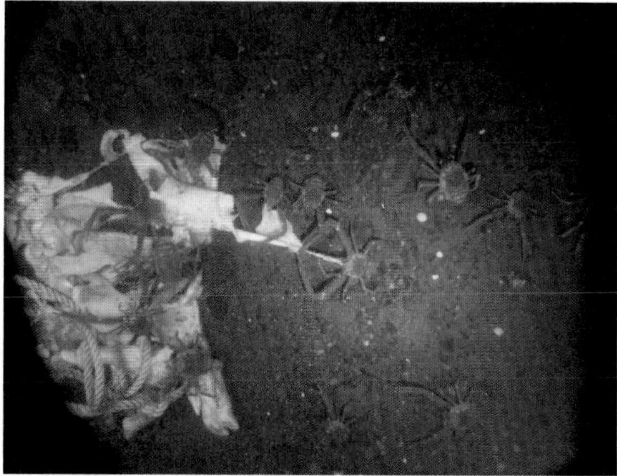

Fig. 12.32 Day 24, 2007. *Munida quadrispina* removing remaining tissue from bones (VENUS Project, University of Victoria)

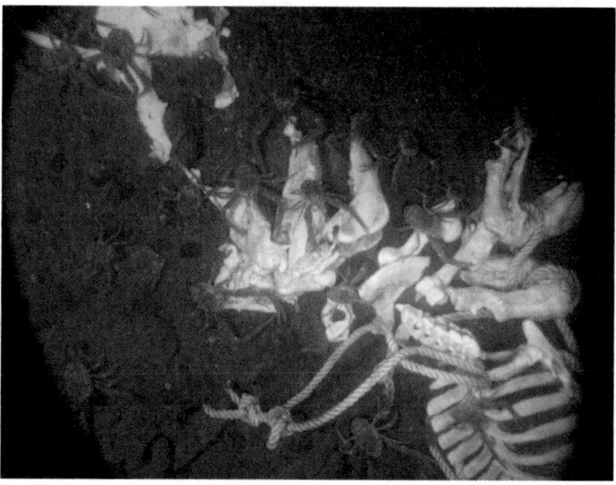

Fig. 12.33 Skin pulled from skull revealing skeletonized skull and mandible (VENUS Project, University of Victoria)

present in low numbers on the carcass, no *C. magister* remained, and once the skin was gone, no *O. obtusa* were noticed. A bacterial mat began to form on the bones after skeletonization. Most of the cartilage had been removed by Day 42. The bones were recovered 5 months after submergence and are undergoing further study.

The levels of dissolved oxygen in the water were much lower in this second study, with levels of 1.4 mL/L at the start of the study, which dropped below 1mL/L by Day 4, and continued to drop to 0.1–0.4 mL/L for the latter part of the study (Fig. 12.34). Temperature rose from 9.3°C to 9.42°C in the first few days then remained at a constant 9.4°C.

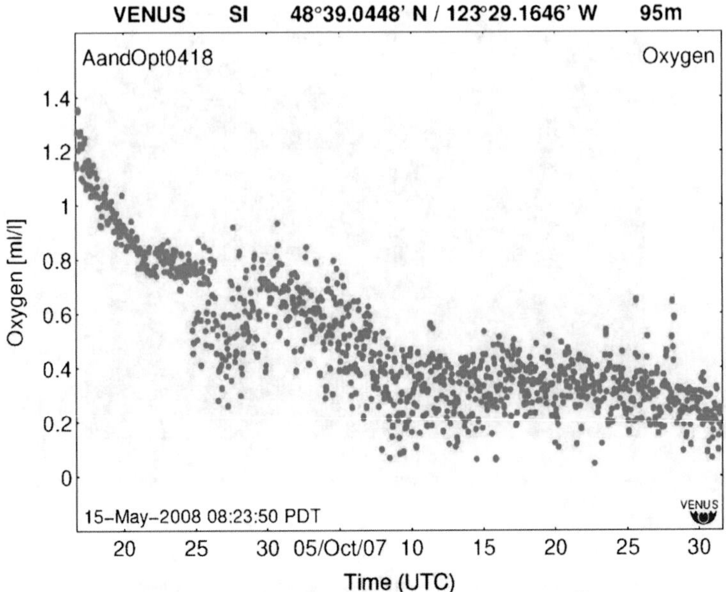

Fig. 12.34 Dissolved oxygen levels during the 2007 VENUS pig study (VENUS Project, University of Victoria)

12.3.3 Third Deployment, 2008

The third carcass was deployed in late September, 6 weeks later in the year than carcass one and 2 weeks later in the year than carcass two. The animal was again euthanized by pin-gun and was weighted as carcass two. The carcass was deployed at a slightly different site, about 65 m from the first site. The substrate was again a thick layer of silt with some cobble over rock but this site was close to a large rock and was 4 m deeper than the earlier site, at a depth of 99 m.

The carcass sank immediately, as in the previous deployments. However, in contrast to past deployments, this carcass was only very slightly attractive to fauna, with only a few *Munida quadrispina* attracted. No *Pandalus platyceros* or *Cancer magister* were attracted during the first few months. This is probably a result of the much lower oxygen levels at the time of deployment (Fig. 12.35). Oxygen levels at time of deployment were ~0.3 mL/L and remained around 0.1–0.5 mL/L until a sudden and rapid increase in late December with oxygen reaching levels of 1–2.5 mL/L. Temperatures remained similar to the previous studies at around 9.2°C until late December when they dropped by 1°C and remained at ~8.2°C until the end of the experiment. The effect of this change in environment on the carcass colonization and taphonomy, particularly the change in oxygen levels, was very clearly seen. The carcass remained completely intact for months. At first, a few squat lobsters were observed picking at the skin, creating small grazed areas, but they were unable

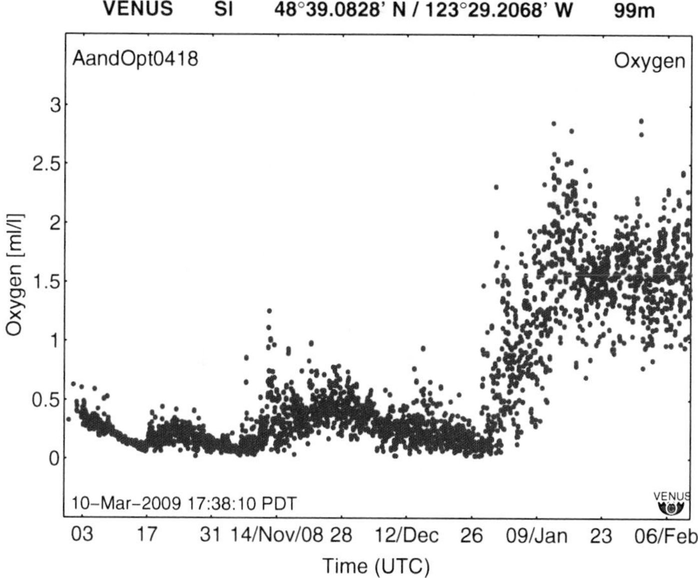

Fig. 12.35 Dissolved oxygen levels during the 2008 VENUS pig study (VENUS Project, University of Victoria)

Fig. 12.36 Day 27, 2008. Small grazed areas seen on thighs and around abdominal line caused by *Munida quadrispina* (VENUS Project, University of Victoria)

to break the skin (Fig. 12.36). After a few weeks, even squat lobsters were no longer observed at the carcass. The carcass became covered in silt and a bacterial mat developed over the entire body (Fig. 12.37). This mat is thought to be a result of the activity of filamentous sulphur bacteria, which occur in low oxygen

Fig. 12.37 Day 79, 2008. Bacterial mat covering entire carcass. Carcass still almost entirely intact (VENUS Project, University of Victoria)

environments where decomposition produces hydrogen sulphide (Juniper 2009; Herlinveaux 1962). This mat was removed in places on several occasions, probably caused by the movement of *Lyopsetta exilis* (slender sole) across the body, removing the silt layer and the biofilm. *Lyopsetta exilis* were seen frequently in the area and were seen to swim over the carcass. Their action may have simply dislodged the accumulated biofilm and silt without actually breaking the skin. During this acute hypoxic period these were virtually the only fish observed.

In late December, almost 3 months post submergence, the oxygen levels rapidly rose to above 2 mL/L and immediately, very large numbers of fish were seen close to the carcass (Fig. 12.38) as well as the first *C. magister*. Numbers of arthropods increased over subsequent days with several *C. magister* and very large numbers of *P. platyceros* feeding on the carcass (Fig. 12.39). Interestingly, much fewer *M. quadrispina* were seen at this time. The water in the area around the carcass became extremely hazy with the sudden and rapid rise of animal activity which is thought to be correlated with the rapid rise in oxygen levels and related to the animal activity, particularly that of the fish (Yahel et al. 2008).

Tissue loss was slow but steady from this point until the carcass was mostly skeletonized by mid February, 135 days after submersion.

12.3.4 Discussion

In the first two VENUS studies, the carcass biomass was removed through animal activity rather than classic decomposition. In contrast, in the shallower experiments, biomass loss was partially due to animal activity, but was primarily due to natural decomposition. In the shallower experiments, the carcasses floated when

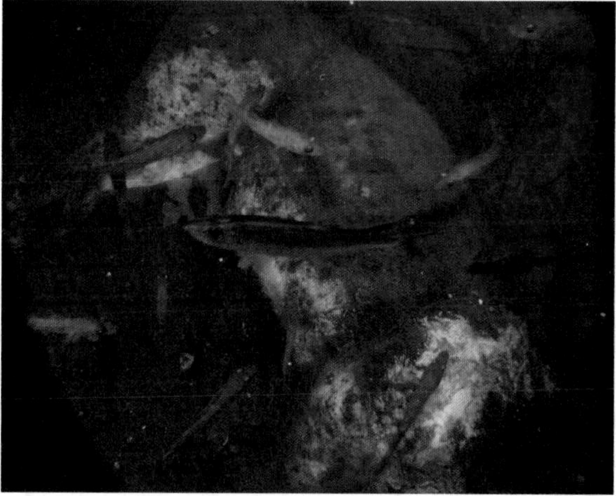

Fig. 12.38 Day 92, 2008. Rapid oxygen renewal in Saanich Inlet resulted in large numbers of fish repopulating the area (VENUS Project, University of Victoria)

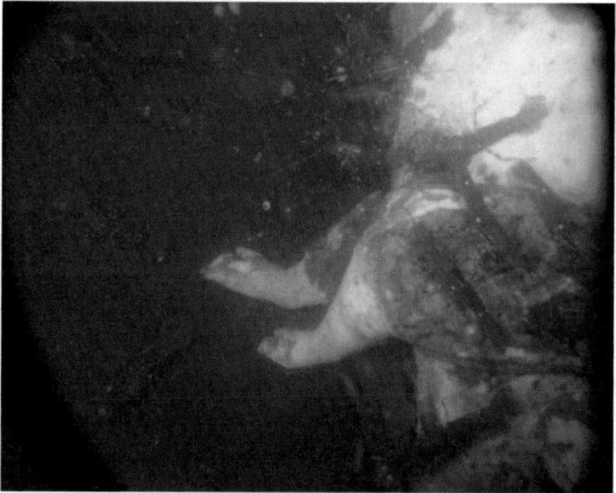

Fig. 12.39 Day 106, 2008. Large numbers of *Pandalus platyceros* on carcass. Note large grazed areas on shoulder and legs (VENUS Project, University of Victoria)

first placed, then sank and later refloated (Anderson and Hobischak 2004, 2002). Whether a body will refloat depends on many factors including but not limited to; timing and type of last meal, water temperature, depth, body mass index and health of the individual and is, therefore, very variable (Teather 1994). Bodies that come to rest below 61 m (200 ft) are not expected to refloat, as cold temperatures and high pressure reduce the volume of gases and also make the gases very soluble in water and tissues (Teather 1994). Therefore, flotation and reflotation were not expected in the deeper studies. In the first two VENUS studies, bacterial decomposition was not seen and no visible signs of adipocere formation or tissue degradation were observed other than that created by animals, although a bacterial mat did start to form on the bones of the second carcass just before recovery. In the third VENUS study, a bacterial mat was seen to form over the entire carcass. In the shallower experiments, tidal action and currents impacted the remains and would have resulted in their removal had they not been tethered. At 94 and 99 m, there was little impact of tides or currents so movement of the carcasses was entirely due to animal action.

Dissolved oxygen levels in the water can have a major impact on the fauna attracted to a carcass. Levels of dissolved oxygen were not directly measured in the shallow experiments in Howe Sound, however, these shallow waters are, in general much more highly oxygenated that that of the Saanich Inlet, where the deeper carcasses were deployed. It is probable that the dissolved oxygen levels at the shallower research sites were at or above 5 mL/L (Tunnicliffe 2006). The Saanich Inlet has restricted water circulation due to a submerged sill at its mouth. The anoxic basin is only flushed when water above the sill is dense enough to run into it, so for much of the year, the basin is anoxic (Jamieson and Pikitch 1988; Tunnicliffe 1981; Anderson and Devol 1973).

The levels of dissolved oxygen in all the deeper coastal studies were low. Levels below 2.0 mL/L are considered hypoxic, with levels below 1.0 mL/L being stressful for most animals (Diaz and Rosenburg 1995). In 2006, the dissolved oxygen levels were hypoxic after the first few days but did not drop below 1 mL/L until part way through the study. In 2007, oxygen remained below 1 mL/L for most of the study. Nevertheless, this did not discourage the fauna, although when oxygen was very low, *M. quadrispina* were the main and sometimes the only invertebrates present (Peters 2007). As the Saanich Inlet is regularly hypoxic at this depth, it is probable that many of the local species are adapted to these conditions (Tunnicliffe 1981). In the Saanich Inlet, Tunnicliffe has reported that even when dissolved oxygen is below 1.0 mL/L the zone between 85 and 100 m did not exhibit low diversity or small populations (Tunnicliffe 1981). Also, although crustaceans are in general not tolerant of hypoxic conditions, *M. quadrispina* are an exception, being present in conditions below 0.2 mL/L (Peters 2007; Burd and Brinkhurst 1985; Farrow et al. 1983). The lack of species diversity may reflect the hypoxic conditions. It does not reflect depth as great faunal diversity can be seen at much greater depths on whale falls (Kemp et al. 2006; Baco and Smith 2003; Smith and Baco 2003; Smith et al. 1989). It was clearly seen that *M. quadrispina* was less affected by the hypoxic conditions, but even *C. magister* was commonly seen on the remains despite very low oxygen levels. Many decapod crustacea, including *Cancer magister*, are capable of surviving for periods of time in acute hypoxic conditions by employing various physiological mechanisms to cope with low oxygen (Bernatis et al. 2007). These include increasing the ventilation of the branchial chambers, exhibiting brachycardia, and altering blood flow, directing blood to tissues requiring more oxygen (Bernatis et al. 2007). They have also been shown to exhibit differing behavioural responses to hypoxic conditions such as entering hypoxic conditions to obtain food but moving the food to a more oxygen rich environment to feed

(Bernatis et al. 2007). However, in these studies, the same crabs, identified by barnacle pattern on their carapace, stayed to feed for days on the carcass, ripping and eating tissue in situ. Bernatis and colleagues' study show that well fed crabs moved to higher oxygen levels and remained at that site for 48 h (Bernatis et al. 2007), however, crabs in these studies were clearly very well fed but remained at the carcass until little remained or oxygen dropped too low.

In the first two VENUS studies, the carcass was deployed when oxygen was at an acceptable level for the three dominant arthropod species. However, the third carcass was deployed when oxygen levels were very low. In such conditions, *Cancer magister* and *Pandalus platyceros* were unable to access the carcass and although low numbers of *Munida quadrispina* were able to reach the carcass, they were unable to break the skin. This left the carcass intact for months until the oxygen levels increased. Clearly, the larger arthropods are required to break the pig skin. As pig skin is so similar to that of humans, it is likely that the same would occur with a human body. In this particular habitat, oxygen was a major driving force for decomposition and animal feeding. In the low oxygen environment, decomposition produced hydrogen sulphide (Herlinveaux 1962).

Temperature can obviously impact both decomposition and faunal colonization (Simpson and Knight 1985; Spitz 1980; Fisher and Petty 1977; Jaffe 1976; Picton 1971; Mant 1960). Temperature was not continuously measured in the earlier experiments as it was during the VENUS studies, but was measured during dives. In the shallow experiments, temperatures ranged from 6–10°C, whereas they held very constant at 9.4–9.8°C during the VENUS studies, except for a 1°C drop in the latter part of the third study. As these temperatures are considered reasonably warm and were very similar, it is unlikely that temperature was a major variable.

The most obvious faunal differences between the experiments at 7.6 and 15.2 m and those done at 94 and 99 m are that, at the greater depths, there was much less species diversity but many more actual animals feeding on the remains. At 94 m, the tissue was lost almost entirely to animal feeding rather than decomposition, with carcasses reduced to bones in a matter of weeks. This is much more in line with the personal experience of public safety divers (MacFarlane 2001; Teather 2000). In the deeper studies, the carcass was dominated by just three species of large crustacean: *C. magister*, *M. quadrispina* and *P. platyceros*, with *M. quadrispina* the most common, being present in 85.7% of the photographs taken in the 2006 study (Peters 2007). Even during the very low oxygen study, the majority of the fauna were still *M. quadrispina*, *C. magister* and *P. platyceros*.

In the Saanich Inlet, *Munida quadrispina* are endemic but occurred in much greater numbers in the presence of a carcass (Peters 2007). *Cancer magister* and *P. platyceros* were not commonly seen on the sea floor prior to the carcass deposition so were directly attracted, although they are endemic to the area. *Pandalus platyceros* is the largest species of shrimp in British Columbia and is usually associated with rocky terrain, although the VENUS studies were all conducted on a substrate consisting of a rock base, thickly covered in a light silt.

Pandalus platyceros is commonly found in the Saanich Inlet, with greatest densities at a depth of 70–85 m (Jamieson and Pikitch 1988). These animals were all actively attracted to the carcass immediately after deposition. It is believed that most animals detect a carcass using chemoreception, although more recent work on *Pandalus* sp. has shown that hydroacoustic stimuli may also be important, either detecting the carcass fall itself, or the feeding activity of other crustacea (Klages et al. 2002). In this case, it may also be the sound of ROPOS as it deposited the carcass. A much greater diversity of fauna were seen on the shallower carcasses, and these fauna were not limited to crustacea but rather included many invertebrate phyla and families. However, the actual number of organisms was low. Invertebrates were clearly attracted to the remains, and invertebrates could be seen moving actively towards the carcasses in both sets of experiments. Although it was not possible to tell whether there was a turnover in the specimens of *M. quadrispina* and *P. platyceros*, some of the specimens of *C. magister* were quite distinctive due to the barnacles adhering to their carapaces. Therefore, it was possible to tell that much of the time, the same crabs were returning to the carcass, or perhaps not leaving the vicinity at all.

A fourth crustacea, the sea louse, *Orchomenella obtusa* was seen on the deeper carcasses in small numbers in 2006 and very large numbers in 2007. Species in this genus have been known to eat large pieces of seal meat bait in less than 24 h and were reported to be collected by the "bucketful" in experiments in the McMurdo Sound area of the Ross Sea in Antarctica (Dearborn 1967 cited in Sorg et al. 1997, p 572). Fish caught in the traps were frequently eaten alive by the time the traps were checked. The temperature of the water in the McMurdo Sound averaged −1.8°C so these amphipods can clearly survive such cold temperatures (Sorg et al. 1997). They have been reported as being common in the Saanich Inlet at depths of 80–210 m and were found in large swarms on dead and dying prawns in a mass anoxic fatality (Jamieson and Pikitch 1988). *Orchomenella obtusa* is in the Family Lysianassidae which are commonly referred to as "sea lice". Teather reported that crustacea, in particular sea lice, have been known to remove almost all the tissue from a body in less than 12 h in some situations, and to commonly partially skeletonize a body in less than a week (Teather 1994). The actual conditions in which this rapid skeletonization occurred were not mentioned, but the majority of Teather's case histories come from British Columbia waters. In the present studies, large numbers of small amphipods were only seen in the 2007 VENUS pig, with smaller numbers seen briefly on the 2006 carcass. These were also not in the numbers described by Teather, who reported that the body could be almost completely obscured by the small shrimp, and that they could become a hazard to the public safety diver when they recover the body, by causing panic (Teather 1994). As well, Teather reported their arrival on humans within 24 h of submergence (Teather 1994) and lysianassid amphipods were found on deployed cetacean carcasses within an hour of deployment (Jones et al. 1998). In the present studies, small unidentified amphipods were seen on the shallow carcasses within 24 h, although not in great numbers and *O. obtusa* was not seen until Day 11 and Day 14 respectively on the deeper carcasses.

In the shallow carcasses, the remains were not skeletonized for many weeks whereas the first two deeper carcasses were skeletonized in less than a month. This is more in line with other studies that suggest that skeletonization usually occurs within a month (Sorg et al. 1997; Teather 1994), although soft tissue can persist for a year or more in certain situations (Sorg et al. 1997). Sorg and colleagues report that cartilage loss was observed in one human case at 10 months but was found in some cases up to 18 months after submergence (Sorg et al. 1997). In these studies, cartilage was removed by Day 42 when oxygen levels were acceptable. A case report from 1987 in waters very close to the VENUS experiments showed that a human body could be at least partially skeletonized with disarticulation of the mandible within 3 weeks of death in these waters (Skinner et al. 1988). This report referred to the historic cases of *Rex v. Sowash* and *Rex v. Charles King* in which two men in 1924 (during American prohibition) were murdered near Vancouver Island, for the liquor they were carrying. Their bodies were weighted down and dropped in the ocean at a depth of approximately 30 m (Skinner et al. 1988). The bodies were not recovered despite extensive searches. As no case at that time had been won in Canada without the *corpus delecti*, prosecuting counsel decided to conduct an experiment to prove that the remains would have been lost within weeks of death, so could not have been expected to be recovered. A quarter of beef was weighted and sunk in the same waters and retrieved after 1 month. The beef had been reduced to bones in that time by crabs and amphipods. This was used in the case to show that the failure of the police to find the bodies was "comparatively immaterial" (p141) and the defendants were convicted (Skinner et al. 1988).

The vertebrate that removed a large piece of the carcass on Day 2 in the first study and half the carcass on Day 18 in the second study is unknown, but suspected to be *Hexanchus griseus*. This is a deep water shark normally feeding primarily on cartilaginous and bony fish (Ebert 1986) although it has also been known to eat marine mammals and invertebrates (Wheeler 1975; Hart 1973, cited by Ebert 1986). This shark rests in deep waters during the day and swims to shallower waters at night to feed (Dumser and Türkay 2008). In a human case, *H. griseus* was seen near the remains, but did not bite it (Dumser and Türkay 2008). In this case, the shark may have been attracted by the large numbers of invertebrates or may have merely been 'tasting'. In both cases, it did not come back to finish the carcass but rather seemed to take a single swipe at it, then leave it alone. Sharks and rays are considered to be first-order scavengers of human bodies in coastal waters (Rathbun and Rathbun 1997; Sorg et al. 1997).

Prior to the large piece of tissue removal from the first VENUS carcass, *C. magister, M. quadrispina* and *P. platyceros* were attracted to the carcass and were found picking at the orifices, skin and silt on the remains, with no one area being more attractive than another. Once the large piece of tissue was removed, the focus of all the scavengers was at the bite site and this continued throughout the duration of the study, with very little attention paid to the head or front half of the animal. In the second VENUS carcass study, despite a head wound, the abdominal area was the most attractive, being opened up by *Cancer magister* very rapidly. This became the focus of subsequent feeding, with again, little activity at the head.

Although few faunal studies have been done on human bodies, or on human models such as pigs in the marine environment, whale carcass communities, or whale falls have been studied sporadically for over 150 years (Smith and Baco 2003). In the past, the accidental discovery of a whale carcass provided a rich opportunity to study the ecology of a very large ephemeral pulse of organic material (Smith and Baco 2003). More recently, carcasses of a variety of cetaceans have been deliberately deployed for study (e.g. Kemp et al. 2006; Witte 1999; Jones et al. 1998).

Such extremely large carcasses (up to 160 t adult body weight (Smith and Baco 2003)) are very different from that of a human and most whale studies have been conducted at depths much greater than the experiments presented here e.g. 4,000–4,800 m (Jones et al. 1998). However, whale falls do provide information that is valuable in understanding the ecology and community structure that builds in relation to a sudden input of nutrients in the ocean. Smith and Baco state that bathyal (pelagic zone 1,000–4,000 m below surface) carcasses provide a massive amount of nutrients to the deep sea floor, with the organic carbon in a 40 t whale being equal to that which normally sinks from the "euphotic zone to a hectare of abyssal sea floor over 100–200 years" (Smith and Baco 2003, p312). Smith and Baco have divided the decomposition and succession process on whale falls into four stages (2003). The first is the "*mobile scavenger stage*" (p 318) during which great numbers of large scavengers remove the soft tissue. These scavengers include sharks, hagfish and invertebrates. This stage can last for several months up to 18 months and in itself contains a temporal succession (Smith and Baco 2003). The increased macrofauna in the vicinity of experimentally placed whale falls had an inverse relationship with the abundance of nematodes in the sediment for at least 30 m from the carcass, although after 18 months, nematode abundance increased, probably due to increased nutrient levels in the sediment (Debenham et al. 2004). The second stage is that of the "*enrichment opportunist stage*" (Smith and Baco 2003, p 319) which can last for months and up to years, where large aggregations of heterotrophic macrofauna colonize the now enriched sediments and the bones. The third stage is the "*sulphophilic stage*" (p 322) which lasts decades. Here, a chemoautotrophic assemblage colonize, as the bones anaerobically produce sulphide during sulphate reduction (Smith and Baco 2003). Those organisms that are able to tolerate sulphur proliferate, including chemoautotrophic and heterotrophic bacteria, some isopods and Galatheid crabs, mytilids, dorvilleid polychaetes, cocculinid limpets, provannid gastropods and columbellid snails, as well as others (Smith and Baco 2003). This is very similar to the fauna found at hot vents (Smith et al. 1989). The final stage is the "*reef stage*" (p 325) which occurs after the organic material has been removed and only mineral material remains. At this time, the remains would be colonized by suspension feeders (Smith and Baco 2003). In the present experiments, due to the much smaller carcasses and the short time of observation, only the mobile scavenger stage was observed, although other stages may well have occurred as sediment below and around the carcass was not sampled and examined. The low oxygen environment of the thuird VENUS carcass study may have been similar to the sulphophilic stage. In further experiments, it would be interesting to examine the bones and sediment for a longer time period.

It has been suggested that sessile marine life such as barnacles and bryozoans could be excellent indicators of elapsed time since submergence (Sorg et al. 1997). Such animals do tend to grow at predictable rates and could be useful in estimating the time that has elapsed since the remains were skeletonized. They have been used in some human cases (Dennison et al. 2004; Sorg et al. 1997; Skinner et al. 1988), although caution must be exercised as growth rates vary with quantity and quality of food as well as water temperature and other physical parameters (Sorg et al. 1997; Bertness et al. 1991). In the first VENUS study, the carcass was removed from camera range within 3 weeks of submergence and no bones were recovered but in the second, the bones were recovered 5 months after submergence and no sessile organisms were found to be attached to them. This was, perhaps, too short a time to assess encrustation by sessile organisms. However, in the shallower studies, the carcasses were observed for some months post submergence, and very few such organisms were recovered from the bones. On Day 47 in the shallow spring experiments, a barnacle appeared to have settled in the left eye but was gone, possibly scavenged, by the next sampling time. Barnacles did settle on a tag used to restrain a carcass at Day 225 in the fall experiment, and mussels were seen in the area but not on the remains, with mussel beds developing nearby (Anderson and Hobischak 2002). Some mollusks were found attached to the bones at final collection (Table 12.4) but these were only observed many months after death. Therefore, the value of sessile organisms may be limited to the later postmortem interval.

When a body is recovered from the water, the remains often exhibit abrasions and other damage. It is important to understand whether this trauma was caused pre or peri-mortem versus post-mortem, as this may have a bearing on the manner of death (Stubblefield 1999). If such trauma can be explained by animal activity, then valuable investigative time is not wasted, and the presence of the post-mortem trauma may help to explain where the remains have been in the water, i.e. depth or habitat. In these studies, marks on the soft tissue by a number of animals were observed and could be mistaken for ante-mortem wounds if not studied carefully. Circular grazed areas were seen here and similar shallow oval or circular holes approximately 10–20 or more mm in diameter were seen in the skin of drowning victims, believed to be caused by crabs in the Japan Sea (Koseki and Yamanouchi 1963) Such trauma has also been seen on hard tissue causing crater-like defects (Sorg et al. 1997; Mottonen and Nuutila 1977).

Bodies are frequently recovered a considerable distance from where they went into the water and taphonomic information may be valuable in determining where the body went into the water and what conditions it has been exposed to in the time of submergence. Sorg et al. relate a case in which human remains were found 4 weeks after death in the clam flats where the Damariscotta River enters the Gulf of Maine (Sorg et al. 1997). The question was whether the remains had entered the water up river or from the ocean. Marine shell fragments and a single spine from the green sea urchin, *Strongylocentrotus droebachiensis,* indicated that the remains originated from the estuary or the ocean but were unlikely to have come from the river (Sorg et al. 1997). In another case, human remains were recovered after 32 years submergence but the condition of the body suggested that the upper and lower halves of the body had been in different environments. The lower body showed little signs of erosion damage, but the upper body exhibited extensive erosion. As well, the upper body showed signs of having been in a well oxygenated environment, with octopus eggs in the clothing, while the lower body exhibited evidence of an anoxic environment. The authors' conclusions were that the remains had been partially buried in the sand (Sorg et al. 1997). These cases illustrate the possibility of determining the conditions under which the body has been by interpreting faunal activity and decompositional changes.

12.4 Conclusions

In all these studies, no classic succession of invertebrate species was observed, in contrast with insect colonization in terrestrial environments. Most invertebrate fauna were opportunistic scavengers and fed on the remains at all times, until no soft tissue was left. Nevertheless, the fauna on the remains, the decomposition and the feeding patterns do suggest that a better understanding of decomposition in the marine environment will be valuable in estimating elapsed time since submergence, identifying the environment in which the body has decomposed and identifying the origin of marks on the remains. The VENUS experiments are ongoing.

Acknowledgments This sort of research cannot be done alone and I owe a tremendous debt of gratitude to the many people who donated hundreds of hours of in-kind support to this work, in the form of divers, boats, hovercrafts and field technicians. I would like to thank the Royal Canadian Mounted Police, in particular, Sgt. Ken Burton, Skipper of the Nadon, and all the RCMP divers that helped in my research; the Canadian Coast Guard and all the many divers, boat and hovercraft operators that assisted; the Vancouver Aquarium Marine Research Centre, in particular Dr. Jeff Marliave and Mr. Jeremy Heyward; and the Canadian Amphibious Search team, in particular, Mr. Tim MacFarlane and all the divers of CAST. I would particularly like to thank the VENUS project; Dr. Verena Tunnicliffe, the Project Director, for inviting me onboard and Mr. Paul Macoun, Dr. Richard Dewey and Mr. Adrian Round for their continuous support. I also wish to thank Ms. Niki Hobischak for her ongoing assistance.

This work was partially funded by the Canadian Police Research Centre, Ottawa, Ontario, Canada and I wish to thank Ms. Julie Graham and Mr. Steve Palmer for their support. Finally, and most importantly, I wish to thank Cpl. Bob Teather, RCMP, CV (deceased), for inspiring all my marine work and for providing advice, support and encouragement throughout. He is greatly missed.

References

Anderson GS (2008) Investigation into the effects of oceanic submergence on carrion decomposition and faunal colonization using a baited camera. Part I. (No. TR-10-2008). Canadian Police Research Centre, Ottawa, ON

Anderson GS (2009) Insect Succession on carrion and its relationship to determining time since death. In: Byrd JH Castner JL (eds) Forensic entomology: the utility of arthropods in legal investigations, 2nd edn. CRC, Boca Raton, FL p 201–250

Anderson JJ, Devol AH (1973) Deep water renewal in Saanich Inlet, an intermittently anoxic basin. Estuar Coast Mar Sci 1:1–10

Anderson GS, Hobischak NR (2002) Determination of time of death for humans discovered in saltwater using aquatic organism succession and decomposition rates (No. TR-09-2002). Canadian Police Research Centre, Technical Report, Ottawa, ON

Anderson GS, Hobischak NR (2004) Decomposition of carrion in the marine environment in British Columbia, Canada. Int J Legal Med 118(4):206–209

Anderson GS, VanLaerhoven SL (1996) Initial studies on insect succession on carrion in southwestern British Columbia. J Forensic Sci 41(4):617–625

Baco AR, Smith CR (2003) High species richness in deep-sea chemoautotrophic whale skeleton communities. Mar Ecol Prog Ser 260:109–114

Bernatis JL, Gersenberger SL, McGraw IJ (2007) Behavioural responses of the Dungeness crab, *Cancer magister*, during feeding and digestion in hypoxic conditions. Mar Biol 150:941–951

Bertness MD, Gaines SD, Bermudez D, Sanford E (1991) Extreme spatial variation in the growth and reproductive output of the acorn barnacle *Semibalalnus balanoides*. Mar Biol Prog Ser 75:91–100

Boyle S, Galloway A, Mason RT (1997) Human aquatic taphonomy in the Monterey Bay area. In: Haglund, WD,Sorg, MH (eds) Forensic taphonomy. The postmortem fate of human remains. CRC, Boca Raton, pp 605–613

Burd BJ, Brinkhurst RO (1985) The effect of oxygen depletion on the galatheid crab Munida quadrispina in Saanich Inlet, British Columbia. In: Gray, JS,Christiansen, ME (eds) Marine biology of polar regions and effects of stress on marine organisms. Wiley, New York, pp 435–443

Catts EP, Goff ML (1992) Forensic entomology in criminal investigations. Ann Rev Entomol 37:253–272

Chorneyko K, Rao C (1996) Aquatic misadventures: water related deaths in the Hamilton–Wentworth region, 1989–1994. Can Soc Forensic Sci J 29(3):165–173

Davis JH (1992) Bodies in water. Solving the puzzle. J Florida Med Assoc 79(9):630–632

Davis JB, Goff ML (2000) Decomposition patterns in terrestrial and intertidal habitats on Oahu Island and Coconut Island, Hawaii. J Forensic Sci 45(4):836–842

Dearborn JH (1967) Stanford University invertebrate studies in the Ross Sea 1958–1961: General account and station list. The Fauna of the Ross Sea, Part 5. New Zealand Department of Science and Industry Research Bulletin 176:31–47

Debenham NJ, Lambshead PJD, Ferreo TJ, Smith CR (2004) The impact of whale falls on nematode abundance in the deep sea. Deep-Sea Res Pt 1, 51:701–706

Dennison KJ, Kieser JA, Buckeridge JS, Bishop PJ (2004) Post mortem cohabitation – shell growth as a measure of elapsed time: a case report. Forensic Sci Int 139(2–3):249–254

Diaz RJ, Rosenburg R (1995) Marine benthic hypoxia: a review of its ecological effects and the behavioural responses of benthic macrofauna. Oceanogr Mar Biol 33:245–303

Dix JD (1987) Missouri's lakes and the disposal of homicide victims. J Forensic Sci 32(3):806–809

Donoghue ER, Minnigerode GC (1977) Human body buoyancy: a study of 98 men. J Forensic Sci 22:573–579

Dumser TK, Türkay M (2008) Postmortem changes of human bodies on the bathyal sea floor – two cases of aircraft accidents above the open sea. J Forensic Sci 53(5):1049–1052

Ebbesmeyer CC, Haglund WD (1994) Drift trajectories of a floating human body simulated in a hydraulic model of puget sound. J Forensic Sci 39(1):231–240

Ebbesmeyer CC, Haglund WD (2002) Floating remains on Pacific Northwest waters. In: Haglund, WD,Sorg, MH (eds) Advances in Forensic Taphonomy. Method, theory and archeological perspectives. CRC, Boca Raton, pp 219–240

Ebert DA (1986) Biological aspects of the sixgill shark, Hexanchus griseus. Copeia 1986:131–135

Farrow EG, Levings CD, Foreman R, Tunnicliffe V (1983) A review of the Strait of Georgia Benthos. Can J Fish Aquat Sci 40:1120–1141

Fisher RS, Petty CS (1977) Forensic pathology: a handbook for pathologists. National Institute of Law Enforcement and Criminal Justice, U. S. Department of Justice, Washington, DC, pp 201

Giertson JC, Morild I (1989) Seafaring bodies. Am J Forensic Med Pathol 10(1):25–27

Haglund WD (1993) Disappearance of soft tissue and the disarticulation of human remains from aqueous environments. J Forensic Sci 38:806–815

Haglund WD, Sorg MH (2002) Human remains in water environments. In: Haglund WD, Sorg MH (eds) Advances in forensic taphonomy. Method, theory and archeological perspectives. CRC, Boca Raton, pp 201–218

Hart JL (1973) Pacific fishes of Canada. Fish Res Bd Can Bull 180:1–740

Herlinveaux RH (1962) Oceanography of Saanich Inlet in Vancouver Island, British Columbia. J Fish Res Bd Can 19:1–37

Hobischak NR (1998) Freshwater invertebrate succession and decompositional studies on carrion in British Columbia (No. TR-10-98). Canadian Police Research Centre, Technical Report, Ottawa, ON

Hobischak NR, Anderson GS (1999) Freshwater-related death investigations in British Columbia in 1995–1996. A review of coroners cases. Can Soc Forensic Sci J 32(2 and 3):97–106

Hobischak NR, Anderson GS (2002) Time of submergence using aquatic invertebrate succession and decompositional changes. J Forensic Sci 47(1):142–151

Jaffe FA (1976) A guide to pathological evidence. The Carswell Co. Ltd, Toronto, ON, pp 174

Jamieson GS, Pikitch EK (1988) Vertical distribution and mass mortality of prawns, Pandalus platyceros, in Saanich Inlet, British Columbia. Fish Bull 86:601–608

Jones EG, Collins MA, Bagley PM, Addison S, Priede IG (1998) The fate of cetacean carcasses in the deep sea: observations on consumption rates and succession of scavenging species in the abyssal north-east Atlantic Ocean. Proc R Soc Lond B 265(1401):1119–1127

Kahana T, Almog J, Levy J, Shmeltzer E, Spier Y, Hiss J (1999) Marine taphonomy; adipocere formation in a series of bodies recovered from a single shipwreck. J Forensic Sci 44(5):897–901

Keh B (1985) Scope and application of forensic entomology. Ann Rev Entomol 30:137–154

Kelly D (1990) Postmortem gastrointestinal gas production in submerged Yucatan micro-pigs. MA Thesis, Colorado State University, Fort Collins, CO

Kemp KM, Jamieson AJ, Bagley PM, McGrath H, Bailey DM, Collins MA, Priede IM (2006) Consumption of a large bathyal food fall, a six month study in the NE Atlantic. Mar Ecol Prog Ser 310:65–76

Klages M, Muyakshin S, Soltwedel T, Arntz WE (2002) Mechanoreception, a possible mechanism for food fall detection in deep-sea scavengers. Deep-Sea Res Pt 1, 49:143–155

Koseki T, Yamanouchi S (1963) The postmortem injury on the drowned bodies inflicted by aquatic animals, especially amphipods. Jpn J Legal Med 18:12–20

Lawlor W (1992) Bodies recovered from water. A personal approach and consideration of difficulties. J Clin Pathol 45:654–659

Lo SJ (2007) Factors influencing adipocere formation. BA Hons Thesis, Simon Fraser University, Burnaby, B.C.

Lord WD, Burger JF (1994) Arthropods associated with Harbor seal (*Phoca vitulina*) carcasses stranded along the New England coast. Int J Entomol 26(3):282–285

Mant AK (1960) Forensic medicine – observation and interpretation. The Year Book Publishers Inc., Chicago

Mellen PFM, Lowry MA, Micozzi MS (1993) Experimental observations on adipocere formation. J Forensic Sci 38:91–93

Mottonen M, Nuutila M (1977) Postmortem injury caused by domestic animals, crustaceans and fish. In: Tedeschi G (ed) Forensic medicine : a study in trauma and environmental hazards, vol II. W.B. Saunders, Philadelphia, PA

O'Brian TG (1994) Human soft-tissue decomposition in an aquatic environment and its transformation into adipocere. University of Tennessee, Knoxville, TN

O'Brien TG (1997) Movement of bodies in Lake Ontario. In: Haglund WD, Sorg MH (eds) Forensic taphonomy. The postmortem fate of human remains. CRC, Boca Raton, pp 559–565

O'Brien TG, Kuehner AC (2007) Waxing grave about adipocere: soft tissue change in an aquatic context. J Forensic Sci 52(2):294–301

Payne JA (1965) A summer carrion study of the baby pig Sus scrofa Linnaeus. Ecology 46:592–602

Payne JA, King EW (1972) Insect succession and decomposition of pig carcasses in water. J Georgia Entomol Soc 73:153–162

Peters KL (2007) Distribution, density and feeding biology of *Munida quadrispina* (Decapoda, Galatheidae), and other scavengers, in Saanich Inlet in relation to food resources and low oxygen conditions. B.Sc. Hons. Thesis, University of Victoria, Victoria, B.C.

Petrik MS, Hobischak NR, Anderson GS (2004) Examination of factors surrounding human decomposition in freshwater: a review of body recoveries and coroner cases in British Columbia. Can Soc Forensic Sci 37(1):9–17

Picton B (1971) Murder, suicide or accident. Robert Hale and Company, London

Polson CJ, Gee DJ (1973) The essentials of forensic medicine, 3rd edn. Pergamen, Toronto, ON

Rathbun TA, Rathbun BC (1997) Human remains recovered from a shark's stomach in South Carolina. In: Haglund WD, Sorg MH (eds) Forensic taphonomy. The postmortem fate of human remains. CRC, Boca Raton, pp 449–458

Simonsen J (1977) Early formation of adipocere in a temperate climate. Med Sci Law 17:53–55

Simpson K, Knight B (1985) Forensic medicine, 9th edn. Edward Arnold (Publishers) Ltd, Baltimore, MA, pp 348

Skinner MF, Duffy J, Symes DB (1988) Method and theory in deciding identity of skeletonized human remains. J Forensic Sci 21(3):138–141

Sloan NA, Robinson SMC (1983) Winter feeding by asteroids on a subtidal sandbed in British Columbia, Canada. Ophelia 22(2):125–142

Smith CR, Baco AR (2003) Ecology of whale falls at the deep-sea floor. Oceanogr Mar Biol Ann Rev 41:311–354

Smith CR, Kukert H, Wheatcroft RA, Jumars PA (1989) Vent fauna on whale remains. Nature 341:27–28

Sorg MH, Dearborn JH, Monahan EI, Ryan HF, Sweeney KG, David E (1997) Forensic taphonomy in marine contexts. In: Haglund WD, Sorg MH (eds) Forensic taphonomy. The postmortem fate of human remains. CRC, Boca Raton, pp 567–604

Spitz WU (1980) Drowning. In: Spitz WU, Fisher RS (eds) Medicolegal investigation of death. Guidelines for the application of pathology to crime investigations, 2nd edn. Charles C. Thomas, Springfield, IL

Stubblefield P M.A. (1999) Homicide or accident off the coast of Florida: trauma analysis of mutilated human remains. J Forensic Sci 44(4):716–719

Teather RG (1994) Encyclopedia of underwater investigations. Best Publishing Company, Flagstaff, AZ, pp 186

Tomita K (1976) On putrefaction and flotation of dead bodies. Hiroshima J Med Sci 25:155–176

Tunnicliffe V (1981) High Species diversity and abundance of the epibenthic community in an oxygen-deficient basin. Nature 294:354–356

Wentworth P, Croal AE, Jentz LA, Eshghabadi M, Pluck G (1993) Water-related deaths in Brant County 1969–1992: a review of fifty-seven cases. Can Soc Forensic Sci J 26(1):1–17

Wheeler A (1975) Fishes of the world. Macmillan, New York, NY

Witte U (1999) Consumption of large carcasses by scavenger assemblages in the deep Arabian Sea: observations by baited camera. Mar Ecol Prog Ser 183:139–147

Yahel G, Yahel R, Katz T, Lazar B, Herut B, Tunnicliffe V (2008) Fish activity: a major mechanism for sediment resuspension and organic matter remineralization in coastal marine sediments. Mar Ecol Prog Ser 371:195–209

Zimmerman KA, Wallace JR (2008) The potential to determine a postmortem submersion interval based on algal/diatom diversity on decomposing mammalian carcasses in brackish ponds in Delaware. J Forensic Sci 53(4):935–941

Juniper K BC Leadership Research Chair in Marine Ecosystems and Global Change, Victoria Experimental Network Under Sea, University of Victoria, B.C. Canada, personal communication.

MacFarlane T President, Canadian Amphibious Search Team, Mission, B.C., personal communication.

Teather RG Cpl. RCMP, Ret. Delta, B.C., personal communication.

Tunnicliffe V Project Director, Canada Research Chair in Deep Oceans, Professor in Biology and in Earth and Ocean Sciences, University of Victoria, Victoria, B.C., personal communication.

Wallace JR Professor, Dept. Biology, Millersville University, Millersville, PA, personal communication.

Chapter 13
The Insects Colonisation of Buried Remains

Emmanuel Gaudry

13.1 Introduction

In our society, burial of a deceased person is a common habit in a normal situation. In opposition, manmade burial by authors of homicide and/or their accomplices to hide the body of their victim is more seldom.

Exhumation of a buried corpse can be ordered by legal decision (second expert conclusion). It can also be accidental or required by authorities for investigation purposes (mass grave) or identification of victims (natural disaster or accident).

Location of an illegal grave is a hard task and the excavation needs to apply adapted techniques and to gather different specialists (forensic investigators, pathologists, dentists, botanists, entomologists, etc.).

Decomposition of corpses is affected by burial environment (conservation, mummification, adipocere formation, etc.) because of climatic, edaphic and biological parameters. It is well known that the decomposition process in soil is vastly different than on the surface, due to different parameters. Accessibility of the necrophagous entomological fauna to the body is either disturbed (arrival delayed) or inhibited. The composition of the insects' population can be strongly modified in addition to the duration of the life cycle. When colonisation is possible, the number of insects and the diversity of population are different than in normal conditions. In this situation, the estimation of the post mortem interval (PMI) is harder because of the lack of information (temperature values), and the lack in the species of forensic interest.

In the last century, studies helped to improve knowledge of the fauna associated with buried carrion. In the last period, some workers focused on a few groups of insects (or arthropods) in order to highlight a predictable pattern of their arrival on a corpse, helping to estimate the PMI. Groups are still neglected, because of the

E. Gaudry
Département Entomologie, Institut de recherche criminelle de la gendarmerie nationale, Rosny-sous-Bois Cedex, France

J. Amendt et al. (eds.), *Current Concepts in Forensic Entomology*,
DOI 10.1007/978-1-4020-9684-6_13,

little motivation for this topic and the lack of taxonomists. However, study of this specific fauna may provide interesting information for investigation purposes.

To better understand the dynamics of insect populations and interpret their processes, it is important to have a better knowledge of the decomposition of corpses when buried. Study of this necrophagous population can give useful information, even if the PMI estimation is less easy to determine. Nevertheless, advances in this particular field of forensic entomology will improve the efficiency of analyses.

13.2 Burial and Exhumation

The first studies dealing with burial fauna were conducted during the nineteenth century (Mégnin 1894; Motter 1898). They mainly described the coffin fauna. The other major data dealing with burial environment were provided by experiments using animal models, generally pig carcasses (Payne et al. 1968; Turner and Wilshire 1999; VanLaerhoven and Anderson 1999), or human cadavers (Rodriguez and Bass 1985).

Voluntary exhumation can be conducted for scientific reasons (mass exhumation in the nineteenth century, knowledge of funeral practices in antique civilization) or investigative purposes, to reconstruct a chain of events, cause and time of death (homicide, war crime, natural disaster, etc.). The discovery of a buried cadaver makes the management of such a crime (or death) scene more complex and information harder to obtain.

It is important to make the difference between an illegal grave excavated after a murder and mass grave exhumation (as in Kosovo, Chyprus, Lebanon, Rwanda, Argentina, etc.). In these latter cases, researches conducted at a large scale by governmental or non-governmental organisations, gathering forensic experts in different fields, helped significantly in the knowledge of burial cadavers, especially in forensic entomology. Many programmes are still ongoing and international prosecution will probably ask for new research. That is why the author decided to devote part of this chapter to this topic.

13.2.1 Burying: Why, Who, When, How?

Burying of people after the death is a very common custom. However, time of burial after death can differ significantly according to culture and religion. In agriculture (animal breeding), burial may be necessary because of health and safety reasons, to stop the spreading of a disease causing the death of animals and thus avoiding contamination by necrophagous insects or other scavengers such as badgers, foxes, and birds of prey.

Blanchard (1915) reported how the technique of burial was used to get rid of cadavers of soldier and carcasses of horses during battles of the First World War.

Mass graves were dug with addition of lime and a thick layer of soil to destroy large numbers of cadavers and carcasses at the same time. In his article the author related the danger represented by 'swarm' of flies attracted by decaying flesh whose proboscis, legs and wings were contaminated by this 'impure contact'.

Burial may also be due to particular circumstances such as natural disaster (earth quake, landslide…). Mass grave had sometimes been dug in dramatic situation caused by natural disaster (Thieren and Guitteau 2000). Some authors of homicide disposed off the corpse of their victim by burying. Manmade burials, except in normal circumstances, are not very common. Haglund and Sorg (2002) reported statistics dealing within the state of Washington during a 9-year period (1981–1990). Murderers used this *modus operandi* in only 1.38% of cases.

In France, elaboration of similar balance sheet is hard to do. Checking the database of the Forensic Entomology Department of the Forensic Science Institute (French Gendarmerie, IRCGN) revealed that since 1992 (creation of this laboratory), 33 referrals of more than 800 listed dealt with totally or partially buried bodies, that is in about 4% of expertise works. Half of them were related to real burial of corpses (2%).

From 1987 to 2008, 33 discoveries of 'buried' cadavers were also recorded in a French national database. Twenty-eight cases were real burial victims.

13.2.2 Exhumation: Why, Who, When, How?

13.2.2.1 Coffin

We wrote previously that time of burial after death can differ according to culture and religion. But within the same culture, habits may have changed over time. Indeed in the last century, a deceased person was laid in a coffin and exposed to his relatives in his house a few days (generally three) before being inhumed, letting enough time for necrophagous flies under favourable conditions to oviposit. Nowadays, dead persons are within a short time transported and kept at a morgue, avoiding insect colonisation Wyss and Cherix (2006).

Exhumation is generally conducted under legal decision to enable a second expert opinion to find the cause of death or to perform a paternity test (in France a young woman claiming that a famous actor was her father obtained the judicial authorities' permission to exhume his body for DNA analyses). Motter (1898) in the United States could study the fauna in 150 exhumed coffins. Bourel et al. (2004) related in their article exhumations requested not only by local authorities but also by insurance companies to clarify the cause of death (disease, accidental death at the workplace).

In other situation, study of the fauna associated with buried carrion helped to improve knowledge in funeral practices in ancient civilizations such as the Egyptians who already knew that the mummies (or corpses) in the process of mummification could be damaged by 'thanatophagous' insects (Huchet 1996).

The initial purpose of exhumation is to collect evidence to help investigation: identification of one (or several) victim(s), determination of the cause of death, determination of the *modus operandi*, and estimation of the time of death. In order to reconstruct the chain of events, it is necessary to study skeleton fragment, teeth, prostheses, hairs, jewellery, clothes, personal belongings, etc.

Study of soil, insects or pollen sampling may indicate the location of death or the season. The presence of a gun, gun shot bullet, or knife, are evidences to infer a potential cause of death. The present list of elements is not exhaustive, and many other samples can also be analysed (toxic, drug products, etc.).

13.2.2.2 Illegal Graves and Mass Graves

Legally speaking a mass grave is a place where at least three or several illegal executed victims, sometimes more than 100 as in Bosnia, Rwanda or Iraq, were buried without being killed during war fights (Kalacska and Bell 2006).

Historically, forensic investigation, after exhumation of corpses from mass graves, has been carried out since the 1980s with the aim to highlight action of torture and illegal execution of victims (as in Argentina). The United Nations International Criminal Tribunals for the former Yugoslavia (ICTY) is one of the main active prosecutions. The main aims of exhumation for ICTY are to (i) confirm witness testimony in case of violation of Human rights, (ii) obtain evidence in order to give argument for prosecution, (iii) establish cause of death and post mortem interval (http://www.un.org/icty). At the end of the Second World War the International Military Tribunal (IMT) was established by the allies. At this period, the element of responsibility of crime was mainly based on testimony or written documents. Thanks to the improvement of sciences, forensic investigations were conducted in Ukraine in 1990, by the Australian Special Investigation Unit to locate Second World War mass graves (Skinner and Sterenberg 2005) and find evidences with modern techniques. More recently, the Commonwealth War Graves Commission (CGWC) started the recovery of Australian and British soldiers who died in one of the most tragic battles of the First World War (Battle of Fromelles in Pheasant Wood, in July 1916) in Northern France. This 1-year project associates several scientists from different disciplines (anthropology, odontology, DNA, etc.) to study in a laboratory set up close to the mass grave site; all kinds of evidences collected to help to identify a maximum number of soldiers and create the first war cemetery in 50 years, 93 years after this battle (http://www.cwgc.org).

With more efficient techniques of recovery and development of forensic science, in addition to influence of public opinion and media (political position, human rights organisations, etc.), many investigations carried out during these past 30 years revealed presence of mass graves all over the world, especially in South America, thanks to the action of agencies[1] for the investigation of mass graves such

[1] The list of the following associations is not exhaustive.

as Argentine Forensic Anthropology Team, Latin American Forensic Anthropology Association, Physicians for Human Rights (PHR), American Association for the Advance in Science and United Nations International Criminal Tribunals for Rwanda in Africa (Skinner and Sterenberg 2005). Other independent organisations are involved in research such as the Institute for International Criminal Investigation (IICI), Inforce Foundation, British Association for Human Identification (BAHID) and Centre for International Forensic Assistance. In Europe, United Nations International Criminal Tribunals for the former Yugoslavia (ICTY) and International Commission for Missing persons from the former Yugoslavia (ICMP) largely contributed to the exhumation in association with different kinds of forensic experts.

However, forensic entomologists seldom took part in such research and the main information came from observations from non-specialists in this field, generating a lack of precision in the description of the development stage, or in the species identification or even mistakes in PMI estimation.

In 2000, a Disaster Victims Identification (DVI) team from Belgium gathering investigators and forensic experts (pathologists, ondontologists, anthropologists and entomologists) was asked to work in Kosovo under the aegis of ICTY (Beauthier et al. 2000; Beauthier 2008). In this part of Europe, the United Nations Mission in Kosovo (UNMIK) still leads exhumation, in association with ICMP (Skinner and Sterenberg 2005). These organisations helped and still participate in collecting different kinds of evidences or material elements (bones fractures, presence of gunshot bullets, marks made with different kinds of tools on bones, etc.) that can help the investigation with aims to punish the authors of such acts. Another reason is less investigative and more humanitarian, by providing information (dental finding, bone elements, prostheses, DNA, jewellery, personal properties, etc.) in order to facilitate identification of victims and enable families to receive the dead body of their relative so that they can go into mourning. In this kind of mission, methodology for investigation and identification should be coordinated. Such work improved significantly the knowledge in forensic taphonomy (Haglund and Sorg 2002), which deals with buried bodies.

13.3 Buried Cadaver: Decomposition Process in Soil

13.3.1 Decomposition Process

Death is characterised by the end of vital activities (so-called negative markers) and positive markers responsible for modifications for morphology and structure of the body (Campobasso et al. 2001). Negative markers, appearing immediately after the stop of activity of the heart, loss of breathing and circulation, are followed by fall of body temperature (*calor mortis*), rigidification (*rigor mortis*), apparition of lividity (*rubor mortis*), dehydration and acidification. Positive markers are rather destructive, with degradation of tissues (organic matter) caused by autolysis, autodigestion and putrefaction, involving action of anaerobic and aerobic bacteria or fungi. Results of this last process can be very different according to the environment. In a dry medium, the decay is rather dry and is called mummification. In wet conditions, or in an aqueous environment, it is called adipocere because of the appearance of waxy substances.

Regarding exposed corpses, most of the authors consider five stages of decomposition: fresh, bloated, putrefied, active putrefied, and dried remains. However, some of them proposed four to six stages of decomposition (Payne 1965). In burial environment, such modifications are hard to observe, unless little holes are dug into exhuming carcasses during the experiment. Continuous control of this process is not easy and observations are less accurate than in open air.

Payne et al. 1968 buried baby pigs at 50–100 cm deep in a clay soil from June to November (5 months). They observed five stages of decomposition: fresh, inflation (5 days after burial), deflation decomposition (+10), disintegration (+25), and skeletonisation (+80). Vanlaerhoven and Anderson (1999) used lambs (33 kg at 30 cm deep, 16-month experiment from June) and proposed five stages too: fresh (until 2 weeks after burial), bloated (6 weeks), active decay (3 months), advanced decay (11 months), and dry remaining stages (16 months). Turner and Wilshire (1999) buried pigs (55 kg) at 10 cm deep from December to April (5 months, cold conditions). They observed an active decay on carcasses 5 months after burial, but no skeletonisation.

Thus, the number and the duration of the decaying stages are not always similar but depend on several parameters, either biotic or abiotic.

13.3.2 Factors Affecting Decomposition Process in Soil

Several studies showed that the rates of decomposition of a cadaver in soil are slower than in open air. Some authors wrote that the rate was eight times reduced (Rodriguez and Bass 1985; Rodriguez 1997), mainly because of the soil environment but also because of the limitation of the wide spreading of decaying odours, and resctriction of the accessibility of the necrophagous fauna (vertebrates and invertebrates). Two types of preservation of soft tissues are generally observed on buried cadaver: mummification and adipocere formation (Forbes 2008). Decomposition depends on the nature of the deceased person (weight, size, nature of death and post mortem treatment, clothes, and coffin type), the environment (temperature, rain, wind, etc.), and the physical parameters of the burial site (depth, soil type, moisture, air content, temperature, etc.): so-called intrinsic and extrinsic factors. Studies too showed that cadavers in coffins decompose more rapidly than those directly buried in soil (Forbes 2008). Indeed, the bodies' dissolution is accelerated in coffins and does not allow adipocere formation.

The main taphonomic factors affecting the decay rate in soil are listed in Table 13.1 (after Haglund and Sorg 2002). Some of them will be discussed below.

Table 13.1 Main taphonomic factors affecting the decomposition in soil (after Haglund and Sorg 2002)

Grave characteristics
• Depth
• Compaction
Inclusions
Temporal factors
• PMI prior to burial
• Duration of burial
Season
Body characteristic
• State of decomposition
• Cause of death
• Body habitus
Presence of clothes, wrapping
Soil characteristics
• pH
• Oxygen contents
• Drainage
• Moisture
• Compaction
• Coarseness and type of soil (clay, sand, silt) composition of soil (C/N etc.)
• Contaminants
Proteolytic activity
Body assemblage characteristics
• Thickness and extent
Position relative to core and perimeter
Other characteristics
• Temperature during pre-burial period
• Post-burial exposure of remains to atmosphere
• Disturbance, treatment (deposit of toxical product: acid, limestone, etc.)
Insects, bacteria (aerobic and anaerobic), plants

13.3.2.1 Temperature

Because soil acts as a shield preventing solar radiation, temperatures are lower and fluctuation is less important. Such fall of temperature has a direct impact on the rate of decomposition by cooling the body, which remains thus during a longer period than in an exposed location. Rodriguez (1997) wrote that at depths greater than 2 ft (about 64 cm), fluctuation of temperature was quite non-existent. Below 4 ft (greater than 1.28 m), a cadaver was very well preserved over 1 year and skeletonisation took 2–3 years.

Studies showed that cadavers buried in the summer demonstrated a greater rate of decomposition than those buried in winter (Rodriguez and Bass 1985). Fresh temperatures tend to inhibit bacterial activity and proliferation. Putrefactive process is highly inhibited below 10°C and above 40°C. A temperature of 37°C is considered as an optimum for having significant impact on bacterial growth and decomposition

(Forbes 2008). Temperatures have a direct impact on insect biology. Activity of flight and oviposition of adults depend on temperature. The development rate of immature stages, from eggs to metamorphosis, is also temperature-dependent and specific (Kamal 1958; Marchenko 2001; Lefebvre and Pasquerault 2003). Donovan et al. (2006) proposed cold tolerance, introducing a hypothesis that there was within the same species of Calliphoridae (*Calliphora vicina* Robineau-Desvoidy) a biogeographical variation in larval growth. Other authors also showed that during the post-embryonic development, influence of cooling may be totally different on the successive developmental stages of a Calliphoridae species, *Protophormia terraenovae* (Robineau-Desvoidy) (Myskowiak and Doums 2002).

13.3.2.2 Depth

Depth has a significant effect on the rate of decomposition. Rodriguez and Bass (1985) reported the following observations on buried human cadavers. Cadavers buried at 1.2 m (4 ft) for 1 year showed a good state of preservation with skeletonisation limited to the head, hand and feet. The body was quite totally covered in adipocere. Cadavers buried at 0.6 m (2 ft) for 6 months showed little decomposition. Bones of the hands and feet were skeletonised. The genital area was decomposed. The body was dark brown in colour and covered with white adipocere (chest and legs). Cadavers buried at 0.3 m (1 ft) for 3 months were decomposed. Skeletonisation of arms and legs was complete with disarticulation at the major long bone joints. The feet were mummified. Small adipocere was visible along the upper chest cavity. Small traces of mould and fungi were observed on trousers.

The same authors highlighted that the duration of burial influences the decomposition rate. Indeed, cadavers buried at 0.3 m (1 ft) for two and a half months exhibited moderate decomposition with muscle visible on hands and legs and adipocere covering slightly the legs. Body colouration was dark pinkish brown. Small patches of fungi on the exposed chest were observed. Cadavers buried at 0.3 m (1 ft) for 1 month showed slight decomposition. Hands and feet were preserved, and bloating was slight in the face and marked in the abdomen. Body colouration was mainly dark pink. Fungi were more visible on trousers.

13.3.2.3 Soil Composition

Studies showed that the type of soil has to be taken into account to interpret the decomposition process and thus insect activity in soil. The decomposition rate of organic matter (cadaveric material) varies according to the characteristics of soil (Carter and Tibett 2008). A soil mainly composed of clay retains moisture and induces covering of the body by adipocere (Turner and Wilshire 1999). Lime burial sites (highly alkaline) delay the decomposition by adipocere formation and by limiting the survival and thriving of destructive bacteria. Forbes (2008) considered a loamy soil as a common soil that can be considered as control for experimentation.

Table 13.2 Stages in the decomposition of buried pigs after Wilson et al. (2007). Moreland (overlying boulder clay and coarse weathered sediments of Millstone Grit), woodland (free-draining loam with degraded Millstone Grit inclusions overlying Millstone Grit), pasture (overlying a silty clay on coal measure)

		Moreland		Woodland		Pasture
	Month	6	12	6	12	24
Fresh stage	Begun at death, includes rigor mortis, post mortem hypostasis and cooling Continues until bloating of the carcass is visible Skin intact, hair firmly anchored					
Primary bloat stage	Accumulation of gases within the body No disarticulation Hair and epidermis losses Soil–skin interface grey Strong odour					
Secondary bloat stage	Body still bloated Disarticulation of limbs Purging Soil–skin interface black Strong odour					
Active decay stage	Deflation of the carcass Disarticulation of the limbs and head Flesh and skin still present Carcass very wet Strong odour					
Advanced decay stage	Collapse of abdomen, rib cage Most or the flesh liquefied or gone Skin, bone, fat and cartilage may remain Carcass very wet Adipocere formation					
Skeletonisation stage	Flesh, skin and cartilage disappear Some adipocere and ligaments may remain					

Wilson et al. (2007) showed varying conditions after exhumation of pig carcasses buried for 6, 12 or 24 months in three contrasting fields (Table 13.2).

13.3.2.4 Oxygen Content

Fungi and aerobic bacteria need a minimum amount of oxygen to thrive (Forbes 2008). Thus, a poor presence of oxygen delays the decomposition process. In other words, preservation of tissue is better when oxygen is low.

13.3.2.5 Moisture

Moisture is present in soil located close to a water environment and in soil mainly composed of clay. Clay retains moisture and helps adipocere formation on the surface of the bodies, helping conversion of triglycerids to free fatty acids. In dried soil the body is rather mummified. The season can be also very important in such a process. Indeed, warm temperature but also very cold temperature (freeze-drying) makes the body desiccate and induce mummification. In a dry environment, adipocere formation may be due to the moisture included in the tissues.

13.3.2.6 pH

Soil pH is important too because high acidity as well as high alkalinity of soil can inhibit bacterial growth and, consequently, the decomposition process. Corpses in sealed coffins with adipocere formation release odour of ammoniac, inducing an alkaline environment. On the contrary, a slight alkaline pH of soil is favourable for the thriving of destructive bacteria (Forbes 2008).

The presence of decaying bodies has two kinds of effect on soil pH, at first increasing alkalinity followed by a fall of pH (Carter and Tibett 2008). Rodriguez and Bass (1985) showed an increase of pH prior to burial and after exhumation whatever the depth. The higher increases were observed at 0.3 m deep after 3 months (+2.1), at 0.6 m after 6 months (+1.1) and at 1.2 m after 1 year (+0.5). However, at a shallow depth of 0.3 m (1 ft), below 2 months and a half, the difference was lower (+0.2 max). Wilson et al. (2007) also observed that 12 months after burial of pig carcasses (*Sus scrofa*) at 0.3 and 0.6 m deep, pH rose significantly at woodland sites (+2.1 and +2.6).

13.3.2.7 Microbial Degradation

Invertebrates, fungi and also saprophagous bacteria (aerobic) participate actively in the decay of organic matters. Indeed, the microbial community produces proteolytic enzymes responsible for the breakdown of polymers in organic matter. Such a process is vital to the local ecosystem. Studies have shown that several parameters have a direct influence on bacteria proliferation and thus on biomass loss. Temperature has a major impact on it. An increase from 10°C to 30/35°C of temperature doubles the microbial activity. In other words, at lower temperature, the microbial activity is lower and hence organic matter is better preserved. On the other hand, Wilson et al. (2007) showed that burial of pigs' carcasses in soil led to higher microbial community. Carter and Tibett (2008) added that repeated burial of cadavers (or carcasses) in the same grave soil simulates microbial proliferation and enhances decomposition. Such information has to be taken into account because of potential impact on the insect fauna.

13.3.2.8 Fauna

Macrofauna and microfauna can be attracted by a decaying corpse causing damages and affecting the decomposition process. Contrary to carcasses in open air, a burial environment prevents attacks from scavengers (birds, rats, foxes, badgers, etc.). Holes of rodents were observed close to buried sheep carcasses at 10 cm depth 1 year after burial, probably because of presence of crevices in soils (Gaudry E, unpublished data). Such holes made by these animals very likely increase the accessibility of the entomological fauna to the cadaveric material and enhance decomposition.

Decaying carcasses and bodies (exposed or buried) provide, during a more or less long period (according to depth, season, etc.), an important quantity of nutritive substrate. Such necromasses can also be used by other organisms such as invertebrates for mating or as shelter. That is why the diversity of arthropods that can be found during collection is important, gathering necrophages, predators, parasites, omnivores and opportunists (Dadour and Harvey 2008). Collection of evidences at a burial death scene is particularly important and requires a specific protocol much more complex than in a crime scene where the body is exposed above the surface.

13.4 Forensic Excavation Technique

Forensic entomologist must not forget that he/she may be included in a global crime scene or cadaver recovery process. Several authors have already described such processes regarding open air cadavers (Catts and Haskell 1990; Byrd and Castner (2001); Amendt et al. 2007).

Detection of graves or mass graves can be very complicated and time-consuming, according to the characteristics of the supposed site of the grave (large or not), its location (accessibility for technicians) and time elapsed since burial. All of these parameters have a direct impact on the type of method and material used for detection. Grave site is a very specific environment because the number of victims (one–two to dozens), and the type of evidence (material elements: weapons, bullets, teeth, bones, jewellery, documents, etc.) are dispersed in a three-dimensional death scene. Thus, collection of these different clues has to be performed even more carefully, because of potential loss or missing of evidences and damages to bones or elements caused by the exhumation techniques (shovel, bulldozer, etc.). Knowledge of such techniques, rather closer to archaeology than the forensic field, is mandatory. In fact, such process requires a mix of archaeology, anthropology and forensic techniques. Some of the different location techniques are listed below. If some of them are very sophisticated and are used with parsimony, other, more basic ones, can be useful for forensic entomologists for common cases dealing with buried remains.

13.4.1 Location Techniques

Detection and recovery of buried bodies have for long concerned anthropological and archaeological studies. It became more recently a focus of forensic investigations. Whatever the context (clandestine graves, mass graves), locations are very complex because of the search area: surface and time elapsed since the burial. Indeed, modification of soil growth of vegetation can mash a burial site several months after. Such a task is generally time-consuming and requires personnel (safeguarding of the scene, recovery with replacement of staff) and material that is more or less sophisticated (detection material, engine and light by night). They include non-intrusive foot search methods such as observation (vegetation, soil) and use of air scent dogs (Killam 1990) or aerial observation (photography and video, infrared). Intrusive ground search methods include techniques that could cause damage of the site by the use of probes, combustible gas vapour detectors and soil analyses: pH, chemical composition, organic content, etc.

Other search methods are based on geophysical prospecting (magnetic surveying, metal detectors, ground-penetrating radar, resistivity or electromagnetic surveying), or remote-sensing. Parapsychological methods have also been used sometimes (Killam 1990).

13.4.1.1 Non-intrusive Ground Search Methods

Existing techniques performed at ground level are more or less sophisticated and dedicated to a relatively reduced surface area. They imply that the site is accessible to personnel.

13.4.1.1.1 Testimonies

Witness or survivor testimonies are a traditional way of grave 'detection' at a different scale used nowadays.

13.4.1.1.2 Basic Observation

Basic observation of vegetation (foliage, plant growth) and soil changes (coloration, sinking or compaction) can also be taken into account for investigators. Rodriguez and Bass (1985) reported that greater burial depths (1.2 m; 0.6 m) caused deeper depression than shallower burials (0.3 m). They observed that because of the redeposition of deeper layer of soil at the surface, coloration of topsoil was different above the grave than in the surroundings. Such method needs trained staff and favourable weather to improve efficiency of search operations.

13.4.1.1.3 Foot Search

In order not to forget any evidences, a common progression pattern can be realised. The search area can be divided into codified zones in which different teams may work. The foot search process must be performed from the outside toward the inside in order to minimize disturbance within the zone. The searcher can thus move in decreasing concentric circles within the zone. The line or strip search consists of a kind of zigzagging between limits of the search area. Search can be conducted by a team of searchers in right line separated at 2 m from one extremity of the delimited area to the other. Another variation consists of separating people in two teams facing each other (interlocking lines), each of them starting from one boundary (natural or man-made) and starting again perpendicularly (Killam 1990). For more efficiency a mix of strip and interlocking lines can be used.

13.4.1.1.4 Cadaver Dogs

Air scent dogs are trained to detect different kinds of elements: drugs, explosives, human scent of live humans. Decaying bodies produce gas, carbon dioxide, hydrogen sulphide and methane that reach the soil surface and are carried by wind. Such gases are water-soluble. Thus rain increases the soil moisture, favouring gas detection by dogs. Each step of the decomposition process thus produces a distinct scent and is discriminated by the dogs trained for it. They are trained to locate not only buried or concealed human remains, but also body fluids. Such a technique is employed for more than 30 years. Cadaver dogs are adaptable to any area without causing any damage or disturbance and can smell a cadaver from 2 weeks until several years after (12–15 years). They may also be efficient for early PMI (>3 h) (Oesterhelweg et al. 2008). This possibility however, needs further studies.

They can be used either in cases of crime investigation or in search of disaster victims. After the 2004 Tsunami in South East Asia, an air scent dog team from the French Gendarmerie worked in the debris of hotels in Phuket Island to find bodies (Fig. 13.1).This team is more commonly required to work through the French territory. Cadaver dogs can be a good help in the location and recovery of scattered body elements in a large area, reducing time and manpower (Komar 1999).

13.4.1.1.5 Aerial Observation/Photography

Aerial observation from helicopter or airplane of relief due to modification of soil compaction can also be useful. Digging modifies the layers of soil disposition and causes differences in colour contrast of soil, as in the case of older graves that show colour contrast of vegetation (type, difference of growth, etc.). Prosecution can also use satellite photography and photographs provided by government secret agencies. Aerial photography may help to cover a large search area but needs aircraft (expensive) and specialists.

Fig. 13.1 Cadaver dog and handler team from the French Gendarmerie (CNICG, Gramat) in the process of detecting buried human corpses after Tsunami in 2004 (Khao-Lak, Thaïland) (photography: E. Gaudry, personal collection, France)

13.4.1.2 Intrusive Ground Search Methods

Probing with a soil-sampling stick can be used to detect differences in soil compaction causing softness. Holes made by insertion of probes in soil may not only be used to measure temperature and pH, but also for soil analysis and combustible gas vapour detection (Killam 1990). This method causes minimal damage, but is time-consuming. Excavations with heavy engines (bulldozers) can also be performed but includes risk of damage.

13.4.1.2.1 Passive Geophysical Prospecting Methods

The following methods listed below are based on the measurement of physical and chemical modification of soil after burying. They are more sophisticated and not commonly used although they do not necessarily cause deterioration to the site.

13.4.1.2.2 Soil Stratigraphy

This discipline takes into account the distinct layers in a non-disturbed soil in comparison with grave backfill mixing the different layers all together and modifying chemical and physical characteristics. It is time-consuming and its efficiency is questionable.

13.4.1.2.3 Detection of Cadaver Scent

Decaying cadavers emanate gases during a defined period (mainly ammonia, hydrogen, methane, sulphur and carbon dioxide). Such scent can be detected to localise the body in soil when external conditions enable it.

13.4.1.2.4 Magnetic Surveying

This technique is commonly used in petroleum search, detecting with a gravimeter micro-variations in the earth's magnetic field caused by burial and backfilling (difference of density). This technique can easily detect metallic objects but can be disturbed by interferences leading to false positives. This passive prospecting method requires trained personnel who are able to interpret signals and is sometimes not adapted to locate a grave.

13.4.1.3 Active Geophysical Prospecting Methods

13.4.1.3.1 Geophysical Resistivity

This sophisticated method is based on the use of conductance of an electrical current through different soil layers. Discrimination between surrounding soils of graves and disturbed soil due to digging and refilling operations can be highlighted by this technique.

13.4.1.3.2 The Ground Penetrating Radar or GPR

The ground-penetrating radar (GPR) detecting changes in soil (Kalacska and Bell 2006) provides data to search and rescue team for locating and measuring distance to trapped victims (disaster situation) but also buried objects prior to digging (Perrot et al. 2007). This material is very sensitive but expensive and requires trained people to use it and interpret signals. Such a tool is regularly used by the Signal, Images and Voices Processing Unit (IRCGN, France) under the aegis of the Criminal Investigation National Unit (Voillot et al. 2006), a crime scene investigation team belonging to the IRCGN (Fig. 13.2). Complementation by other techniques is necessary to increase significantly the efficiency of search.

Fig. 13.2 Use of Ground-Penetrating Radar (GPR) to locate illegal graves (photography, IRCGN, France)

13.4.1.3.3 Metal Detector

This material can detect conductive metal by generation of an electromagnetic field transmitted to the soil surface. When the field meets a metallic object, an eddy current is created, thus generating a loss of the power signalled by the detector. Such material is easy to use and convenient whatever the condition. But it requires trained people and can detect a metallic object at shallow depth only.

13.4.1.3.4 Remote Sensing

When a search covers a larger area, recent studies show that optical remote sensing applying the process of interaction of light with a material can be a powerful tool. Hyperspectral imagery acquired from aircraft is a powerful detection material that can provide specialists analysing such data accurate information regarding the location of a burial site simultaneously or retrospectively (Kalacska and Bell 2006).

13.4.2 On the Scale of the Forensic Entomologist

Forensic entomologists can be involved in the search process and participate actively in it. When the site is accessible, observation of soil compaction can be done. But when time has elapsed, such compaction is not so easy to discriminate.

At shallow depth, decomposition of organic matter produces decaying fluids. They impregnate the soil and provide rich nutrients sought by roots of plants (Rodriguez and Bass 1985). That is why difference of growth of plant and contrast of colour can be observed above the grave. These authors added that changes of foliage, due to previous burial, are signs of potential digging. In contrast, a deep grave modifies the layer of soil disposition and also does not permit roots to reach the soil imbibed of rich nutrients because of the digging. Such a situation generates retardation of surface plant growth (Rodriguez and Bass 1985) and creates an artificial but visible difference in the vegetation aspect. Moreover, disturbance to the soil is shortly linked to the importance of digging but also reduces the above-grave vegetation development. Such observations can be good indicators for forensic investigators (including forensic entomologists) to locate a burial site.

Rodriguez and Bass (1985) proposed to use methods employed by archaeologists to locate buried sites and funeral pits by inserting probes (pH, soil conductivity, temperature) into the soil of expected graves and observing changes in comparison to natural soil (experiments performed between 0.3 to 1.2 m). However, such methods require to have already defined a site with potential graves and do not apply to large perimeters. At shallow depth, the presence of flies flying and landing on soils may also indicate the presence of decaying organic matter.

The Forensic Entomology Department (IRCGN, France) has initiated since 2008 cooperation with the cadaver dog and handlers' investigation team specialised in cadaver search (CNICG – France) and the *Signal, Images and Voices Processing Unit* (IRCGN) investigating in the same field with GPR. Usually, the local forensic pathologist and local crime scene technicians (CST) are involved in such recovery operations. The aim of this approach is to propose to investigators and magistrates a complementary experts team from the location of a corpse until the sampling operation. When a search of cadaver operation is known and planned, each team can alert one another (CNICG is located 600 km from IRCGN) and can in this way be available in a short duration. Experience shown that investigators focus on identification elements and evidence that could provide information on the cause of death (gun, bullet, knife, string, bone element, etc.) which may be forgotten. The presence of a forensic entomologist within the team decreases the risk of forgetting insect collection. For instance, such an approach may provide promising results and satisfy the different personnel involved (judge, investigators, CST, etc.).

13.4.3 Recovery

As written previously, investigating to locate buried corpses requires a complementary team of specialists and material adapted to the case. Recovery of cadavers also needs the assistance of traditional specialists: CST and forensic pathologist. As the time elapsed after the events can be more or less long, a lot of parameters may be different between burial and discovery such as climate and other environmental conditions. Several protocols concerning exhumation of clandestine graves and a strict methodology exist (Byrd and Castner 2001) that recommend the digging of a second, deeper grave beside the primary one, helping to excavate more carefully the cadaver and exploring the grave surroundings in a better way Beauthier et al. (2000).

In recent articles, different authors propose to better understand events surrounding death (crime scene or disaster) and to consider each grave as a potential crime scene, and include forensic archaeologists (Menez 2005) to carry out careful excavation and mapping and to define limits of the grave (Skinner and Sterenberg 2005), as well as forensic anthropologists: experts in the identification and determination of altered bone elements, forensic odontologists, ballistic specialist, botanists, and entomologists. It is obvious that such a multidisciplinary team is not easy to gather in every situation dealing with a buried cadaver. Material of excavation has also to be adapted to the type of situation: shovel, sieve, bulldozer, metal detector. Particular installation has to be set up sometimes: tent, chain of treatment to sort elements from soil of tables for corpse examination Beauthier et al. (2000).

Coexistence of different specialists (discipline, structure, status) may lead to a problem of management. That is why defining limits for each of them is mandatory as also the presence of a scene manager (Skinner and Sterenberg 2005).

Excavation can start with bulldozers to remove the first layers of soil. If not, soil is manually removed with a shovel in order to reach remains or skeletons. Exhumation is performed very carefully not to lose evidences or deteriorate corpses or bones. This crime scene is managed in the usual way: evidences are identified, photographed, collected and inventoried.

Once exhumation has ended, the remaining soil is carefully sorted with sieve or a more sophisticated system (chain). Corpses are submitted to radiography and medico-legal expertise (external and internal exams, DNA sampling, fingerprint sampling, entomological sampling, etc.).

13.5 Actual Knowledge of Necrophagous Entomofauna on Buried Remians

The diversity of arthropods collected can be important, gathering necrophagous, predators, parasites, omnivorous and opportunists (Dadour and Harvey 2008). Arthropods found in the surroundings of buried bodies are described in two different classes: Arachnida (Acari, Spiders, Millipedes) and Insecta (Collembola, Thysanura, Blattodea, Hymenoptera or Ants, Dermaptera, Diptera and Coleoptera). Within this population, few taxa (mainly Diptera and some Coleoptera) have a forensic interest in estimating the PMI. One cannot exclude the impact of groups such as predators (Spiders, Coleoptera but also pseudoscorpions or Dermaptera) that play a role in determining the size of others groups. Among opportunists are Blattodea (more active at night) and Thysanura, which feed on stored product or detritus. Both are observed on carrion during the later steps of decay. Hymenopterans gather taxa with a wide diversity of diet: necrophagous (bee and wasps), predators of larvae and pupae of Diptera and Coloeoptera (Vespidae) but also parasites of Diptera during larval and pupal stage (Chalcididae, Braconidae, etc.).

Acari or Mites are small specimens almost never used in estimating PMI. Some of them feed on eggs or larvae of others insects; others that feed directly on fluids of decaying tissues (so as Lepidoptera) are detritivorous or fungivorous. Megnin (1894) associated them within the sixth wave, during the later stages of decay. Goff (1989) described four gamasid families known to use carrion insects as carriers: Diptera (Muscidae: *Musca domestica*) and Coleoptera (Silphidae or Carabidae). His studies showed changes over time in the acari population which could be linked with PMI indication. They can thus be interesting forensic indicators.

Lepidoptera were also listed in the chronological succession pattern proposed by Megnin; Pyralidae and Tineidae were respectively observed during fats rancid and dry remains stages. Several other species have been observed on buried animal carcasses (Payne et al. 1968; Vanlaerhoven and Anderson 1999).

13.5.1 Necrophagous Insects Found in the Surroundings of Buried Bodies

Within the microfauna, necrophagous insects are said to be the most important factors after temperature, impacting the decomposition rate (Mann et al. 1990). Burial significantly inhibits access of scavengers to the cadaver. Necrophagous insects are attracted by decaying odours, which are masked by the soil covering above the body. Some of them can reach the decaying body through cracks in the soil created either by bloating of the body (Rodriguez and Bass 1985), crevices caused by dry weather or holes dug by rodents or other predators (Gaudry E, unpublished observation). Adult females can then lay eggs directly on the surface of the body. In some species, females lay eggs on the surface of the soil. After hatching, young first instars migrate into the soil to reach the body, feed and develop (Smith 1986; Rodriguez and Bass 1985).

Rodriguez and Bass (1985) reported the following observations. Cadavers buried at 1.2 m (4 ft) for 1 year showed a good state of preservation. No carrion insect activity was observed on corpses buried at 0.6 m (2 ft) for 6 months either. Cadavers buried at 0.3 m (1 ft) for 3 months was decomposed and numerous larvae of Diptera, pupae, puparium and adults were observed. Their identification revealed that they belong to the Calliphoridae and Sarcophagidae family (Table 13.4). Cadavers buried at 0.3 m (1 ft) for two and a half months and for 1 month exhibited few larvae of Diptera.

In this study, authors observed that the adults of Calliphoridae not only try to reach the cadaver through small cracks and crevices in the soil, but also by oviposition in the soil cracks following heavy rain.

Cold and anoxic conditions favour preservation of a carcass even at shallow depth but at the same time may significantly inhibit insect colonization. Turner and Wilshire (1999) started a 5-month experiment in winter and observed only Calliphoridae after 4 months: *Calliphora vomitoria*, *Lucilia* sp.; Muscidae: *Phaonia* sp.; Phoridae: *Conicera tibialis* and Silphidae: *Nicrophorus investigator* (Table 13.4).

13.5.2 Insects Found in Coffins

In Europe, mass exhumations carried out during the nineteenth century provide basic knowledge with respect to necrophagous insects developing on buried bodies (Table 13.4).

In France, Mégnin (1894) described a succession on buried cadavers with predominance of Dipterans (Muscidae: *Ophyra* spp., Phoridae undetermined) during the first year and Coleopterans (Staphylinidae: *Philonthus* sp., Rhizophagidae) during the second year. Motter (1898) reported a study of 150 disinterments in the vicinity of Washington. Among the population of arthropods collected, he identified many species actually considered forensically interesting such as Dipteran (Phoridae: *Phora clavata* Loew, *Conicera* sp.; Muscidae: *Compsomyia macellaria* Fabricius; *Lucilia caesar*[2] Linné, Anthomyidae: *Homalomyia* sp., *Ophyra leucostoma*[3] Wiedemann, Sepsidae: *Piophila casei*[4] Linné and Stratiomyidae), Coleopterans (mainly Staphylinidae: *Atheta* sp., *Actobius poederoides* Lec., *Homalota* sp., *Lathrobium simile* Lec., *Staphylinus cinnamopterus* Gravenhorst, *Poederus littorarius* Gravenhorst, *Philonthus* sp., *Eleusis pallida* Lec., *Actobius umbripennis* Lec.). Hymenopterans and Acari were also collected during this study.

Other studies on human bodies had been carried out. Bourel et al. published in 2004 observations on 22 exhumed cadavers in the Lille area (North of France). Such exhumations were requested by legal authorities in order to have the conclusion of a second expert in case of suspected death. On burial ranging between 2 and 29 months at about 2 m deep, authors identified 10 species of insects dominated by Dipterans, especially Muscidae (*Hydrotaea capensis*) Phoridae (*Conicera tibialis, Leptocera caenosa, Megaselia rufipes, Triphleba hyalinata*) and Fanniidae (*Fannia manicata* and *F. scalaris*). Regarding Coleopterans, Staphylinidae, individuals of *Philonthus* sp., also described by Mégnin and Leclercq, and *Omalium rivulare* were identified. A single species of open air cadaver was observed (*Calliphora vicina*) on a corpse exhumed after 3 months. Although the presence of some of these species may help to confirm a confinement of the body, authors logically confessed that it was hard to estimate the period of burial or PMI with these studies (definition of generation, end of life cycle, temperature prediction).

Merritt et al. (2007) reported a cold case with a cadaver that was exhumed 28 years after death from a cemetery in Michigan. The body had been embalmed and buried at 6 ft (unsealed casket and unsealed cement vault). Authors confirmed that the vault rested 1.8 m below at the base and had 1.2–1.5 m of soil on top. The predominant group identified belonged to collembolan (thousands of specimens). Workers wrote that the moist soil conditions at this depth explained their abundance. A Diptera species, *Conicera tibialis* (Phoridae), known to produce several generations

---

[2] Calliphoridae

[3] Muscidae

[4] Piophilidae

and mate inside the coffin, was identified. A large number of Acari (Glycyphagidae) were collected too.

13.5.3 Insects Found in Caves

The fauna of the cave may be treated in opposition the open air insect population. A cave or abyss can be considered a medium for delaying or inhibiting locations of remains by insects, and thus their colonization. However, some authors reported surprising observations.

Turchetto et al. (2008) collected puparia in the surroundings of bones of a German soldier captured and executed during the liberation of a village in the province of La Spezia (NW Italy). Remains were found in a cave or *foiba* located 240 m above sea level (647 m deep with a 15 m wide opening). Puparia belonged to Calliphoridae (one puparium of *Calliphora* sp.), Fannidae and Heleomyzidae. Vanin and Vernier (2005) updated a list of species of Diptera collected in a cave (*Grotta della Guerra*, Vicenza Province, Italy) including Heleomyzidae*: Heleomyza serrata* and *Heteromyza atricornis*. Four other species were identified: *Limonia rubeculosa* (Limoniidae), *Psychoda parthogenetica* (Psychodidae), *Culex pipiens* (Culicidae) and *Leptocera nigra* (Sphaeroceridae).

Wyss and Chérix (2006) tried to recontruct the chain of events after the discovery of a dead hill-walker in an abyss 18 days after being declared missing. They placed a bait containing pig meat (muscle and liver) 10 m above the ground at the bottom of the abyss in the dark for 17 days in August. Twelve days after the disposal of the bait, they observed oviposition of *Calliphora vicina* and *Calliphora vomitoria*. Adults of *Neoleria inscripta* (Heleomyzidae) and *Stearibia nigriceps* (Piophilidae) were identified too. In the same abyss, they had previously collected larvae of *Calliphora vomitoria* (Calliphoridae) and pupae of *Neoleria inscripta* and *Megaselia rufipes* (Phoridae) on a roe deer carcass 6.5 m deep. Living Carabidae, Catopidae (*Catops picipes*) and Staphylidae represent Coleoptera. A cave or abyss is considered a confined medium that significantly delays or forbids the arrival of insects on remains and thus their colonization. According to the geographical location and altitude, ambient conditions may be strongly different from open air conditions. Cave and abyss are mediums presenting a strong and obvious similitary with regard to the burial environment.

13.5.4 Necrophagous Insects of Forensic Importance

The faunal inventory on exposed carcasses has already been documented in various geographical locations. The diversity of insects and arthropod populations colonizing corpses and animal carcasses is limited in a burial environment. Different authors (Mégnin 1894; Leclercq 1978; Smith 1986; and others) have highlighted the modification of insect populations between exposed and buried remains.

Amendt et al. (2004) and Gennard (2007) focused on the main families and genus of insects of forensic importance (Table 13.3); therefore Diptera and Coleoptera are the most represented orders on exposed cadavers.

Diptera order showed a great diversity of population. Among the Diptera, Calliphoridae (genus *Calliphora*, *Chrysomya*, *Cochliomyia*, *Lucilia*, *Phormia*) and Muscidae (*Hydrotaea*, *Musca*, *Muscina* and *Ophyra*) are predominant families.

Among the Coleoptera, Silphidae (genus *Necrodes*, *Nicrophorus*, *Silpha*), Staphylinidae, Dermestidae and Histeridae are well represented.

In order to propose a (non-exhaustive) list of insects of forensic importance from buried remains, we compared inventories proposed by Amendt et al. (2004) and Gennard (2007) with different studies (Table 13.4).

The composition of populations might be divided into five main categories of depths: very shallow, shallow (10–30 cm deep), deep (40–60 cm deep), very deep grave (from 90 cm) and coffin (2 m).

Table 13.3 Listing of insects of forensic importance (after Amendt et al. 2004 and Gennard 2007)

Amendt		Gennard	Gennard(additional species)
Order/family	Important genera		
Diptera			
Calliphoridae	*Calliphora*,*Chrysomya*, *Cochliomyia*, *Lucilia*, *Phormia*	Listed	
Drosophilidae	*Drosophila*	Not mentioned	
Ephydridae	*Discomyza*	Not mentioned	
Fanniidae	*Fannia*	Listed	
Heleomyzidae	*Heleomyza*, *Neoleria*	Not mentioned	
Muscidae	*Hydrotaea*, *Musca*, *Muscina*, *Ophyra*	Listed	
Phoridae	*Conicera*, *Megaselia*	Listed	
Piophilidae	*Piophila*, *Stearibia*	Listed	
Sarcophagidae	*Liopygia*, *Sarcophaga*	Listed	
Sepsidae	*Nemopoda*, *Themira*	Listed	
Sphaeroceridae	*Leptocera*	Listed	
Stratiomyidae	*Hermetia*, *Sargus*	Not mentioned	
Trichoceridae	*Trichocera*	Listed	
Coleoptera			
Cleridae	*Necrobia*	Listed	Nitidulidae
Dermestidae	*Attagenus*, *Dermestes*	Listed	Carabidae
Geotrupidae	*Geotrupes*	Listed	Scarabaeidae
Histeridae	*Hister*, *Saprinus*	Listed	Trogidae
Silphidae	*Necrodes*, *Nicrophorus*, *Silpha*	Listed	Tenebrionidae
Staphylinidae	*Aleochara*, *Creophilus*	Listed	
Lepidoptera			
Tineidae	*Tineola*		
Hymenoptera			
Ichneumonidae	*Alysia*		
Pteromalidae	*Nasonia*, *Muscidifurax*		

Table 13.4 Examples of insects (and arthropods) of forensic importance in burial environment. Bou: Bourel et al. 2004; Gau: Gaudry et al. 2006; Mot: Motter 1898; Meg 1:Megnin 1894 (Ivry) and Meg 2: Megnin 1884 (St Nazaire); Mer: Merritt et al. 2007; Pay: Payne et al. 1968; Smi Smith 1986; Tur: Turner and Wilshire 1999; Van 1: CMH zone Ex; Van 2: SBS zone Ex (After VanLaerhoven and Anderson 1999)

	Type of grave	<10 cm		Shallow (~10–30 cm)					Deep (~40–60 cm)		Very deep (≥90 cm)		Coffin (2 m)					
Order/family	Important genera/author	Lun	Tur	Lun	Van1	Van2	Rod	Gau	Lun	Pay	Pay	Gau	Bou	Mer	Mot	Smi	Meg	Meg2
Diptera																		
Calliphoridae	*Calliphora, Chrysomya, Cochliomyia, Lucilia, Phormia*		×		×	×	×						×		×	×	×	
Drosophilidae	*Drosophila*	×	×	×	×	×		×		×	×	×	×		×	×	×	×
Ephydridae	*Discomyza*				×	×		×				×	×					
Fanniidae	*Fannia*					×		×										
Heleomyzidae	*Heleomyza, Neoleria*																	
Muscidae	*Hydrotaea, Musca, Muscina, Ophyra*																	
Phoridae	*Conicera, Megaselia*	×	×	×	×	×	×	×	×	×	×	×	×	×	×	×	×	×
Piophilidae	*Piophila, Stearibia*				×			×		×	×	×			×			
Sarcophagidae	*Liopygia, Sarcophaga*				×			×				×			×			
Sepsidae	*Nemopoda, Themira*				×													
Sphaeroceridae	*Leptocera*																	
Stratiomyidae	*Hermetia, Sargus*																	
Trichoceridae	*Trichocera*																	
Coleoptera																		
Cleridae	*Necrobia*	×	×	×	×	×		×	×	×	×	×	×		×	×	×	×
Dermestidae	*Attagenus, Dermestes*	×		×	×	×		×	×			×						
Geotrupidae	*Geotrupes*	×		×	×	×			×									
Histeridae	*Hister, Saprinus*																	

(continued)

Table 13.4 (continued)

	Type of grave	<10 cm		Shallow (~10–30 cm)					Deep (~40–60 cm)		Very deep (≥90 cm)	Coffin (2 m)						
Order/family	Important genera/author	Lun	Tur	Lun	Van1	Van2	Rod	Gau	Lun	Pay	Pay	Gau	Bou	Mer	Mot	Smi	Meg	Meg2
Silphidae	*Necrodes, Nicrophorus, Silpha*																	
Staphylinidae	*Aleochara, Creophilus, Philonthus*																	
Hymenoptera																		
Formicidae					×					×	×				×			
Pteromalidae					×													
Others		×		×					×						×			
Acari		×		×					×	×	×				×			
Collembola		×		×					×	×	×				×			

In a shallow grave, the diversity of Arthropods is logically greater, the with presence of different families of Diptera, Coleoptera and Hymenoptera as well as Acari and Collembola. In a shallow grave and within the coffin fauna, Calliphoridae are well represented. In the second medium (coffin), cultural habits may have a direct impact of this observation due to the exposition of the deceased before being placed in a casket. Predominance of Muscidae and Phoridae is obvious after burial. In deep and very deep graves, insect activity is poor, with very few families within each order. Data are rare in deeper graves (>90 cm).

Among Coleoptera, Silphidae, Staphylinidae and Histeridae are well represented in shallow graves, whereas at greater depth and witinin the coffin, occurrence of Staphylinidae is important. In the studies considered it is the only family of Coleoptera collected.

Burial modifies the hierarchy established at the surface, discriminating insects of primary importance and those of secondary importance, because of the lack of the former under the soil surface depending on the depth. Time of death, period of burial, exposition of the dead body or carcass before burial have an impact on the insect composition. Time spent before exhumation is important because it has an impact on the insect collection. Indeed, a long period enables an easier insect colonization.

13.5.4.1 Dipterans Found on Buried Corpses

13.5.4.1.1 Calliphoridae

Calliphoridae, considered a major family of forensic importance, is generally collected from a very shallow grave. Lundt (1964) reported that a layer of soil was sufficient to prevent Calliphoridae colonization (*Calliphora* and *Lucilia*). Their presence at a greater depth may indicate that the victim has been exposed prior to being buried. If not, no Calliphoridae specimens colonize remains when the soil layer above the cadaver is superior to 40 cm thick. Thus, estimation of reliable PMI is hardly possible.

13.5.4.1.2 Fannidae

Species belonging to this family prefer a moist habitat. According to Vanlaerhoven and Anderson (1999) such behaviour may explain the presence of immature stages of *Fannia canicularis* at early stages of decomposition. The same species was identified by Gaudry et al. (2006) at 30 cm depth with *F. scalaris*. *Fannia manicata* and *F. scalaris* were identified by Bourel et al. (2004) in coffins.

13.5.4.1.3 Heleomyzidae

Turner and Wiltshire reported that a PMI estimation was based upon a *Tephoclamys rufiventris* (Heleomyzidae). Heleomyzidae are regularly collected among the population of unburied (Smith 1986) and buried cadaver (Vanlaerhoven and Anderson

1999). Specimens belonging to Heleomyzidae have been collected at 90 cm depth, 1 year after the burial of lamb carcasses (Gaudry et al. 2006). A lack of identification and/or development data on species of this family does not enable it to be associated to insect population of forensic importance.

Such a situation can be compared to the Stratiomyidae (soldier flies) collected by Motter (1898) and other workers (see below).

13.5.4.1.4 Muscidae

Lundt (1964) highlighted the predominance of Muscidae in a shallow grave (2.5 and 10 cm deep) represented by two species: *Muscina pabulorum* Fallen and *Ophyra leucostoma* Wiedemann, whose females lay eggs on the soil surface above the buried substrate (flesh). Vanlaerhoven and Anderson (1999) observed a chronological succession pattern on buried carrion bteween Calliphoridae (*Calliphora vomitoria*) and Muscidae (*Hydrotaea* sp., *Morellia* sp. and *Ophyra leucostoma*) 6 weeks after death, as previously reported by Smith (1986). *Muscina* and *Ophyra* were predominant genus in a shallow grave (Gaudry et al. 2006). From very deep graves, the authors collected individuals of *Muscina* sp. *Ophyra capensis* was identified by Bourel et al. (2004) in coffins.

13.5.4.1.5 Phoridae

Scuttle flies are described in many experiments of exhumation of bodies. As previously mentioned for Muscidae, this family is the most represented in a burial environment. Some authors (Disney 1994) suggested that some Phoridae collected on buried carrion may play a similar fonction to Calliphoridae in exposed ones.

Lundt (1964) found adults of *Metopina* sp. and *Conicera* sp. at a depth of 25–50 cm 4 days after, burial in Germany.

Payne et al. (1968) collected *Dohrniphora incisuralis* (Loew) and *Metopina subarcuata* Borgmeier from pig carcasses buried 50–100 cm deep, on the third day and actively feeding on the seventh day.

This family was also present in Coastal Western Hemlock zone and Sub-boreal Spruce (British Columbia). Vanlaerhoven and Anderson (1999) collected in this area specimen of *Dohrniphora* sp. but only at the adult stage on pig carcass buried 30 cm below the surface. *Conicera tibialis* Schmitz specimens have been collected at 90 cm depth, 1 year after the burial of lamb carcasses (Gaudry et al. 2006).

Bodies buried may be revealed by the presence of coffin flies, *Conicera tibialis*, whose presence on the soil surface could indicate the location of a buried body (Leclercq 1999, Gennard 2007). Disney (1994) reported too frequent occurence of this species on buried corpses .

They are generally found at pupal stage on hair of exhumed deceased persons or peal and wool of animal carcasses (Fig. 13.3). Their adults are supposed to be able to move to a significant depth (1 m and more when found in coffins at 2 m deep) through the ground (Disney 1994). Other authors (Colyer 1954a,b) suggested a 'one-way traffic' from the body to the ground level.

Fig. 13.3 Pupae of Phoridae found in hair of a buried cadaver in France (2008) (photography: E. Gaudry, IRCGN, France)

13.5.4.1.6 Piophilidae

Motter collected *Piophila casei* Linné (named as Sepsidae) among the fauna of many coffins studied. *Stearibia nigriceps* (Meigen) was collected by Vanlaerhoven and Anderson (1999) but only at the adult stage on pig carcasses (see above).

13.5.4.1.7 Stratiomyidae

Main Stratiomyidae (Hermetiinae, Sarginae) are more opportunistic than necrophagous. That is why they can be classified at first as species of secondary forensic importance. Motter (1898) found specimen of Stratiomyidae in coffins. *Hermetia illlucens* (Linné) (Stratiomyidae) has been described on buried human remains in the United States and Hawaï (Lord et al. 1994). Studies showed that this species colonizes remains during later stages of decay (drier, post-decayed stages).

13.5.4.2 Coleopterans Found on Buried Corpses: Staphylinidae and Rhizophagidae

Coleoptera are present in every kind of medium: surface, ground, on carrion, vegetation, etc. Many specimens belonging to this order are found on exposed-above ground as well as buried bodies. If the diversity is very important in open-air corpses, Coleoptera are less predominant underground (grave and coffin). Such situations may be partially explained by their trophism. Many of them are not carrion feeders but predators or fungi feeders (see below).

13.5.5 Staphylinidae and Rhizophagidae

Several workers who published on burial fauna often observed two families within Coleoptera: Staphylinidae and Rhizophagidae. Staphylinidae family is almost always represented in a burial environment. Genus *Philonthus* sp. was identified by Megnin (1894). Vanlaerhoven and Anderson (1999) listed on average three different species within the same genus. *Philonthus* sp. was mainly identified at 10 and 30 cm deep by Gaudry et al. (2006). At deeper layers of soil (25–50 cm), Lundt (1964) reported that the genus *Atheta* sp. dominated. *Rhizophagus paralellocolis* (Gyllenhal) (Rhizophagidae) is placed in Mégnin's fourth wave 2 years after the death (exposed above-ground corpses). Smith (1986) associated the genus *Philonthus* sp. (Staphylinidae) and *Rhizophagus* sp. (Rhizophagidae) with buried human corpses, although these small beetles are rather predators of larvae of Diptera. Other authors linked the presence of *Rhizophagus paralellocolis* (Gyllenhal) to the presence of fungi present on cadavers (Hakbijl 2000). Such findings may be interesting for further studies associating development data of fungi to those of this species to indicate a period for time of death.

13.5.6 Silphidae

Authors who worked on coffins did not collect individuals of Silphidae. In very deep graves no Silphidae were found either. They are only described in shallow or very shallow graves (Lundt 1964; Turner and Wilshire 1999; Vanlaerhoven and Anderson 1999). Shubeck (1968) observed that a shallow layer of sand (2–3 cm) significantly inhibited attraction of *Necrophila americana* (Linné) to buried foetal pig carrion in Maryland (USA), as well as other Silphidae. He observed no individuals below 3 cm and few specimens at 3 cm depth, suggesting that ability for carrion beetles to reach carrion under shallow layers of sand was limited. Previously, Lundt (1964) reported collection of Staphylinidae (larvae and adults) from 2.5 to 50 cm depth.

13.5.7 Histeridae

In very deep graves (90 cm), Gaudry et al. (2006) found Histeridae specimens (genus and species unknown). In shallow graves they identified *Margarinotus carbonarius* (Hoffmann). At very shallow depth, they collected *Margarinotus brunneus* (Fabricius) and *M. ruficornis* (Grimm). Vanlaerhoven and Anderson (1999) collected *Hister depurator* Say and other specimens (genus and species unknown). Lundt (1964) collected Histeridae specimens in shallow graves: *Saprinus semistratus* Scriba 2.5 cm underground and *Hister cadaverinus* Hoffmann at 10 cm depth – no Histeridae specimens have been collected in coffins.

Cleridae and *Dermestidae* families are rarely observed underground or in coffins. Champollion, a French Egyptologist, observed adults of *Corynetes glabra* Champollion, a species similar to *Necrobia rufipes* De Geer (Huchet 1996) between fingers of mummies.

13.5.7.1 Hymenopterans Found on Buried Corpses: Ants and Parasitic Wasps

The order Hymenoptera includes few species having a direct forensic importance. Being opportunists, they can either feed on the flesh of the corpse or be predators of immature stages of others insects (eggs, larvae). However, these Hymenopterans are not considered primary forensic indicators, unlike other species belonging to the same order.

Ants (Formicidae, Myrmicidae) are found during the entire decaying process and then regularly collected even in burial conditions (Bornemissza 1957; Vanlaerhoven and Anderson 1999). Motter observed *Crematogaster lineolata* Say, *Aphenogaster* sp. and *Myrmycina latreilli* André, *monomorium minutum* Mayer (Myrmicidae) but also *Brachymyrmex heeri* Forel , *Lasius americanus* Emery, *L. flavius* De Geer, *Ponera contracta* Latreille and *Camponotus melleus* Say (Formicidae). Payne et al. (1968) found *Prenolepis imparis* Say (Formicidae) at early stages of decay. Species of Formicidae (*Anoplolepsis longipes*) have already been studied to estimate a PMI on skeletal remains found in a metal toll in Hawaï by Goff and Bani (1997). Payne et al. (1968) observed that *Prenolepis imparis* Say (Hymenoptera, Formicidae) was predominant on buried pigs.

Parasitic wasps such as *Nasonia vitripennis* (Hymenoptera, Pteromalidae) are regularly found on buried carcasses (Vanlaerhoven and Anderson 1999). Lundt (1964) collected Braconidae and Proctotrupidae at 50 cm depth. Cynipids and Diaprides were also found by Payne et al. 1968 from 50 to 100 cm depth.

13.5.7.2 Other Insects Found on Buried Corpses: Collembola (Spring Tails)

Merritt et al. (2007) collected a species of Collembola (*Sinella tenebricosa* Folsom) from a casket containing a 28-year-old body that was exhumed. Collembola species had already been identified from exhumation (Motter 1898; Payne et al. 1968). In his paper, Meritt did not estimate a PMI by the case study but suggested a potential (forensic) utility of these specimens to confirm a post mortem transportation of the corpse during a given season thanks to the knowledge of specific findings linked to the season or specificity of a soil type.

13.5.7.3 Other Arthropods Found on Buried Corpses: Acari (Mites)

Insects as well as other arthropods are involved and collected during the decaying process of organic matter. Acari (Gamasidae) were collected by Motter (1898) or

Merritt et al. (2007) (Glycyphagidae) from buried cadavers in classical graves. Goff (1989) observed in Hawaii Island changes in the population of Acari that could be associated with post mortem intervals (if PMI > 17 days) considering four Gamasid families (Macrochelidae, Pachylaelapidae, Parasitae and Uropididae). Mégnin (1894) and Smith (1986) listed Acari in the sixth waves on exposed corpses, but provided no information about their presence on buried ones. Macrochelidae are phoretic on *Nicrophorus* species (Silphidae, Coleoptera). Disney (1994) summarized different phoretic mites on Scuttle fly (Phoridae), suggesting host specificity (to be confirmed by further studies).

13.6 Post Mortem Interval Estimation

13.6.1 Temperature

Temperature is the most critical parameter taken into account for PMI estimation (accumulation degree day and accumulation degree hours method: ADD and ADH (Marchenko 2001)). While such parameters are rather easy to collect by own measurement and/or collection of weather station data, the situation is different for burial scenes. Data between underground, surface and ambient temperatures (weather station: 1.2 m hight) are different and depend on several parameters. Fluctuations are reduced underground in comparison with the surface. It is difficult to consider which temperature is most representative of the condition in which the corpses decayed because data are rarely available.

Rodriguez and Bass (1985) observed increase of cadaver temperature over soil temperature recorded at the same depth. They highlighted that the mean temperature differential was directly proportional to the depth of burial: +3.4°C at 1.2 m; +5°C at 0.6 m; +7–10°C at 0.3 m.

Vanlaerhoven and Anderson (1999) suggested that soil temperature was the best indicator to estimate the PMI with insects that have developed on a buried corpse, better than weather station data.

13.6.2 Insects

13.6.2.1 Calliphoridae

Calliphoridae may be collected from shallow graves. Their study can help to estimate a minimum PMI, because delay caused by burial cannot be determined. Moreover, colonization of the corpse before burial cannot be excluded. In this situation the estimate is more accurate. Turner and Wilshire (1999) suggested that cold and anoxic conditions in a heavy clay soil had preserved the carcasses and significantly delayed the arrival of *Calliphora vomitoria*. The use of this species would have led to a wrong estimation of PMI.

Wyss and Chérix (2006) observed oviposition of *Calliphora vicina* and *Calliphora vomitoria* 12 days after the disposal of a bait at 10 cm depth at the bottom of an abyss in the dark. The average temperature was 5°C. In this situation, the location of the substrate delayed attraction of insects but did not stop it. They concludes that death of the hill-walker occurred just after his disappearance. The absence of light is not a sufficient criterion for considering egg-laying to be a nocturnal phenomenon. Authors did not observe in which period of the day they occurred (probably a diurnal period). Indeed, photoperiod has a hormonal and maternal determinism. Modifications in the circadian rhythm are known to be observed in the following generation and not in the actual one (oviposition).

Absence of Calliphoridae in the samples associated to the study of environmental conditions (weather station) of the scene of death may indicate (if conditions are favourable) that time elapsed between death and burial was shorter.

Forensic entomologists may have difficulties in estimating PMI when species that are hard to identify are collected, such as Sarcophagidae and Phoridae (3,000 species in the world). The main consequence in this situation is also the lack of development data of some species in these families.

13.6.2.2 Muscidae

Megnin described *Muscina stabulans* within the first wave on buried carrion. The chronological succession pattern observed by Vanlaerhoven and Anderson (1999) between Calliphoridae (*Calliphora vomitoria*) and Muscidae (*Hydrotaea* sp., *Morellia* sp. and *Ophyra leucostoma*) suggested interesting developments to estimate PMI. In shallow graves genus *Ophyra* and *Muscina* were described as predominant (Lundt 1964). If Calliphoridae are not observed, development data of species belonging to these genuses may be interesting to estimate the minimal period of insect colonisation after burial.

13.6.2.3 Phoridae

Their study does not necessarily provide an accurate estimate of the PMI as these flies do not mandatorily appear at the surface after the first generation. Smith (1986) also suggested that many generations could breed in the coffin, as many adults and puparia could be found in many exhumations. However, it is possible to estimate a minimum PMI of burial basing the analysis upon puparia that are found, considering that the period of emergence can at least occur at the moment of collection. Some authors (Disney 1994) suggested that some Phoridae collected on buried carrion may play a similar fonction to Calliphoridae in exposed ones.

Greenberg and Wells (1997), Leclercq (1999) wrote that Phoridae adults are known to arrive at a body with a delay in comparison with Calliphoridae or Sarcophagidae, observed in early stages of decay, suggesting they could not be necessarily associated as insects of secondary forensic importance. During the winter period, Manlove and Disney (2007) estimated in a case example a PMI with *Megaselia abdita* (Schmitz), whereas Calliphoridae were collected. At the temperature taken into account by authors, the Phoridae provide a longer life cycle and therefore provide a better PMI than the Calliphorids usually analysed.

13.6.2.4 Stratiomyidae

Hermetia illucens (Linné) (Stratiomyidae) is a widespread Diptera (Leclercq 1997). Lord et al. (1994) collected *H. illlucens* in a corpse from a grave site and used its development time to estimate the PMI. Lord et al. mentioned that *H. illlucens* was frequently predominant on remains in the drier post-decay stage of decomposition above ground. A late arrival (colonization of exposed corpses 20–30 days post mortem), associated with its life cycle duration, may provide longer estimates according to the authors. It could be a potential PMI indicator whose development can be used in combination with data of other arthropod species. However, further studies (Tomberlin et al. 2005) in the United States did not confirm a late arrival but rather an early one (6 days post mortem) as confirmed by Pujol-Luz et al. (2008) in Brazil (3 days).

13.6.2.5 Coleoptera

The Coleoptera rarely colonize a carcass at early stages, as suggested by the authors (Shubeck 1968), because of a lack of efficiency in detecting carrion for some of them. Thus their forensic importance is lesser than Diptera. Staphylinidae, Histeridae and Silphidae are mainly collected on buried carcasses or corpses (Lundt 1964; Smith 1986; Leclercq 1978; VanLaerhoven and Anderson 1999), mainly from shallow graves (<30 cm deep). In coffins, Staphylinidae seems to be the only family observed (Mégnin 1894; Motter 1898; Smith 1986; Bourel et al. 2004; Merritt et al. 2007) as in deep graves (about 90 cm deep) (Payne et al. 1968). Moreover, many of them are generally predators or immature stages of Dipterans. Their arrival is not directly in relation with the presence of the corpses. For all these reasons Coleopterans collected on a buried cadaver and its surroundings cannot be considered as reliable forensic indicators, and are almost never used to estimate PMI.

The relation highlighted by Hakbijl (2000) between the occurrences of *Rhizophagus paralellocolis* (Gyllenhal) and the presence of fungi present on the cadavers may be interesting. The association of growing data of fungi in addition to thermo-dependent development time of this species may help to estimate PMI.

13.6.2.6 Hymenoptera

Females of *Nasonia vitripennis* usually lay eggs in their host (pupae of Calliphoridae, Muscidae) 1 day after the pupation when the skin of the larva has separated from the inner pupal cuticle. In this situation the forensic entomologist can only assess

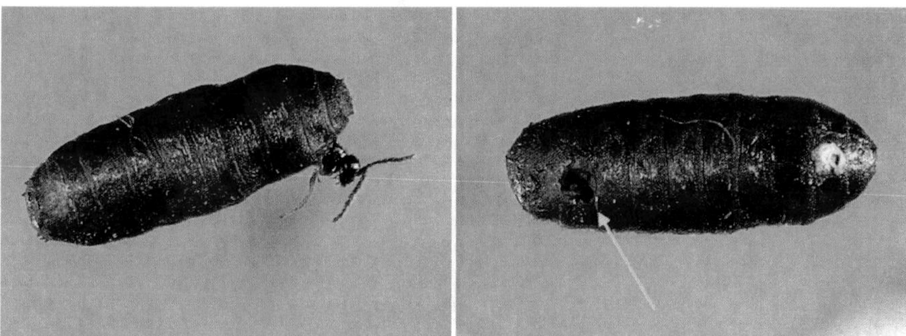

Fig. 13.4 Parasitic wasp emergence from a pupae of *Lucilia sericata* (Meigen); circular hole made by the wasp (photography: E. Gaudry, IRCGN, France)

damage in the rearing: no adult flies' eclosion but pupae with little circular holes in the cuticle (Fig. 13.4). Thus, PMI estimation is hardly possible. Moreover, the tiny adults of parasitic wasps observed in the incubator are susceptible to contaminate other rearing boxes and compromise other analyses. Grassberger and Frank (2003) studied the temperature-related development of *N. vitripennis* using host of pupae of *Protophormia terraenovae* (Robineau-Desvoidy) (Diptera, Calliphoridae) in order to allow estimating an extended PMI. They confirmed a development time inversely related to the temperature for this species and defined a lower threshold and ADD summation from 15°C to 30°C, thus suggesting to use this as a reliable forensic indicator for estimating PMI. This approach is promising but further research is necessary.

13.7 Conclusion

During Springfield, we (Department of Forensic Entomology – IRCGN, France) were called for entomological sampling on a corpse buried at about 80 cm depth in the West of France (Fig. 13.5). The soil was mainly composed by clay. A layer of farm quicklime (about 20 cm) covered the corpse from 40 cm depth onward. The victim was supposed to be killed 6 months ago. We collected larvae, many pupae and empty pupae of Heleomyzidae, some pupae and empty pupae of *Megaselia* sp. (Phoridae) and adults of Sphaeroceridae and Aleocharinae sp. Heleomyzidae and Sphaeroceridae were found in the wax layer covering the body between 40 and 60 cm depth. Phoridae were collected from the hairs of the victim at about 90 cm depth (Fig. 13.3). Phoridae were collected in large quantity. Several generations in burial environment may have succeeded without being able to precise more. Pupae of Heleomyzidae were hard to identify until the species' taxonomic level. Presence of adults of Sphaeroceridae and Aleocharinae sp. was not very helpful in the expertise work. In that situation, estimation of PMI was hard to make.

Fig. 13.5 Illegal grave with a body initially buried at 80 cm depth (Photography: E. Gaudry, IRCGN, France)

However, the absence of Calliphoridae or Sarcophagidae may indicate burial a short while after death (the time of homicide was supposed to occur in early autumn under good condition of temperature). In this case, burial has very likely inhibited post-burial colonisation.

In the West of France, during summer (at the end of July), a cadaver was found buried at 20 cm depth in an uncultivated field. The body was wrapped in a carpet (Fig. 13.6). A large population of insects was collected:

- Sarcophagidae (empty pupae)
- Muscidae: *Muscina* sp. (empty pupae), *Ophyra ignava* (Harris) (empty pupae)
- Phoridae (empty pupae)
- Nitidulidae: *Carpophilus* sp. (adult), unidentified (larvae)
- Staphylinidae: *Creophilus maxillosus* (Linné) (adult), *Gauropterus* sp. (adult), Aleocharinae (adult), Staphylininae (adult)
- Histeridae: *Saprinus semistriatus* (Scriba) (adult)
- Scarabaeidae: *Onthophagus ovatus* Linné (adult)

The analysis was based upon the hypothesis of time of emergence on the day of sampling. With such a hypothesis, a minimum PMI was estimated at least 6 weeks before. Recent information from investigators revealed that the time of death was established at the middle of May, which was 10 weeks ago. In the present case study, burial and wrapping delayed the arrival and colonisation of the body by insects but did not

Fig. 13.6 Illegal grave with a body initially buried at 20 cm depth. (Photography: Gendarmerie Nationale, France)

inhibit it helping the PMI estimation. This case study can be listed in the shallow grave category. Such estimation provided to investigators was then very important to identify the victim.

Motter (1898) argued about the notion of seasonality in the presence of certain species (e.g. presence of Muscidae indicating burial in summer and not in winter). He rather preferred to state that such presence may be due to exposure of the bodies to more favourable temperatures for the post-embryonic development of these species. He observed, as occurred for exposed cadavers, a chronological succession arrival of the various species colonising the bodies within distinct waves at specific periods of decay.

Lerclercq reported in 1993 a case dating from 1974 where a cadaver was found buried at 50–60 cm depth in a claying soil during the autumn season. On the remains that appeared at the surface of soil, entomologists identified *Limosina (paracollinella) curvinervis* (Diptera, Sphaeroceridae), *Brachyopa* sp. (Diptera, Syrphidae) and *Coprophilus* sp. (Coleoptera, Staphylinidae). Lerclercq wrote in his article that the victim was deceased 1 year ago without specifying further the implication of the entomological evidence in this expertise work.

These non-exhaustive cases reveal that the study of entomofauna associated with buried carrion needs further experiment and observation. Better knowledge of the impact of intrinsic and extrinsic factors such environment (temperature, rain, wind, etc.), and physical parameters of the burial site (depth, soil type, moisture, air content, temperature, etc.) are needed too.

Among the necrophagous population collected on buried corpse and its surroundings, whose composition may be significantly different than on the surface, species initially considered as groups of secondary importance may have a more important role in estimating PMI using ADD or ADH. Interaction between arthropod species (predation, parasitism, succession) needs further studies and can be an interesting option just as interaction between insects and fungi.

The discovery of a buried cadaver (criminal case, mass grave, disaster) is uncommon and often becomes a media event. Criminal investigations are no more confined only to the competence of magistrates and investigators. The public demands information, and the media need topics for reporting. Magistrates and investigators need to solve cases in such conditions. Forensic sciences are still developing and their importance in investigation is still growing. Techniques are becoming more and more efficient.

Accidental exhumation of buried corpses, but also requirement by authorities for investigating purposes, or second expert opinion or identification of victims (natural disaster or accident) may demand growing implication of forensic specialists such as forensic entomologists.

Study of this necrophagous population can give useful information, even if the PMI estimation is less easy to determine. Progress in the last 50 years has been significant thanks to several authors. Advances in this particular field of forensic entomology need improvement in the accuracy and reliability of the analyses in.

Acknowledgements Jens Amendt, Institute of legal Medicine, Frankfurt am Main - Germany. Jean-Pol Beauthier, Laboratoire de médecine légale et d'anthropologie médico-légale, Bruxelles –Belgique. Laurent Chartier, Département Signal image parole, IRCGN, gendarmerie nationale – France. Jérôme Carlier, Département Signal image parole, IRCGN, gendarmerie nationale – France. Laurent Dourel, Département Entomologie, IRCGN, gendarmerie nationale – France. Fabrice Galli, Centre national d'instruction cynophile de la gendarmerie nationale – France. Dorothy Gennard, Lincoln University, Lincoln – United Kingdom. Jean–Bernard Huchet, université de Bordeaux 1, Bordeaux – France. Hans Huijbregts, National Museum Natural History, Leiden – Nederland. Yvan Malgorn, IRCGN, gendarmerie nationale – France. Guy Marchesi, Service central de renseignements judiciaires et de documentation, gendarmerie nationale – France. Jean-Jacques Menier, National History Museum, Paris - France. Jean-Bernard Myskowiak, Département Thanotologie anthropologie odontologie, IRCGN, gendarmerie nationale – France. Thierry Pasquerault, Département Entomologie, IRCGN, gendarmerie nationale – France. Patrick Perrot, Département Signal image parole, IRCGN, gendarmerie nationale – France. Henri Plouchart, Direction Centrale de la Police Judiciaire – France. Géraldine Salerio, Institut de médecine légale de Montpellier – France. Jean Richebe, Département Thanotologie anthropologie odontologie, IRCGN, gendarmerie nationale – France. Stefano Vanin, Università di Padova, Padova– Italia. Benoît Vincent, Département Entomologie, IRCGN, gendarmerie nationale – France.

References

Amendt J, Krettek R, Zehner R (2004) Forensic Entomology. Naturwissenschaften 91:51–65

Amendt J, Campobasso CP, Gaudry E, Reiter C, LeBlanc HN, Hall MJR (2007) Best practice in forensic entomology – standards and guidelines. Int J Legal Med 121(4):90–104

Beauthier JP, Boxho P, Crèvecoeur JM, Leclercq M, Lefevre P, Vogels L (2000) Mission du Team Belge au Kosovo: science et justice à la rencontre du drame humain. Premiers résultats. Biométrie humaine et anthropologie 18(1–2):43–48

Beauthier JP (2008) Traité de médecine légale. Ed. De Boeck, Bruxelles, 837p

Blanchard R (1915) La destruction des cadavres en temps de guerre. Mode opératoire des Japonais en Mandchourie, vol 23(6). Revue d'Hygiène et de Police Sanitaire, Masson et Cie, Paris, pp 563–577

Bornemissza GF (1957) An analysis of arthropod succession in carrion and the effect of its decomposition on the soil fauna. Aus J Zool 5:1–12

Bourel B, Tournel G, Hedouin V, Gosset D (2004) Entomofauna of buried bodies in northern France. Int J Legal Med 118:215–220

Byrd JH, Castner JL (eds) (2001) Forensic entomology: the utility of arthropods in legal investigations. CRC, Boca Raton, FL, p 418

Campobasso CP, Di Vella G, Introna F (2001) Factors affecting decomposition and Diptera colonisation. Forensic Sci Int 120:18–27

Carter DO, Tibett M (2008) Does repeated burial of skeletal muscle tissue (*Ovis aries*) in soil affect subsequent decomposition? Appl Soil Ecol 40:529–535

Catts EP, Haskell NH (1990) Entomology and death/a procedural guide. Joyce's Print Shop, Inc., Clemson, South Carolina, p 182

Colyer C (1954a). More about the "coffin" fly Conicera tibialis Schmitz (Diptera, Phoridae). The Entomologist 87: pp 129–132

Colyer C (1954b). Further emergences of Conicera tibialis, the "coffin fly" (Diptera, Phoridae). The Entomologist 87: 234

Dadour IR, Harvey ML (2008) The role of invertebrates in terrestrial decomposition: forensic application. In: Tibett M, Carter DO (eds) Soil analysis in forensic taphonomy. Chemical and biological effects of buried human remains CRC Press, Boca Raton

Disney RHL (1994) Scuttle flies: the Phoridae. Chapman & Hall Ed, London, p 467

Donovan SE, Hall MJR, Turner BD, Moncrieff CB (2006) Larval growth rates of the blowfly *Calliphora vicina* over a range of temperatures. Med Vet Entomol 20:106–114

Forbes SL (2008) Potential determinants of postmortem and post burial interval of buried remains. In: Tibett M, Carter DO (eds) Soil analysis in forensic taphonomy. Chemical and biological effects of buried human remains. pp 225–246

Gaudry E, Dourel L, Chauvet C, Vincent B, Pasquerault T, Lefebvre F (2006) Burial of lamb carcasses at 3 different depths: impact on the colonization by necrophagous insects. Proceedings of the fourth meeting of the European Association for Forensic Entomology. Bari, Italy

Gennard DE (2007) Forensic entomology: an introduction, Wileyth edn. Chichester, England, p 224

Goff ML (1989) Gamasid mites as potential indicators of post mortem interval. In: Channabasavanna GP, Viraktamath CA (eds) Progres in acarology. Oxford & IBH, New Delhi, India, pp 443–450

Goff ML, Bani HW (1997) Estimation of postmortem interval based on colony development time for *Anoplolepsis longipes* (Hymenoptera: Formicidae). J Forensic Sci 42(6):1176–1179

Grassberger M, Frank C (2003) Temperature-related development of the parasitoid wasp Nasonia vitripennis as forensic indicator. Med Vet Entomol 17:257–262

Greenberg B, Wells JD (1997) Forensic Use of *Megaselia abdita* and *M. Scalaris* (Phoridae: Diptera): case studies, development, rates, and egg structure. Entomol Soc Am 35(3):205–209

Haglund DH, Sorg MH (2002) Advances in froensic taphonomy. In: Haglund, Sorg (ed) Method, theory, and archaeological perpectives. CRC, Boca Raton, Florida, p 507

Hakbijl T (2000) Arthropod remains as indicators for taphonomic processes: an assemblage from 19th century burials, Broerenkerk, Zwolle, The Netherlans. In Huntley J.P. and Stallibrass S Eds, Taphonomy and Interpretation. Symposia for the Association for Environmental Archaeology, N°14. Oxbow Books, Oxford, 95–96

Huchet J-B (1996) L'archéoentomologie funéraire: une approche originale dans l'interprétation des sépultures. Bulletin et Mémoire de la Société d'Anthropologie de Paris 8(3–4):299–311

Kalacska M, Bell LS (2006) Remote sensing as a tool for the detection of clandestine mass graves. Can Soc Forensic Sci 39:1–13

Kamal AS (1958) Comparative study of *sarcosaprophagous Calliphoridae* and *Sarcophagidae* (Diptera). Ann Entomol Soc 51:261–271

Killam EW (1990) The detection of human remains. Thomas CC (ed) Springfield, Illinois, p 263

Komar D (1999) The use of cadaver dogs in locating scattered, scavenged human remains: preliminary field test results. J Forensic Sci 44(2):405–408

Leclercq M (1978) Entomologie et Médecine légale – Datation de la mort, Collection de Médecine Légale et de Toxicologie Médicale N°108, éditions. Masson, Paris, p 100

Leclercq M (1997) A propos de *Hermetia illucens* (Linnaeus, 1758) ("soldier fly") (Diptera Stratiomyidae: Hermetiinae). Bulletin des Annales de la Société Royale Belge Entomologique 133:275–282

Leclercq M (1999) Entomologie et médecine légale: importance des Phorides sur les cadavres humains. Annales de la Société Entomologique de France 35(Supp):566–568

Lefebvre F, Pasquerault T (2003) Temperature-dependent development of *Ophyra aenescens* (Wiedemann, 1830) and *Ophyra capensis* (Wiedemann, 1818) (Diptera, Muscidae). Forensic Sci Int 139:75–79

Lord WD, Goff ML, Adkins TR, Haskell NH (1994) The Black Soldier Fly *Hermetia illucens* (Diptera: Stratiomyidae) as a potential measure of human postmortem interval: observations and case histories. J Forensic Sci 39(1):215–222

Lundt H (1964) Okologische Untersuchungen über die tierische Besiedlung von Aas im Boden. Pedobiologia 4:158–180

Manlove JD, Disney RHL (2007) The use of *Megaselia abdita* (Diptera: Phoridae) in forensic entomology. Forensic Sci Int 175(1):83–84

Mann RW, Bass WM, Meadows L (1990) Time since death and decomposition of the human body: variables and observations in case and experimental field studies. J Forensic Sci 35:103–111

Marchenko MI (2001) Medicolegal relevance of cadaver entomofauna for the determination of the time of death. Forensic Sci Int 120:89–109

Mégnin P.(1894) La faune des cadavres: application de l'entomologie à la médecine légale. Masson, Paris, 214 p

Menez LL (2005) The place of a forensic archaeologist at a crime scene involving a buried body. Forensic Sci Int 152:311–315

Merritt RW, Snider R, De Jong JL, Benbow ME, Kimbirauskas RK, Kolar RE (2007) Collembola of the grave: a cold case history involving arthropods 28 years after death. J Forensic Sci 52(6): 1359–1361

Motter MG (1898) A contribution to the study of the fauna of the grave. A study of one hundred and fifty disinterments, with some additional experimental observations. N Y Entomol Soc VI(4):201–230

Myskowiak JB, Doums C (2002) Effects of refrigeration on the biometry and development of *Protophormia terraenovae* (Robineau-Desvoidy) (Diptera: Calliphoridae) and its consequences in estimating post-mortem interval in forensic investigations. Forensic Sci Int 125:254–261

Oesterhelweg L, Kröber S, Rottmann K, Willhöft J, Braun C, Thies N, Püschel K, Silkenath J, Gehl A (2008) Cadaver dogs – a study on detection of contaminated carpet squares. Forensic Sci Int 174:35–39

Payne JA (1965) A carrion study of the baby pig Sus scrofa Linnaeus. Ecology 46(5):591–602

Payne JA, King EW, Beinhart G (1968) Arthropod succession and decomposition of buried pigs. Nature 219:1180–1181

Perrot P, Chartier L, Carlier J (2007) The role of ground penetrating radar in forensic sciences. Proceedings of the International Crime Science Conference, London

Pujol-Luz JR, da Costa A, Francez P, Uruahy-Rodrigues A, Constantino R (2008) The Black Soldier-fly, *Hermetia illucens* (Diptera: Stratiomyidae) used to estimate the post mortem interval in a case in Amapà State, Brazil. J Forensic Sci 53(2):476–478

Rodriguez WC III (1997) Decomposition of buried and submerged bodies, pp 459–467. In: Haglund WD, Sorg MH (ed) Forensic taphonomy. The postmortem fate of human remains. CRC, Boca Raton, FL, p 636

Rodriguez WCIII, Bass WM (1985) Decomposition of buried bodies and methods that may aid their location. J Forensic Sci 30(3):836–852

Shubeck PP (1968) Orientation of carrion beetles to carrion: random or not random. J N Y Entomol Soc 76:253–265

Skinner M, Sterenberg J (2005) Turf wars: authority and responsibility for the investigation of mass graves. Forensic Sci Int 151:221–232

Smith KGV (1986) A manual of forensic entomology. Trustees of the British Museum, Natural History and Cornell University Press, London, p 205

Thieren M, Guitteau R (2000) Disasters preparedness and mitigation, issue no. 80. Pan American Health Organization (PAHO), p 8

Tomberlin J, Sheppard C, Joyce JA (2005) Black Soldier fly (Diptera: Stratiomyidae) colonization of pig carrion in South Georgia. J Forensic Sci 50(1):152–153

Turchetto M, Lafisca A, Borini M, Vanin S (2008) A study of the entomofauna on some 2nd world war skeletons from a "foiba" in NW-Italy. Proceedings of the sixth meeting of the European Association for Forensic Entomology, Kolymbari (Crete)

Turner B, Wilshire P (1999) Experimental validation of forensic evidence: a study of the decomposition of buried pigs in a heavy clay soil. Forensic Sci Int 101:113–122

Vanin S, Vernier E (2005) Segnalazione di Penicillida dufourii (Diptera, Nycteribiidae) ectoparassita di chirotteri vespertilionidi nella "grotta della guerra" (Ialia, Veneto). Lav Soc Ven Sc Nat 30: pp 9–11

VanLaerhoven SL, Anderson G (1999) Insect succession on buried carrion in two biogeoclimatic zones of British Columbia. J Forensic Sci 44(1):32–43

Voillot J-F, Touron P, Servettaz J, Hebrard J (2006) The national serious crime scene unit of the national gendarmerie. Proceedings of the meeting of the European Association of Forensic Sciences. Helsinki

Wilson AS, Janaway RD, Holland AD, Dodson HI, Baran E, Pollard AM, Desmond JD (2007) Modelling the buried body environment in upland climes using three contrasting field sites. Forensic Sci Int 169:6–18

Wyss C, Chérix D (2006) Traité d'entomologie légale. Les insectes sur la scène de crime. Presses polytechniques et universitaires romandes, Lausanne, p 317

Chapter 14
Forensic Implications of Myiasis

M. Lee Goff, Carlo P. Campobasso, and Mirella Gherardi

14.1 Introduction

Myiasis has been defined variously by numerous different authors over the years. The term itself was first coined by Hope in his 1840 paper entitled "On insects and their larvae occasionally found in the human body" although there were some earlier accounts by other authors. Subsequently there were additional treatments but not with equal restrictions. Possibly the most enduring and practical definition is that of Zumpt in his 1965 work entitled "Myiasis in Man and Animals in the Old World." In this work myiasis is defined as: "the infestation of live human and vertebrate animals with dipterous larvae, which, at least for a certain period feed on the host's dead or living tissue, liquid body-substances, or ingested food." A similar definition was followed by Guimarães and Papavero in their 1999 work on myiasis in the Neotropics. Based on the system of classifying parasites developed by Patton (1922), Zumpt divides these myiasis-causing larvae into Obligatory Parasites and Facultative Parasites. Diptera larvae within the obligatory group develop in the living tissues of the host and this is, in fact, a necessary part of their life cycle. By contrast, the facultative group includes species that are normally free-living, feeding on decaying material, such as animal carcasses, fecal material, and even decaying vegetable materials. Under some While some species included in this category, such as *Phaenicia sericata*, may frequently act as parasites for all or part of their larval development, more commonly, species in this classification are associated with dead tissues present in a wound and do not actually feed on living tissues. Another situation, termed "Pseudomyiasis or Accidental Myiasis," occurs when Diptera larvae are accidentally ingested with

M. L. Goff(✉)
Forensic Sciences Program, Chaminade University, Hawaii, USA

C. P. Campobasso
Department of Health Sciences, University of Molise, Italy

M. Gherardi
Forensic Pathologist, Private Practice, Milano, Italy

J. Amendt et al. (eds.), *Current Concepts in Forensic Entomology*,
DOI 10.1007/978-1-4020-9684-6_14, © Springer Science+Business Media B.V. 2010

food materials and pass through the digestive tract. Keep in mind that this passage is most often passive and may result in the death of the larva. Their presence in the gut of the animal will often trigger various gastric problems, as noted by (e.g. Kenny, 1945). These infestations are not to be confused with the obligatory infestations of mammal digestive tracts by species in the subfamily Gasterophilinae (Colwell et al. 2006).

As a result of the medical/veterinary implications of myiasis, for clinical and practical reasons, there has been an additional subdivision into several different grouping, based primarily on location of the infestation. A comprehensive treatment of these is given in Hall and Farkas (2000). Additional, more recent, accounts include: sanguinivorous myiasis, cutaneous myiasis (Suite and Polson 2007), Auricular myiasis (Rohela et al. 2006), oral myiasis (Droma et al. 2007; Kamboj et al. 2007), traumatic myiasis (Franza et al. 2006), nasopharyngeal myiasis (Gelardi et al. 2009), umbilical myiasis (Duro et al. 2007), intestinal myiasis and urogenital myiasis (Makarov et al. 2006). In many respects these subdivisions do not have any relevance to the type of myiasis or the Diptera species involved, only to the location of the infestation, although the majority would be classified as traumatic myiasis.

From an evolutionary standpoint, myiasis presents an interesting situation. There have been two routes proposed for the origin of the lifestyle by Zumpt (1965): Saprophagous and Sanguinivorous. The saprophagous route appears to begin with a relatively unspecialized larva feeding on carrion or in a benign manner on necrotic tissues associated with wounds. This habit progresses to feeding on healthy tissues adjacent to the necrotic tissues and finally results in an obligatory feeding on healthy tissues with a detrimental effect. In like manner, there appear to be three steps in the evolution of myiasis via the sanguinivorus route. In this route, the larvae begin as blood-feeding predators of other arthropods feeding on fecal material or carrion. From this the habit shifts to feeding on the vertebrate as a blood feeder and finally to the obligatory condition. A similar pattern has been suggested for the parasitic mites in the families Laelaipdae and Macronyssidae by Radovsky (1967). In this pattern, the nest is viewed as a mechanism for bringing the various elements (predators, necrophages, parasites, and hosts) together. A similar situation has not been observed for myiasis in mammals. Most species involved in forensic concerns have probably originated through the saprophagous route.

In the forensic context, myiasis is most frequently associated with the facultative parasites in the Families Calliphoridae, Sarcophagidae and Muscidae. However, some use may be made of species demonstrating the obligatory state, such as *Dermatobia hominis* (Cutrebridae). If not fully appreciated, myiasis can be a significant point of confusion for the forensic entomologist, appearing to give an estimate of the postmortem interval far longer than the actual period of time since death. In other instances, particularly in cases involving the living, an understanding of myiasis may prove to be a significant factor in resolving the case. In this chapter, several different situations involving myiasis will be demonstrated using case studies. Although myiasis is involved in both human and veterinary aspects of forensic investigations, in this treatment, only human involvement is presented.

14.2 Case Studies

14.2.1 Case # 1. Increased Postmortem Interval/Period of Neglect

Specimens were submitted from the remains of a female, 58 years of age, who had been in home care on the island of Oahu, Hawaii, in an extended family situation (two daughters, one son-in-law, and five grandchildren). The decedent had a history of stroke with right side paralysis and only minimal contact with health care professionals. She was described as "difficult" and often remained in a wheelchair for extended periods of time, refusing to speak to family members for periods of several days. Specimens were collected at 0930 on 10 July and submitted to the laboratory for analysis. Submitted specimens consisted of 3rd instar larvae of *Phaenicia sericata* and egg masses of *Chrysomya megacephala*. The 3rd instar larvae of *P. sericata* indicated a mode developmental time of 50 h, based on conditions inside the house. Based on the idea that the species showing the greatest period of residence on the body is indicative of the minimum period of time since death, this indicated an onset of insect activity at approximately 0800 on 8 July. By contrast, the eggs of *Chrysomya megacephala* hatched at 1,400 on 10 July, indicating they had been deposited at approximately 0600 on 10 July. The family members stated that the decedent had last been seen alive at 0100 on 10 July. Examination of the body during autopsy revealed the presence of a large necrotic area on the lower back, penetrating into the abdominal cavity. Maggots of *P. sericata* were restricted to this area, while the egg masses of *C. megacephala* were recovered only from the nasal cavities. The prosecutor in this case felt that this supported the account of the family as to the possible time of death. Although the presence of the 3rd instar larvae of *P. sericata* indicated an instance of neglect and a general lack of care for the decedent, charges were not filed by the prosecutor. Lacking the data concerning distribution of the maggots with respect to the wound and the involvement of *P. sericata* in myiasis in Hawaii, the estimated postmortem interval would have been considerably longer than was actually the case. Too often, specimens are submitted by law enforcement agencies as a single collection from the remains, with no indication of location on the remains of infestations. If pre-existing infested wounds are not noted by those individuals making collections from the remains, the estimated period of time since death could be significantly in error.

14.2.2 Case # 2. Possible Sexual Assault

Specimens were submitted from the vaginal vault of an unidentified female found in Paoha Stream, Oahu, Hawaii, by hikers. At the time the individual was discovered, she was nude and had no recollection of events for the previous 8 days. Examination in the emergency room indicated some trauma to the genital area consistent with a sexual assault. The specimens submitted were identified as 3rd instar larvae of *Chrysomya megacephala*, measuring 15 mm in total length. Adjusting for normal body temperature, it was determined that it would require approximately 3.7 days to reach that stage of development. The degree of healing of the wound and the stage of development of the maggots indicated that the assault may have taken place 4 days prior to her discovery. Unfortunately, although the victim was subsequently identified, her memory of the 8-day period in question remains blank and no resolution was reached with respect to the possible sexual assault.

14.2.3 Case # 3. Period of Neglect

A 16-month-old female was discovered on the edge of Lake Wilson on the island of Oahu, Hawaii. The child was in a clear area surrounded by heavy vegetation (Goff et al. 1991). When found, she was suffering from dehydration, bruising and had numerous insect bites. Initially, the period of exposure was estimated as being 2 days. Given the state of dehydration, a pediatrician suggested that the period was longer and that the child would most probably have died within the next 24 h. The child was clothed in a sweatshirt, t-shirt, a pair of pants and disposable diapers. On the front of the pants, from the waistband to a point below the knees, there were eggs masses of a Calliphoridae (Fig. 14.1). When the clothing was removed, numerous 1st instar larvae (measuring 3–4 mm total length) and fewer 2nd instar larvae ,(measuring 5 mm total length) of *Chrysomya megacephala* were discovered in the diapers and pants (Fig. 14.2). Additional 1st and 2nd instar larvae of this same species were recovered from the vagina and rectum of the child and appeared to be feeding on tissues at those sites. Rearing data for this species from controlled studies conducted by Goff (unpublished data) at 26ºC and 28ºC indicated the most mature larvae would have required 39 and 36 h, respectively, to reach the stage of development represented. Using ADH calculations without a base temperature to adjust for normal body temperature, it was estimated that it would have required 23.5 h to reach the most mature stage of development for the specimens collected from inside the diapers.

In this instance, the time interval estimated by the insect activity did not account for the entire period of abandonment for the child. This was later found to have been approximately 36 h. The fly species, *C. megacaphala*, is a common early invader of decomposing remains in the Hawaiian Islands, typically arriving within minutes following death. Although this species has been implicated in myiasis in other parts of the world (Zumpt 1965; James 1947), it has not been previously implicated in cases of human myiasis in Hawaii. This species is strongly attracted to fecal material and this was probably the initial attraction for the flies in this case as well as Case 1 above. It should further be noted that, in Hawaii, this species has not been observed ovipositing on a moving animal. This would indicate that the child was relatively motionless at the time of oviposition and this would fit well with the more extended period of abandonment in this case. As *C. megacephala* also tends to seek out darker areas for oviposition, it might be assumed that, at the

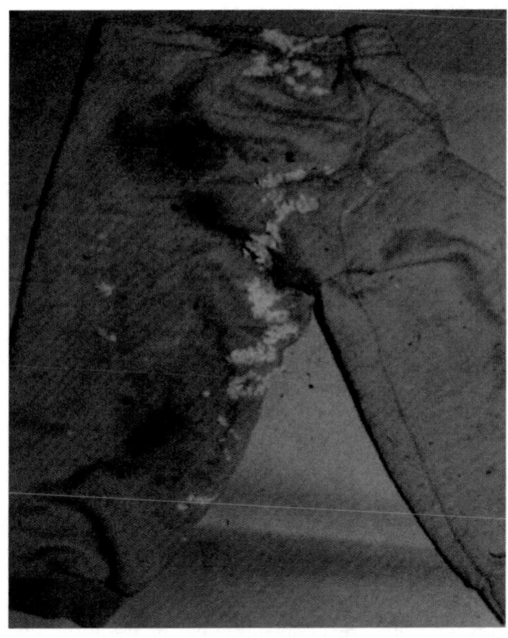

Fig. 14.1 Pants of child showing egg masses

Fig. 14.2 Second instar larva of *C. megacaphala*

time of oviposition, the child was lying face down. This would be indicated by the location of the egg masses on the front of the child's pants.

In this case, the entomological evidence in combination with the report from the pediatrician was used to establish the period of abandonment. The child's mother was charged with attempted murder and subsequently convicted at a trial held in Honolulu in 1990. Also introduced during the trial was evidence of previous abuse, including old bruising of the back and shoulders of the child.

A similar situation was reported by Benecke and Lessig (2001) for a case in Germany. In this instance, however, the child died and a post mortem interval of 6–8 days was established. Three species of flies were represented on the body: *Calliphora vomitoria, Fania canicularis*, and *Muscina stabulans*. While *C. vomitoria* was collected from the face of the child, *F. canicularis* and *M. stabulans* were present only in the genital area. Based on developmental stages present, the authors concluded that the child had not been cleaned for a period of approximately 14 days prior to discovery of the body.

14.2.4 Case # 4. Period of Neglect

One day at the end of June, in Milan (Northern Italy), at almost 06:30 a.m., a 50-year-old white female used an Emergency Call Service Number to request assistance for her 89-year-old mother (Gherardi and Lambiase, 2006). Arriving at the home, the physicians ascertained that the elderly woman was already dead. On the body, they observed some large defects of the skin colonized by Diptera larvae actively crawling and they interpreted them as postmortem changes. The defects were mainly located at the right iliac crest and in the perineal region where some faecal material was present and attached inside the diaper. Skin defects and faeces were heavily colonized by maggots. Based on this evidence, the physicians stated that the death could have occurred few days before body recovery. This estimated time since death was not consistent with the circumstantial data collected from the daughter.

The daughter of the decedent told to the investigators that the previous evening she had prepared the supper and helped feed her mother. A few minutes later, she realized that her mother was breathing in a strange way. This did not alarm her, since this had already happened many times in the past and it was previously naturally resolved. Therefore, she administered to her mother the hypnotic drug that she used to take for sleeping and, in her turn, she went to sleep. The following morning she woke up around 06:00 and found her mother in bed in her usual position. She called her several times, but she never replied. She attempted to waken her by shaking, but was not successful. Finally she decided to ask for medical assistance. With respect to the health of her mother prior to death, she said that the elderly woman had been confined in bed since 2001 when she felt accidentally at home breaking a leg. The long immobility in the bed served to increase pre-existing circulatory problems in the inferior limbs (deep venous thrombosis) causing an ingravescence foot deformation (equine foot with claw toes). Therefore, the elderly woman required daily assistance and was cared for by her daughter. She normally took only anti-inflammatory drugs along with some hypnotic drugs for sleeping.

Some neighbours of the same condominium where the two ladies lived were aware of the incapacitated status of the elderly woman and they stated that they had not seen her since her fall in 2001. Some of them also knew that the two ladies lived alone in their apartment resembling some kind of social isolation.

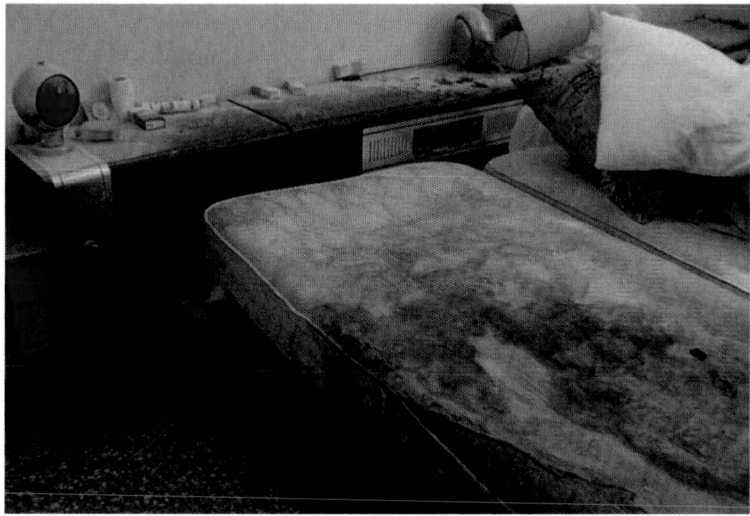

Fig. 14.3 Victim's mattress showing large blackened stains

The neighbours also told the investigators that, on several occasions in the past, they had noted bad odours coming from the apartment. They believed this was because the ladies failed to remove rubbish from their apartment. The physicians as well as the investigators confirmed that the apartment, located at the sixth floor of the condominium, was exposed to sunlight all day long, and was characterized by very poor hygienic conditions and extreme squalor. The mattress on which the body was found lying on was largely covered by brown-blackish stains (Fig. 14.3). This was probably a mixture of dried feces and urine, indicative of poor personal hygiene and suggesting a form of senile-neglect.

Based on the discrepancies between the circumstantial data collected and the physical evidence observed by the physicians, the State Attorney decided to ask for a more detailed cadaver inspection which occurred a few hours following body recovery, at approximately 12:00 the same day. The postmortem changes were consistent with a PMI of 6–12 h according with the report provided by the daughter. Body rectal temperature was 33.7°C, while the room temperature recorded before body removal, at 10:30, was 30°C. Lividity was not fixed and the rigor was in development, present only at small joints like the mandible and fingers. No other signs of putrefaction were observed, such as green or blackish discolouration areas of the skin and/or marbling of veins. At external examination, the body showed poor conditions of nutrition with dystrophic integument along the inferior limbs and muscular masses hardly hypotrophic. The body was dressed in nightclothing and a diaper dirtied with fecal material. The body showed numerous large bedsores at the dorsal region (decubitus lesions caused by supine position), partly covered by sticking plasters, soiled with a brownish mix of feces and urine (Fig. 14.4). Specifically, such decubitus lesions were located in the right

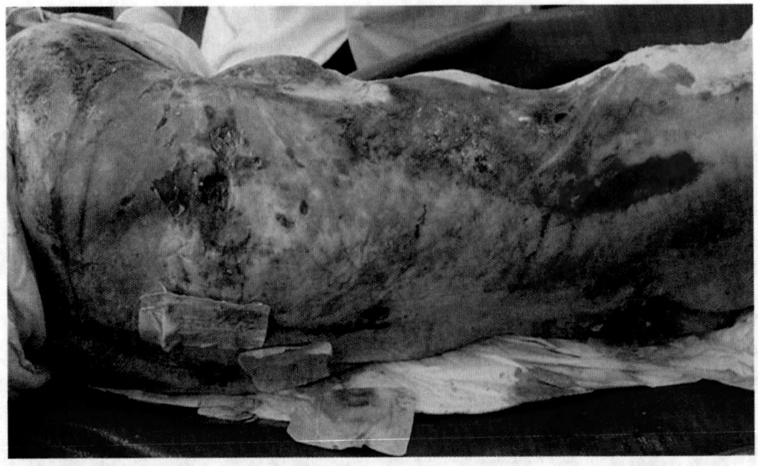

Fig. 14.4 Body of victim showing bedsores covered partially by sticking plasters

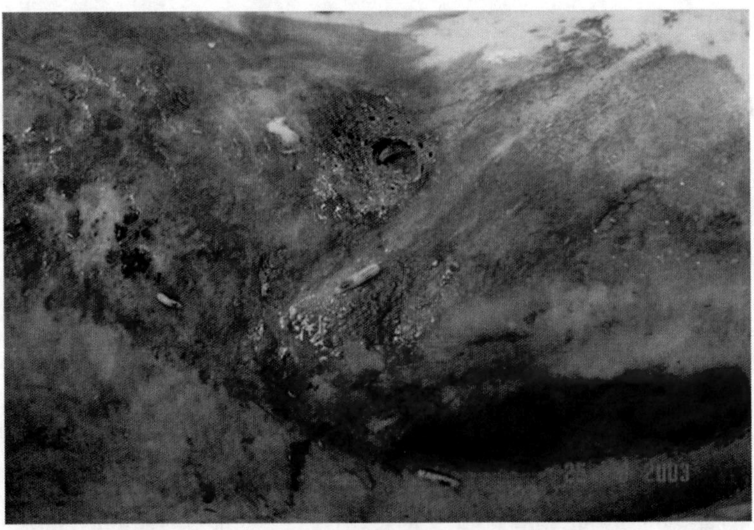

Fig. 14.5 Necrotic tissue at right iliac crest showing infestation by maggots

and left scapular regions, along the dorsal line and in sacral area and also at the right parieto-occipital region, into the sub-mammary folds and in left peri-areolar region. On the right side of the abdomen, close to the iliac crest, there was an additional bedsore defect, round in shape, approximately 1.5 cm in diameter, with irregular necrotic borders reddish in colour (Fig. 14.5). The surrounding integument showed multiple small holes, 0.3 cm in diameter, produced by feeding activity of Diptera larvae. Also the skin of the perineal region, after removal of the faecal material attached, showed large inflammatory reaction mixed with some necrotic areas colonized by Diptera larvae.

However, the stage of development of Diptera larvae collected from the skin defects was not in agreement with the above tanatological findings and the brief PMI. The oldest maggots collected were 3rd instar larvae 15 mm long. The maggots were identified as *Lucilia sericata* specimens no more than 4–5 days of age based on temperatures provided from the nearest meteorological station. During the 7 days prior the body recovery, the MAX Ts recorded in Milan ranged from 33.1°C to 36.6°C (mean MAX T: 35°C) while the MIN Ts ranged between 20.2°C and 25.3°C (mean MIN T: 23°C). These temperatures were used for calculations as these were the only temperature data available to investigators. Although it is well documented that there can be significant differences between temperatures at the scene and those from established weather stations, in this case, the judicial authority did not allow experts to record temperature data at the site in order to compensate for these differences.

In this case the maggots did not show the minimum PMI interval as assumed by the physicians. In fact, Diptera colonization did not occur after death but when the elderly lady was still alive, colonizing the necrotic tissues associated with the bedsores. The period of insect activity (4–5 days) in this case of myiasis showed how long the elderly woman was without any assistance. It was a case of neglect of the elderly by the caretaker, the victim's daughter, who was indicted only for abandonment of incapacitated people (art. 591 Italian Penal Code). The entomological specimens provided enough evidence of such a violation for at least a few days prior to the death. In fact, development of bedsores on the victim may have required less than the 4–5 days in the estimate based on development at ambient temperatures, but certainly not indicative of the 6–12 h estimated postmortem interval presented based on postmortem changes observed to the body. The daughter was found guilty because she did not take care of the multiple decubitus lesions found on the body and/or she never asked for help to any physicians. There was no indictment for homicide since the death was ruled as due to natural causes.

A similar situation was reported by Benecke et al. (2004) for a case from Germany. In this instance, an elderly woman was found dead in her apartment with her foot wrapped in a plastic bag. Based on postmortem changes, the PMI was estimated as being 2 days prior to discovery of the body. Inside the plastic bag were numerous larvae of *Lucilia sericata*, approximately 4 days of age. Judging from the deep tissue loss, it appeared that the maggots had been feeding on the woman for approximately 1 week while she was still alive. Some had already left the body to pupariate elsewhere.

14.2.5 Case # 5. Geographic Movement

A female 22 years of age returned from an archaeology field school held in Belize during the months of June through July 2000. She had been working at a relatively remote although secure site in the jungle for a period of 4 weeks. During this period

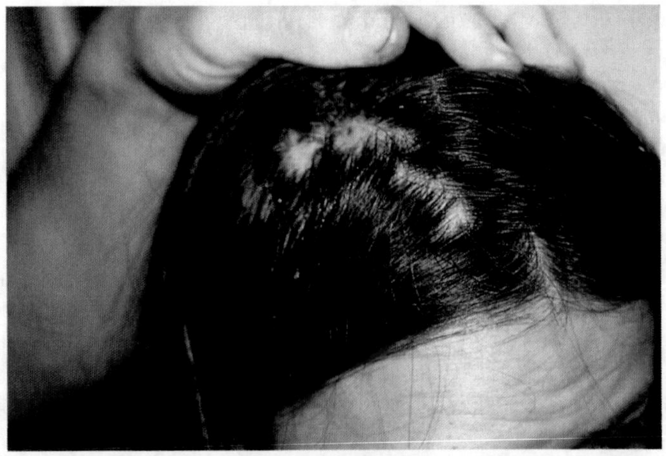

Fig. 14.6 Scalp of student showing locations of two *D. hominis* larvae

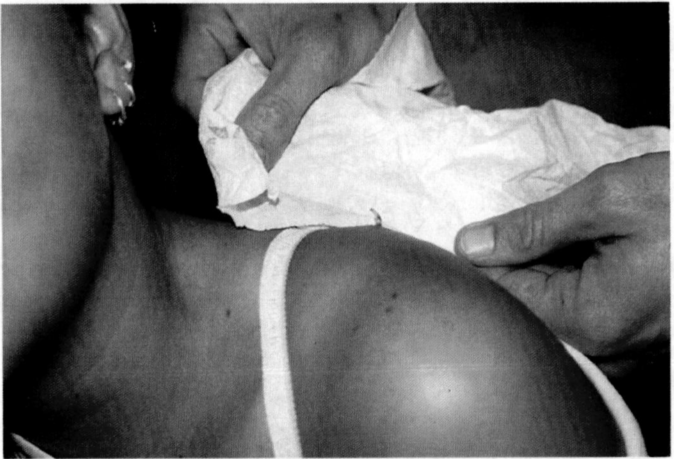

Fig. 14.7 Left shoulder of student showing partially expressed larva of *D. hominis*

of time, she was living is somewhat primitive conditions. Although the students were aware of the presence of *Dermatobia hominis* and the possibility of infestation, a number did acquire the bots. On her return to California, the student was aware of 2 larvae in her scalp (Fig. 14.6) and one on her left shoulder (Fig. 14.7). Initially she was not disturbed but as the larvae matured, she became more aware of their feeding activities and requested that they be removed. The larvae were expressed with no adverse effects to the student (Fig. 14.8).

The life cycle of *D. hominis* is quite unique and deserves some brief comment. The adult fly is somewhat similar in appearance to a bluebottle fly and is a forest-dwelling insect. The adults are non-feeding. The adult female does not oviposit directly on the host animal. Instead, another arthropod is captured and

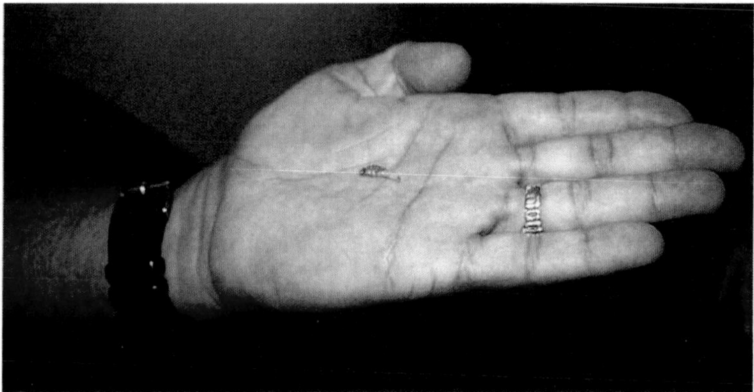

Fig. 14.8 *D. hominis* larva removed from left shoulder of student

eggs deposited on its external surface. Guimarães and Papavero (1999) list 55 species of arthropods in nine families as vectors for *D. hominis* eggs. While the majority are obligate blood-feeding species, several are not. When the egg vector comes in contact with the vertebrate host, the eggs hatch and the emergent larva falls onto the host skin. These larvae may penetrate the skin at the point of first contact or migrate to another part of the body. Once the skin is penetrated, the larva encysts and remains at that site. There is no migration through the host, characteristic of many myiasis causing species, although not those typically encountered in forensic situations. The site of penetration will take the form of a boil, with an opening to the outside maintained to enable the larva to breath. Development to the puparial stage varies, depending on the host and can range from 5 to 10 weeks (James 1947). Once the larva has matured, it leaves the site and drops to the soil to pupariate. The emergence most frequently takes place during the early morning hours. Guimarães and Papavero (1999) indicate that the adult life span is variable, ranging from 2 to 9 days, depending on temperature.

Guimarães and Papavero (1999) list 17 different countries in which *D. hominis* larvae have been recovered from individuals returning from trips to areas of Central and South America where *D. hominis* is present. While none of these reports has a specific forensic nature, given the current ability for travel and transport, *D. hominis* and other myisasis-causing Diptera have the potential to serve as evidence for geographic movement of individuals in criminal cases.

14.3 Discussion

The cases presented above serve to illustrate the potential for Diptera larvae to provide valuable information concerning periods of abandonment and/or neglect, particularly in cases involving the very young and the elderly. While the cases presented here have dealt with the living or individuals only recently deceased, as illustrated by the case documented by Benecke and Lessig (2001) in Germany, it is also possible for information also to be obtained concerning deceased individuals having a more extended postmortem interval. While the problems associated with neglect and abuse of children have been reported over the years, the current trend for an aging population has seen an increase in reports of cases of abuse/neglect in that growing population.

It is crucial that investigators are aware that the period of insect activity (PIA) does not always correspond to the PMI (Amendt et al. 2007). This is because the PIA could either be shorter than the PMI due to delay in access of insects to the body (e.g. initial burial or concealment of the body in a sealed room or a freezer), or longer, as in the cases involving myiasis (larvae feeding initially on necrotic tissues and continuing to feed following death). If the individual dies, as in 2.4, and collections submitted for analyses did not include the sites of infestations, the estimate would be made based on the most mature larvae as indicators of the minimum period of time since death. This would have been significantly longer than was actually the case. A similar situation would have existed in 2.3, had the child died.

One inherent problem in the use of myiasis causing larvae to establish periods of neglect/abandonment lies in the ability of those dealing with the problem to recognize the potential of such evidence. Even among the medical examiner community, there is still a tendency to want to remove maggots from a body as quickly as possible. This frequently results in loss of valuable data for use in the investigation. The situation is even more pronounced among social workers, pediatricians and emergency responders. All too often the first response to a wound infested by maggots is to remove and discard the specimens as quickly as possible and then clean and treat the wound. More recently, during 2007 in Hawaii, a child was seen suffering from dehydration and abuse, including necrotic wounds with maggots visible. The presence of maggots in the wounds was noted but no collections or further documentation was made. As a result, potentially valuable data concerning the period of abuse/neglect were lost.

The examination of remains or victims by an entomologist when insect evidence is present would serve to increase the utility of such evidence. This may not be possible in many instances for a variety of reasons, including legal ramifications in cases involving the living and simply geographic proximity in death investigations. Given that examination of the remains by an entomologist is most often not practical, first responders, emergency room personnel, and death investigators should be trained in proper techniques of recognition, collection and preservation of insect evidence.

References

Amendt J, Campobasso CP, Gaudry E, Reiter C, LeBlanc HN, Hall MJR (2007) Best practice in forensic entomology - standards and guidelines. Int J Legal Med 121:90–104

Benecke M, Lessig R (2001) Child neglect and forensic entomology. Forensic Sci Int 120:155–159

Benecke M, Josephi E, Zweihoff R (2004) Neglect of the elderly: forensic entomology cases and considerations. Forensic Sci Inter 146:195–199

Colwell DD, Hall MJR, Scholl PJ (eds) (2006) The Oestrid Flies: biology, host-parasite relationships, impact and management. CABI, Cambridge

Droma EB, Wilamowski A, Schnur H, Yarom N, Scheuer E, Schwartz E (2007) Oral myiasis: a case report and literature review. Oral Surg Oral Med Oral Pathol Oral Radiol Endod 103(1):92–96

Duro EA, Mariluis JC, Mulieri PR (2007) Umbilical myiasis in a human newborn. J Perinatol 27(4):250–251

Franza R, Leo L, Minerva T, Sanapo F (2006) Myiasis of the tracheostomy wound: case report. Acta Otorhinolaryngol Ital 26(4):222–222

Gherardi M, Lambiase S (2006) Miasi ed entomologia forense: segnalazione casistica. Riv It Med Leg, 3, pp. 617–628

Gelardi M, Fiorella ML, Tarasco E, Passalacqua G, Porcelli F (2009) Blowing a nose black and blue. Lancet 373(28)780

Goff ML, Charbonneau S, Sullivan W (1991) Presence of fecal material in diapers as a potential source of error in estimations of postmortem interval using arthropod developmental rates. J Forensic Sci 36:1603–1606

Guimaråes JH, Papavero N (1999) Myiasis in man and animals in the neotropical region: bibliographic database. Fundaçåo de Amparo a Paesquisa do Estado de Såo Paulo, Brazil

Hall MJR, Farkas R (2000) Traumatic myiasis of humans and animals. In: Papp L, Darvas B (eds) Contributions to a manual of Palearctic Diptera. Science Herald, Budapest

James MT (1947) The flies that cause myiasis in man. US Dept of Agriculture Misc Publ, Washington, DC, p 631

Kamboj M, Mahajan S, Boaz K (2007) Oral myiasis misinterpreted as salivary gland adenoma. J Clin Pathol 60(7):848

Kenny M (1945) Experimental intestinal myiasis in man. Proc Exp Biol Med 60:235–237

Makarov DV, Bagga H, Gonzalgo ML (2006) Genitourinary myiasis (maggot infestation). Urology 68(4):v889

Patton WS (1922) Notes on the myiasis-producing Diptera of man and animals. Bull Entomol Res 12:239–261

Radovsky FJ (1967) The Macronyssidae and Laelapidae (Acarina-Mesostigmata) parasitic on bats. Univ Calif Publ Ent 46

Rohela M, Jamaiah I, Amir L, Nissapatorn V (2006) A case of auricular myiasis in Malaysia. Southeast Asian J Trop Med Public Health 37(Suppl 3):91–94

Suite M, Polson K (2007) Cutaneous human myiasis due to Dermatobia hominis. West Indian Med J 56(5):466–468

Zumpt F (1965) Myiasis in man and animals in the old world: a textbook for physicians, veterinarians and zoologists. Butterworth, London

Chapter 15
Climate Change and Forensic Entomology

Margherita Turchetto† and Stefano Vanin

15.1 Introduction

We believe it is necessary to have a substantial section on climatology with the intent of providing a comprehensive picture that can be useful for the entomologist to explain the discovery of species or to observe communities that differ from those classically described. The most recent theories explaining the rise in global temperature, the consequences for flora and fauna, and the predictions concerning the climatic state in the forthcoming years should, in our view, become part of the knowledge of each forensic entomologist, who is often the first to pick up on and to indicate alien species or species that are changing their distributional areas, or phenology. Forensic entomologists, doctors, and veterinarians have more opportunities for chance encounters with species that are not included in the local checklists because the corpses and animal remains serve as substantial bait, attracting and concentrating in small areas rare species that previously had been scattered throughout the environment.

Many explanations of discoveries, along with the apparent incoherence in the series of surges proposed in the classic table of the colonization of corpses, can be justified in the light of environmental variations induced by actual climatic changes, which are occurring with incredible velocity.

15.2 Causes and Evidence of Global Warming

Before presenting the main evidence for global warming and discussing its probable effects on our specific arguments – behavior and geographic distribution changes of arthropods of forensic interest under climate change – we introduce some of the

M. Turchetto and S. Vanin (✉)
Department of Biology, University of Padova, Italy

†Prof. Margherita Turchetto deceased on July 25th 2009.

J. Amendt et al. (eds.), *Current Concepts in Forensic Entomology*,
DOI 10.1007/978-1-4020-9684-6_15, © Springer Science+Business Media B.V. 2010

most quoted theories on this subject. Global warming is undoubtedly a process that is truly underway; and the theory of the "greenhouse effect," based on knowledge and study of a single big phenomenon, can be a valid approach to study and can aid our predictions of trends in global climate changes.

A planet's climate is decided by its mass, its distance from the sun, and the composition of its atmosphere. The phenomena that mainly affect the changes of the Earth's climate can be classified under the following categories:

- Astronomic effects
- Geological variations
- Human contribution

The effect of these causes, leading to a global warming, can be stated as:

- Increased atmospheric gas content (CO_2, methane, and N_2O)
- Warming of the troposphere
- Increased global average surface heart temperature of about 0.6°C (~1°F) during the twentieth century
- Increased ocean water temperature, especially in tropical areas (+1–2°C) over the past 100 years but also in closed seas and freshwater basins
- Rise in ocean and sea levels and the drift of streams
- Increased precipitation over tropical and Northern Hemisphere areas
- Significant weather events, such as the El Niño/southern oscillation (ENSO), monsoons, hurricanes, and typhoons

The best sources of scientific information on global warming are the Intergovernmental Panel on Climate Change (IPCC) reports. In 1988, the IPCC was set up by the United Nations Environmental Program and the World Meteorological Organization to examine the most current scientific knowledge on climate change. More than 2,500 scientists from about 100 countries have been involved in the program. Scientists, economists, and risk experts along with other interest groups have reviewed all the previous data and scientific information from the last few years to assess what is actually known about the climate: if global warming is already happening; how and why it will predominantly affect the world; what it will mean for people and the environment; and exactly what can be done about it.

The last IPCC reports (*Climate Change 2001: The Third Assessment Report* and *Climate Change 2007: The Fourth Assessment Report*) are the most comprehensive and revised evaluations of global climate change. IPCC reports testify that global warming is occurring on the basis of current climatic, physical, and ecological changes, some of the most important of which are as follows:

- Melting of snow cover, sea ice extent, polar ice caps, and mountain glacier retreat, along with thawing of permafrost
- Extended dry climate and desertification
- Shift of plants and animals toward poles or higher altitudes
- Decline of some plant and animal species; reduction of species' geographic areas, sometimes to niche species: loss of biodiversity

- Migration of nonnative species to new biota and altered composition of previous ecological communities, in both marine and terrestrial environments
- Modification of primary and secondary marine production, affecting planktonic organisms, fish larvae, and large fish and, consequently, the migration of pelagic fish and marine mammals
- Changes in the timing of holoartic species' spring activities, such as earlier shooting and flowering of plants, earlier arrival of migrant animals, earlier breeding and egg-laying of birds, and earlier emergence of insects

Even though these listed effects confirm that changes in the climate system are trending upward, leading to an increase in the average world temperature, it is impossible to foresee how they can interact and modify both weather and climate together in single regions and in such a short time frame.

The alteration of the Earth's physical conditions along with the other abiotic factors that have a variety of effects on local climates may affect environmental assessments and organisms' lives, starting from the vegetal composition and distribution and extending to animal biodiversity in the sea and on land. Consequently, human life and health could also be affected by the altered distribution of organisms, mainly in terms of food availability and vector-borne diseases.

Fossil evidence clearly demonstrates that Earth's climate has repeatedly shifted abruptly in the past and that a dramatic climate change that happened within a decade can have persisted for centuries. The climate shifts do not necessary have global effects: even as the Earth as a whole continues to warm gradually, large regions may experience a precipitous and dramatic shift into a colder climate. The dynamics of the climate has more than one mode of operating, and each mode produces different climate patterns. Earth's history shows that a climate system has a sensitive threshold. Pushed past the threshold, the system can jump quickly from one stable operating mode to a complete different one, as stated by the U.S. National Academy of Sciences report (US National Academy of Sciences 2002).

15.2.1 Great Buffer Systems of the Earth: Atmospheric "Greenhouse Effect", Oceans and Ocean Currents

The Earth has a natural temperature control system: the atmosphere and the oceans constitute the two most important components, regulating the global climate jointly.

The Earth's atmosphere is a mixture of gases, vapors, and aerosols (78% nitrogen, 21% oxygen, 1% other gases; carbon dioxide accounts for just 0.03% to 0.04%, and water vapor varies in amount from 0 to 2%). These atmospheric gases are critical to, and are known as, "greenhouse gases." Carbon dioxide and some other minor gases present in the atmosphere absorb some of the thermal radiation leaving the earth's surface and then emit radiation from much higher and colder levels out to space. These radioactive gases are known as "greenhouse gases" because they act as a partial blanket for the thermal radiation from the earth's surface and thus enable it to be substantially warmer than it would otherwise be – analogous to the effect of a greenhouse. This blanketing is known as the natural greenhouse effect. Without these greenhouse gases, the Earth's average temperature would be roughly 0°C. The "greenhouse" effect is the natural phenomenon that keeps life on the Earth. It is due to the sun's energy, which arrives at the atmosphere and Earth's surface, is partly absorbed warming the earth, and then is partly radiated back through various phenomena. In this delicate balance, allowing the organisms' life, the gas composition of the atmosphere plays fundamental roles as both an insulating layer and a heat-trapping layer (Figs. 15.1 and 15.2). Current terrestrial life is allowed at an average temperature of 15°C (59°F), which depends on the gas composition of the troposphere. The troposphere is the thin layer (10,000 m) of atmosphere that is directly at contact with the Earth's surface and in which high-altitude habitants and flying organisms can survive.

The oceans, the planet's largest reservoir of water, play a fundamental role in the distribution and availability of water throughout the earth. Evaporation from oceans transfers huge amounts of water vapor to the atmosphere, where it travels aloft until it cools, condenses, and precipitates in the form of rain or snow. These weather phenomena can also occur far from the evaporation sites: the equatorial sun warms the oceanic surface and enhances evaporation in tropical areas; winds then blow and drive water vapor in the form of clouds, creating the ocean/atmosphere dynamics that regulate Earth's climate. The oceanic evaporation leaves the tropical water saltier: salt concentration, together water temperature, creates the big currents, collectively known as the "Ocean Conveyor." The currents distribute vast quantities of heat around the planet, playing a fundamental role in governing climate: this oceanic pump is the

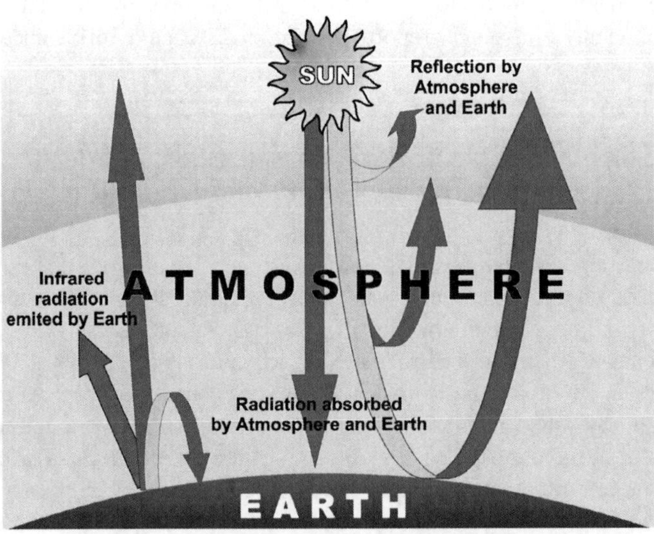

Fig. 15.1 Energy balance between the sun and earth's surface throughout the atmosphere layer

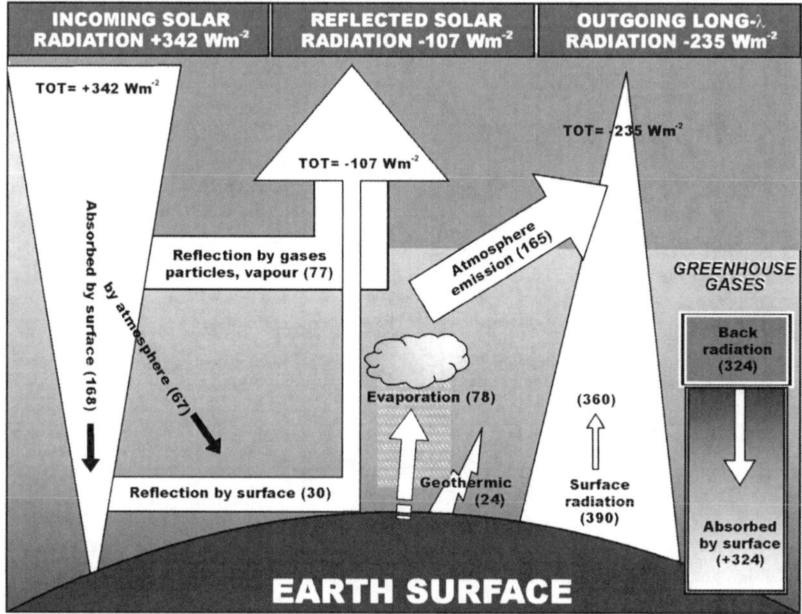

Fig. 15.2 Balance of radiations (W m^{-2}) incoming to or outgoing from the whole earth over 24 h. Greenhouse gases are mixed throughout the atmosphere (depicted here as a light layer) (the source of the major part of records are from the IPCC (Intergovernmental Panel on Climate Change) 1st (1996a, b) and 3rd (2001) reports)

most important mechanism for reducing equator-to-pole temperature differences. The Gulf Stream, for example, moderates climate in the North Atlantic region, increasing the northward transport of warmer waters by about 50%. Oceanic waters then release their heat to the atmosphere at colder northern latitudes, especially during the winter, so the ocean and ocean/atmosphere gradients increase. The Conveyor warms North Atlantic regions by as much as 5°C (Gagosian 2003). Beginning cooler, the waters are also denser than warm waters and sink to the abyss, forming the cool Labrador Stream, which flows into the South Atlantic. This enormous ring, circulating throughout the oceans (Fig. 15.3), is the major factor that affects global climate patterns.

Changes in the oceans' circulation or water properties can disrupt this hydrological cycle on a global scale, causing flooding and long-term droughts in various regions. The El Niño phenomenon demonstrated a few years ago how an oceanic change can dramatically induce precipitation falls or hurricanes in different parts of the planet.

The records of past climate collected from a variety of sources, such as deep sea sediments and polar ice cores, show that the Conveyor has slowed and shut down several times in the past. This shutdown reduced heat delivery to the North Atlantic and caused substantial cooling throughout the region.

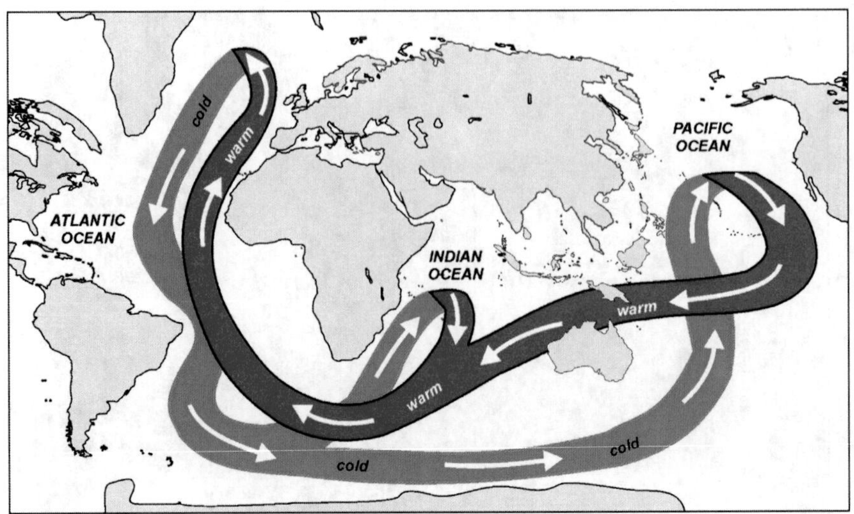

Fig. 15.3 Global conveyor belt thermohaline circulation. The warm water from the equatorial ocean surface is carried toward northerrn Europe. In the North Atlantic, water releases heat into the atmosphere, becomes colder, and begins to sink

15.2.2 Evidence of Global Warming During the Last Hundred Years

15.2.2.1 Climate Changes

Periods of earth warming and cooling occur in cycles. This is well understood, as is the fact that small-scale cycles last 20,000 years. Earth's climate was in a cool period from AD 1350 to about AD 1860, dubbed the "Little Ice Age" (Fig. 15.4). The decline in global temperature was only 0.5°C on average, but its effects were more pronounced at higher latitudes, dramatically characterized by drier climate, harsh winters, shorter growing seasons, and crop failures. Today, the global temperature has warmed to the levels of the former "Medieval Warm Period," which existed approximately AD 1000–1350 (Fig. 15.4). Over the last 400,000 years, the Earth's climate has been unstable, with significant temperature changes, ranging from a warm climate to an ice age as rapidly as within a few decades. These rapid changes suggest that the climate may be sensitive to internal or external climate forcing and feedbacks. As can be seen from the dashed-line curve in Fig. 15.5, temperatures have been less variable during the last 100,000 years. Based on the incomplete evidence available, it is unlikely that global mean temperatures have varied by more than 1°C in a single century during this period. The information presented in this graph indicates a strong correlation between CO_2 content in the atmosphere and temperature. Here is a possible scenario: anthropogenic emissions of greenhouse gases (GHGs) could bring the climate to a state where it reverts to

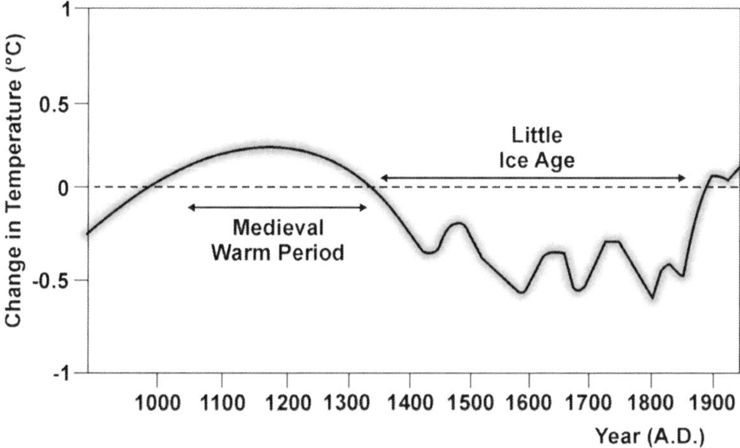

Fig. 15.4 Variations in surface air temperature during the past 1,000 years (from the IPCC third report 2001). The recent "Little Ice Age," from about 1,350 to 1,860, shows modest global cooling (0.5°C average) but heavy effects at higher latitudes

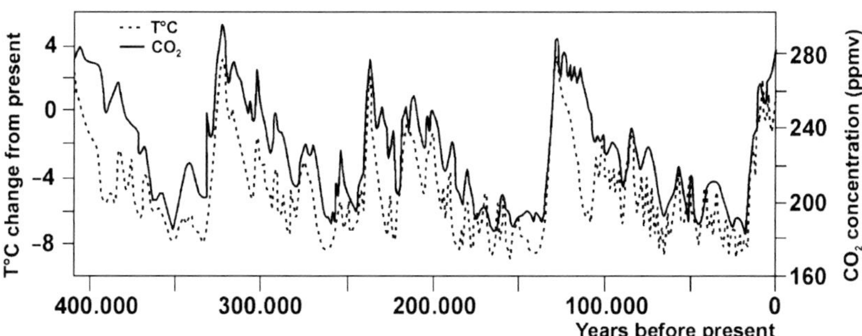

Fig. 15.5 Temperature and CO_2 levels in the atmosphere over the past 400,000 years. The geological records clearly show a correlation between CO_2 (from air bubbles from the Antarctic ice core) and temperature trough glacial/interglacial cycles

the highly unstable climate of the pre-Ice Age period. Rather than a linear evolution, the climate follows a nonlinear path with sudden and dramatic effects when GHG levels reach an as yet unknown trigger point.

15.2.2.2 Ocean Warming: Effects on Coastal Environments

Over the last 100 years, the global sea level has risen by about 10 to 25 cm. Sea level change is difficult to measure, and relative sea level changes have been calculated mainly from tide-gauge data. In the conventional tide-gauge system, the sea level is measured relative to a land-based tide-gauge benchmark. The major problem is that the land experiences vertical movements (e.g., from isostatic effects, tectonic movements and sedimentation), and these movements become incorporated into the measurements. However, improved methods of filtering out the effects of long-term vertical land movements, as well as greater reliance on the longest tide-gauge records for estimating trends, have provided greater confidence that the volume of ocean water has indeed been increasing, causing the sea level to rise within the given range.

It is likely that much of the rise in sea level has been related to the concurrent rise in global temperature over the last 100 years. Within this time scale, the warming and consequent thermal expansion of the oceans may account for about 2 to 7 cm of the observed sea level rise, and the observed retreat of glaciers and ice caps may account for about 2 to 5 cm. Other factors are more difficult to quantify. The rate of observed rise of sea level suggests that there has been a net positive contribution from the huge ice sheets of Greenland and Antarctica, but observations of these ice sheets do not yet allow meaningful quantitative estimates of their separate contributions. The ice sheets remain a major source of uncertainty when accounting for past changes in sea level due to insufficient data about these ice sheets over the last 100 years.

15.2.2.3 Polar Ice Melting

Melting glaciers and land-based ice sheets also contribute to rising sea levels, lowering sea temperature and salinity and threatening low-lying areas around the globe with beach erosion, coastal flooding, and contamination of freshwater supplies. Each year, a volume of water equivalent to the upper 7 mm of all of our planet's oceans falls as snow on the Antarctic ice sheet, and a roughly equivalent amount of ice slips into the sea via glaciers. However, the ice sheet is seldom in a state of balance. From one Ice Age to the next, and from one season to the next, the amount of snow falling differs from the amount of ice leaving, causing parts of Antarctica's frozen reservoir to be alternately drained and replenished. When the ice sheet grows, the global sea levels fall; when it shrinks, sea levels rise. Sea-ice draft is the thickness of the part of the ice submerged under the sea. Sea-ice data acquired in the Arctic pole by submarine researchers have shown a similar melting trend between the periods 1958–1976 and 1993–1997: the mean ice draft at the end of the melt season has decreased by about 1.3 m in most of the deep water of the Arctic Ocean, from 3.1 m during 1958–1976 to 1.8 m during the 1990s. The ice draft is therefore more than a meter thinner than two to four decades earlier (Fig. 15.6). The main draft has decreased more than 3 m under 2 m, and its volume has decreased by some 40%. Given that Antarctica is exposed to the atmosphere and oceans, sudden changes in the Earth's climate can also alter this balance. Because Antarctica is a largely frozen environment, however, its ice was expected

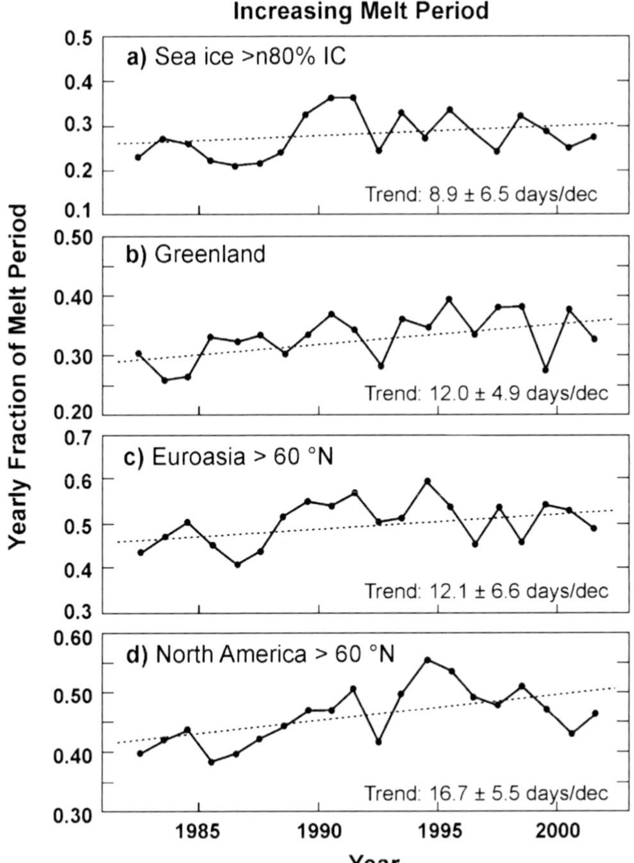

Fig. 15.6 Increasing melt period since the end of the "Little Ice Period" to 2004 in the northern Atlantic, Greenland, Paleartic, and Neartic Regions (data are from the United Nations Environment Programme (UNEP))

to respond only slowly to the recent increase in global temperatures, which have climbed 10 times faster during the twentieth century than at any other time over the last 1,000 years. Now, however, satellite observations tell a different story. Coastal glaciers are retreating and thinning at various places around the entire continent, and it seems that the vast quantity of heat delivered by the Earth's gradually warming oceans may be to blame. The first cracks in Antarctica's icy armor appeared during the mid-1990s. Massive ice shelves floating in bays around the Antarctic Peninsula, a narrow mountain chain that extends northward toward South America, began to disintegrate.

15.2.2.4 Altitudinal Changes

One of the best places to look for changes in plant and animal life that may be caused by a climate change is in the mountains. As the globe warms up, mountaintops get warmer too. Trees start growing at higher altitudes than before, and the tree line shifts upward. The alpine timberline zone is highly sensitive to climate variability (Fig. 15.7). The rise of temperatures during the vegetation period over long periods induces also a rise of the tree line with higher forest stand density. Temperature reductions, however, lead to less dense forests and a drop of the timberline. The Alps records from the last 80–100 years show that plants have been migrating upward at a rate of about 4 m every decade. Researchers of the Institute for Geography at the University of Innsbruck have conducted dendrochronology analyses in the Tyrolean Central Alps where the timberline region is dominated by Swiss stone pine, demonstrating that the distribution area of adult trees and regeneration has moved upward during the last 150 years. In Nevada (USA) the Engleman spruce had moved its habitat upslope 650 feet in just 9 years. Organisms that live on mountains may face a grim future because mountaintops are in many ways like islands. They are isolated clearings that poke up above the tree line (Nicolussi and Patzelt 2006).

Loss of glacier volume has been more or less continuous since the nineteenth century, but it is not a simple adjustment to the end of an "anomalous" Little Ice Age. The data collected by Dyurgerov and Meier (2000) during the 1961–1997 period show trends that are highly variable with time as well as within regions; trends in the Arctic are consistent with global averages but are quantitatively smaller. The averaged annual volume loss is 147 mm year^{-1} in water equivalent, totalling 3.7 × 103 km^3 over 37 years. The time series shows a shift during the mid-1970s, followed by more rapid loss of ice volume and further acceleration during the last decade; this is consistent with climatologic data. Perhaps most significant is an increase in annual accumulation along with a rise in melting; these changes produce a marked increase in the annual turnover or amplitude. The warming of air temperature suggested by the temperature sensitivities of glaciers in cold regions is somewhat greater than the global average temperature rise derived largely from low-altitude gauges, and the warming is accelerating (Dyurgerov and Meier 2000).

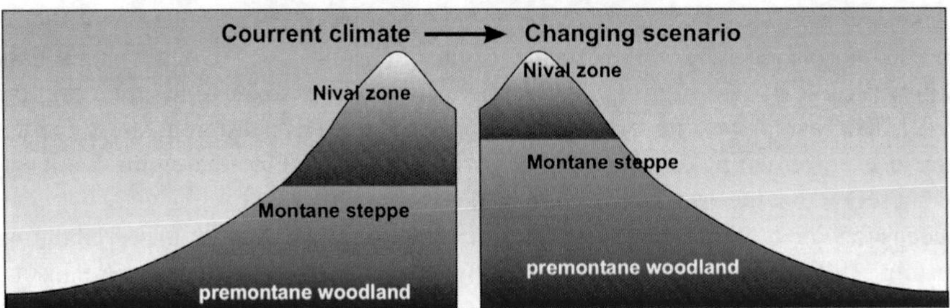

Fig. 15.7 Impact of climate on mountain vegetation belts. (*Left*) current vegetation. (*Right*) simulated vegetation belts under warming and drying climate

15.3 Ecological Responses to Climate Changes

15.3.1 Natural Shift and Strategies of Insects in Rising Temperatures

The effects of global warming on the phenology and distribution of plants and animals were recently well summarized and analyzed by Parmesan (2006). The available studies have demonstrated that climate change has had a strong impact on marine, freshwater, and terrestrial groups in various regions of the world (North and South Poles, temperate and tropical regions, and oceans) and with wide-ranging consequences (i.e., shifts in the distribution of species, range contraction, extinction, community reorganization, and alteration of predator–prey and plant–insect interactions).

Polar and mountaintop species have experienced severe range contractions and are the first groups in which entire species have become extinct due to recent changes in climate. In southern France, populations of the cool-adapted Apollo butterfly (*Parnassius apollo*) have become extinct over the past 40 years in mountainous areas less than 850 m high but have remained healthy in higher regions in the station at elevations greater than 900 m asl. An evident relation between the shift of altitudinal species and the rise in temperatures was documented in Spain, where the lower altitudinal limits of 16 species of butterfly have risen an average of 212 m in 30 years, concurrent with a 1.3°C rise in mean annual temperatures. Plant, bird, and marine life in Antarctica have exhibited profound responses to climate change. The extensive decline in sea-ice extent appears to have stimulated a trophic cascade effect in biological systems. Declines in sea-ice have apparently reduced the abundance of ice algae, which in turn has led to declines in krill, the primary food resource for many fish, seabirds, and marine mammals, with direct consequences in their distribution and population. The extreme effects of global warming on marine communities is well documented at all ecological levels; and climatic variables are in fact responsible for the distribution and dynamics of marine plankton and fish. Phytoplankton and zooplankton communities have experienced strong shifts in tune with regional oceanic climate regime shifts, as well as expected poleward range shifts and changes in the timing of peak biomass. Particularly the sea-ice-dependent Adélie and emperor penguins (*Pygoscelis adeliae* and *Aptenodytes forsteri*, respectively), have almost disappeared from their northernmost sites around Antarctica since 1970.

Changes in climate have also had a dramatic impact on the distribution of terrestrial species in temperate regions. Evident shifts in temperate species of the Northern Hemisphere were well documented in Great Britain, where various taxa were examined. In this area, northward shifts of 18.9 km over a 20-year period were observed in 12 bird species (Thomas and Lennon 1999). The same results were also obtained by studying the distribution of various insect groups (Lepidoptera, Odonata, Orthoptera). Several species of butterfly that were mobile and habitat generalists were found to have expanded north or, in smaller portions, had contracted southward (Parmesan 2006). The distribution change closely matched changes in population densities. It was demonstrated that the impact of global climate change also involved the distribution of species living in tropical areas, resulting in movements or outbreaks of tropical species into more-temperate areas. This was observed for the rufous hummingbird, which has undergone an evident shift in its winter range during the last 30 years from Mexico to the southern United States. Tropical insects have also expanded their distribution e.g., Odonata in the southern United States and the black soldier fly (Diptera, Stratyiomidae) all over the word.

It seems that in the environment will soon prevail on grasslands and herbal ecosystems than forestall biomes, with an enlargement of desert in tropical zones. These new conditions favor herbivorous insects not linked to specific plant–host relationships or induce one host plant's insects to feed on several more widespread plants, as demonstrated for some British butterflies (Hill et al. 2002; Thomas et al. 2001). Highly mobile species, whose breeding habitat is widespread, are likely to track climate changes, when sedentary species have few opportunities to shift their ranges. The speed of expanding species is increased by the emergence of dispersive phenotypes exhibiting increased flight ability. This was demonstrated by Thomas and coworkers (2001) for two species of bush crickets from Britain and by Nimela and Spence (1991) in the ground beetle (Carabidae) invading Canada. In both cases these are populations with wing length polymorphism, and in the marginal areas the long-wing specimens were prevalent (Zera and Denno 1997). In other cases, reported by Hill et al. (1999a, b) for butterflies, thorax size and thoracic flight muscles increased because of a changed during of larval stages. Other similar modification of the behavior and the phenotype led us to think that in many cases evolutionary changes could be a more rapid, convenient response to the rapid climate change than migration to nearby habitats.

15.3.2 Prey–Predator Relationships

The effects of global climate change have also affected ecosystem interactions, altering prey–predator and parasite–host relationships. In temperate regions, the photoperiod is the cue for seasonal synchrony: the interaction between temperature and photoperiod determine the phenology. In The Netherlands, studies have demonstrated the shift in synchrony between plants and plant feeders, which has had serious consequences for functioning of the ecosystem due to the tight multitrophic interactions involved in the timing of reproduction and growth. Visser and Holleman (2001) highlighted that in an ecological system that evolved with strong selection on synchronization, such as with the winter moth's (*Operophtera brumata*) egg hatching and the oak's (*Quercus robur*) bud burst, there was poor synchrony in recent springs due to an increase in spring temperatures without a decrease in the incidences of freezing spells during the winter. The timing of egg hatching in relation to bud burst has had major health consequences for caterpillars. If the eggs hatch prior to bud burst the caterpillars starve, whereas if hatching occurs after bud burst the caterpillars have to eat less digestible leaves due to the increased tannin concentrations. This may lead to lower weight at pupation or to a longer larval period, resulting in a higher probability of being preyed upon or parasitized.

15.4 Globalization and Human Interferences on Animal Distribution

The presence of adventive (= nonindigenous) species in new areas and regions can be due not only to natural shift but also increased by human accidental or conscious transport. In these regions, the allochtonous species can proliferate, spread, and persist to the detriment of the environment and human safety. The invasive species threaten native biodiversity, disrupt ecological processes, and cause significant economic loss. In a world without borders, few if any zones remain sheltered from these immigrations. In fact, in every area reached by humans, new species have been imported from other regions – even in the polar Antarctica continent. In this continent and in the Sub-Antarctic islands a large number of alien microbes, fungi, plants, and animals arrived over approximately the last two centuries, coincident with human activity in the region. Most of the alien species are European in origin. Introduction routes have varied but are largely associated with movement of people and cargo in connection with industrial, national scientific program, and tourism operations (Frenot et al. 2005). Many detrimental nonindigenous organisms have been accidentally introduced through commercial foreign trade, as for example the Western Conifer Seed Bug *Leptoglossus occidentalis* recently introduced in Europe, probably with trees imported from North America (Vanin et al. 2005) and the Red Palm Weevil (*Rhynchophorus ferrugineus*) with palms imported from Asia. The frequency of the introductions through commercial pathways and the subsequent potential for establishment and spread has increased with expanding international trade and globalization. Methods of packing and shipping may play an important role in determining arrival and survival rates of nonindigenous species. For example, 19% of insects intercepted on maritime cargo entering New Zealand were alive upon reaching port destinations, whereas estimates of insect survivorship in air cargo entering Hawaii were much lower. Moreover, insects and other organisms introduced in refrigerated containers may survive at relatively high rates, increasing the risk associated with this pathway, because containers are kept at constant, nonlethal temperatures throughout transport (Work et al. 2005).

Not only expanding global trade is responsible of human alien organism introduction but also the use of nonindigenous parasites and predators for biological control of pests, the introduction of livestock and domestic animals (e.g., Australia), and the trade and abandonment of nonindigenous pets. For example, in southern Florida the proportions of adventive species of fishes, amphibians, reptiles, birds, and mammals range from 16% to more than 42%; the proportion of adventive plants is about 26%, and the proportion of insects is 8%. Almost all the vertebrates were introduced as captive pets but escaped or were released into the wild and established breeding populations; few arrived as immigrants. Almost all the plants were introduced as well, with a few arriving as contaminants of shipments of seeds or other cargo. In contrast, only 42 insect species (0.3%) were introduced for biological control of pests. The remaining 946 species arrived as undocumented immigrants, some of them as fly-ins but many as contaminants of cargo. Most of the major insect pests of agriculture, horticulture, manmade structures, and the environment arrived as contaminants of, and stowaways in, cargo, especially plants (Frank and McCoy 1994). Few species survive in the new climatic and ecological condition, and only a small fraction becomes naturalized, although some naturalized species become invasive. There are several potential reasons why some immigrant species prosper: some escape the constraints of their native predator or parasite; other are aided by human-caused disturbances that disrupt a native community as, for example, landscape fragmentation, cultivation, and husbandry. Moreover, the new climatic conditions allow the survival of several species introduced in northern regions.

15.4.1 Human Health and Global Warming

Earth climate warming and globalization have had a direct impact on human health. Infections involve pathogens, parasites, hosts, and the environment. Some pathogens are carried by vectors or require intermediate hosts to complete their life cycle. The climate can influence pathogens, vectors, host defenses, and habitat. A range of vector-borne diseases are geographically and temporarily limited by variations in climatic variables, such as temperature, humidity, and rainfall. The direct impact of the climate on infectious diseases can occur in three principal ways: effects on human behavior that can have a significant impact on disease transmission patterns; direct effects on disease pathogens; and effects on disease vectors.

For infectious diseases caused by pathogens that have part of their life in the environment or in an intermediate host or vector, climatic factors can have a direct impact on the developmental rate. Most viruses, bacteria, and parasites do not complete their life cycles if the temperature is below a certain threshold (e.g., 18°C for the malaria pathogen *Plasmodium falciparum*). Increases in ambient temperature above this threshold shorten the time needed for the development of the pathogen and thus increase reproduction rates, whereas temperatures in excess of the tolerance range of the pathogen may increase mortality rates.

The geographical distribution and population dynamics of insect vectors are closely related to patterns of the temperature, rainfall, and humidity. A rise in temperature accelerates the metabolic rate of insects and increases egg production, along with the frequency of blood feeds. Rainfall establishes the relatively wet conditions that create favorable insect habitats and in this way increases the geographical distribution and seasonal abundance of disease vectors.

Diseases carried by mosquito vectors are particularly sensitive to meteorological conditions. Excessive heat kills mosquitoes; but within their survivable range, warmer temperatures increase their reproduction rate and biting activity and the rate at which pathogens mature within them. Freezing kills *Aedes* eggs, larvae, and adults, whereas warm winters favor insect survival; thus, expanding tropical conditions can therefore enlarge the ranges and extend the season, with conditions allowing the transmission of pathogens. For some disease vectors, an evident northward shift and population increase related to global warming was demonstrated and simulated with statistical models for ticks belonging to *Ixodes* genus in Neartic (e.g., *Ixodes scapularis*) and Palaearctic (e.g., *Ixodes ricinus*) regions (Brownstein et al. 2005; Lindgren and Gustafson 2001). These species, vectors of several pathogens (e.g., Lyme borreliosis, human babesiosis, anaplasmosis) could become a public health challenge for states in the Northern Hemisphere, such as Canada and the Scandinavian countries.

15.5 Insects of Forensic Interest: Evidence of Latitudinal and Altitudinal Shift Because the Warming Climate Changes

As previously discussed, focusing on insects, alterations in climate can affect abundance (Bale 2002; McLaughlin et al. 2002), species' distribution (Hill et al. 2002; Hughes et al. 2003; Maistrello et al. 2006; Roy andAsher 2003; Thomas et al. 2001; Warren et al. 2001), physiology (Bradshaw and Holzapfel 2001; Musolin and Numata 2003, 2004), synchrony, and relationships with hosts (Kamata et al. 2002; Visser and Holleman 2001) ultimately affecting the structure of whole communities (Harrington et al. 1999; Stenseth et al. 2002; Thomas et al. 2004; Walther et al. 2002; Wuethrich 2000). Carrion-breeding insects are displaying this phenomena with relevant consequences for the postmortem interval (PMI) evaluation as well as for other forensic deductions (e.g., evaluation of the season of death, body displacement).

To evaluate the effects of global climate change on the forensic entomological evidence evaluations, we focused our attention on the available literature and on original data collected in northern Italy in recent years. Global climate change affects the fauna of forensic importance on different scales, both structural and temporal. In fact, both the composition of the carrion community and the arrival and departure times of carrion insects are influenced by the new climatic conditions, with direct consequences for entomological evidence evaluations. In temperate regions, new species, particularly those with originally more southern distributions, have been collected during recent years from both human bodies and animal carrion or traps with meat bait used for forensic entomology experimentation (Grassberger and Frank 2004; Grassberger et al. 2003; Turchetto and Vanin 2004a, b, c; Vanin et al. 2007). Similar reports have come from both America and Europe for different zoological groups, but they are particularly important for Diptera, from a forensic entomology point of view.

The composition of carrion-breeding fauna in a region can be modified by the arrival of new species transported by humans that survive and stabilize thanks to the new climatic conditions or by spontaneous colonization. This has implications not only for the decreased value of certain species in estimating the place of death (in that they are no longer exclusive to southern regions) but also for new temporal and structural relationships in the carrion-breeding composition in a region. This is the case in which the new arrivals are not only saprophagous but also predators of other necrophagous insects.

The calliphorid *Chrysomya albiceps* (Diptera: Calliphoridae) is generally described as a tropical and subtropical species that can be found in Africa, southern Europe, Arabia, India, and Central and South America (Hall and Smith 1993); in recent years, however, as demonstrated by Grassberger and colleagues (Grassberger and Frank 2004; Grassberger et al. 2003), it has shifted to central Europe from the southern regions (Fig. 15.8) and in Italy from southern to northern regions (Vanin, et al. 2009). The species was recorded just recently throughout the years 1996 and 1997 by Arnaldos et al. (2001) close to the city of Murcia and by Martìnez-Sánchez and coworkers (2000) in the Salamanca province. It was collected by Introna et al. (1998) in southern Italy from a partly skeletonized, decaying body during September in a suburban area of the city of Bari. Since 1999, however, it was reported also from Austria, near Vienna; from Romania near Bucharest, where *C. albiceps* constituted 15.6% of all trapped Calliphoridae and 11.3% of all trapped flies; from Croatia (Island of Krk); and in 1995 it was captured on a dead body in

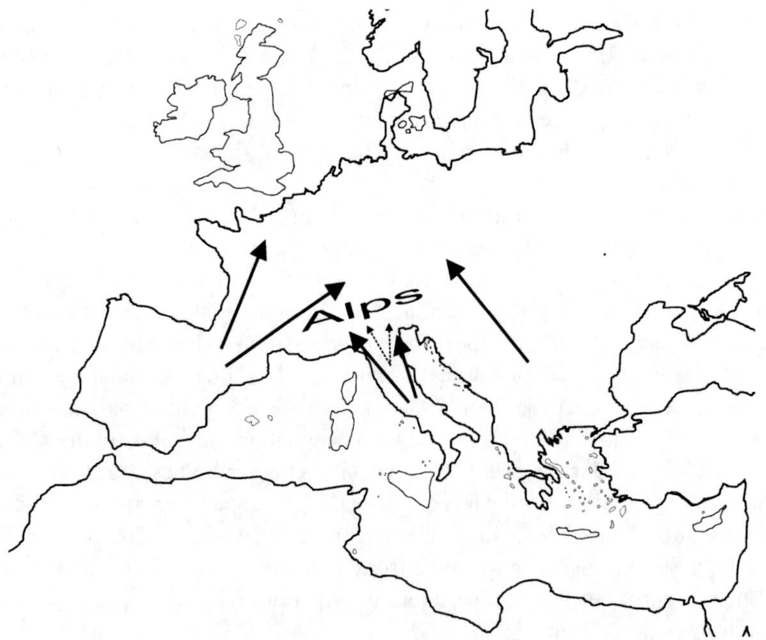

Fig. 15.8 Shift of carrion flies to central Europe from the southern regions and to alpine valleys from the Mediterranean areas. The black arrows indicate the possible migration routes of *C. albiceps*; the dashed arrows indicate the migration of *H. illucens* and *M. scalaris*

an apartment in Zurich, Switzerland (Rognes 1997). Apart from some records of the French gendarmerie in 1996 (Erzinclioglu 2000), reports of *C. albiceps* from the Palearctic region north of the Alps were scarce. The expansion of its range to north of Paris and along the French channel coast during hot summers was probably facilitated by the mild climatic influence of the Gulf Stream (Grassberger and Frank 2004; Grassberger et al. 2003). The second and third instar larvae of this species demonstrate aggressive feeding behavior on local carrion-breeding larvae and, by eating all the early arrivals, could reset the postmortem insect clock.

Similar predatory behavior was also demonstrated by the larvae (Fig. 15.9) of the southern American black soldier fly (*Hermetia illucens*, Diptera: Stratiomyidae). This fly (Fig. 15.10) has been shown to be a ubiquitous inhabitant of human remains throughout the southern, central, and western United States and Hawaii. This species has also been reported from Australia and from Europe (since 1956 from Italy and doubtful from Malta in 1930) (Turchetto 2000; Venturi 1956), but in recent years it has become frequent and common in the northern region and in mountainous areas. Specimens were collected during summer in human remains in plain regions and on traps activated with meat in a Dolomitic Valley at 1330 m asl. The effect of this species on the other carrion-feeding larvae depend not only on its

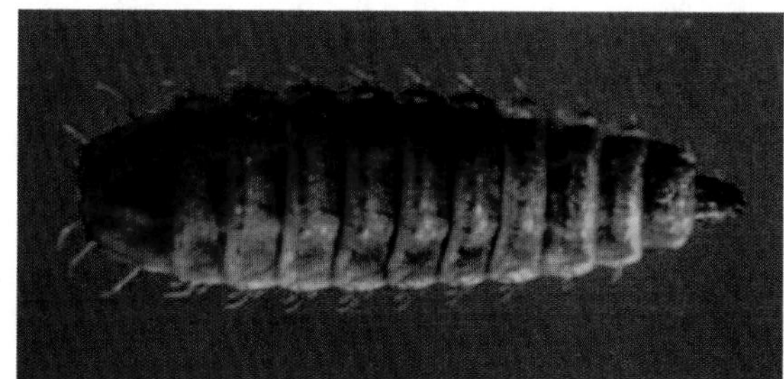

Fig. 15.9 Larva of *Hermetia illucens*

Fig. 15.10 Black soldier fly *Hermetia illucens*

aggressive behavior but also on its adaptability to elevation and resistance to various chemicals.

It is worth mentioning that within the zoosaprophagous Calliphoridae two new introduced species have been reported from Spain: *Chrysomya megacephala* and *Protophormia terranovae* (Martínez-Sánchez et al. 2007a, b; Gobbi et al. 2008; Velásquez et al. 2008). The presence of *P. terranovae* in Spain seem to be related to its use as commercial live bait ("asticot") (Martínez-Sánchez et al. 2007b).

The arrival or shift in new fly species involved in carrion decomposition was reported not only for flies breeding on exposed corpses but also recently for scuttle flies. These flies are able to find buried remains at different depths and are commonly found in graves. For this reason they are aptly named "coffin flies."

Megaselia scalaris (Fig. 15.11) was collected from these flies during forensic entomology experiments in a wild valley in the Alps, in both grassland and a deciduous forest. *Megaselia scalaris* is a polyphagous saprophage species that has been transported around the world by humans (Disney 1994). It is essentially a warm climate species and is common in the lowlands bordering the Mediterranean. In cooler climates it tends to breed in buildings or in places where it can escape frost. It has just recently been reported in Belgium and Japan (Campobasso et al. 2004). Moreover, this species has become more common in northern Italian houses and in mountainous areas, whereas until few years ago it was reported only from the central and southern regions. This is clear evidence of a northern shift of this species.

Not only new necrophagous or predatory dipterans can affect the PMI evaluation but also new fly parasitoids, particularly Hymenoptera (Fig. 15.12). As described above for the predators, parasitoids can affect the body colonization dynamics

Fig. 15.11 The Phorid *Megaselia scalaris*

Fig. 15.12 Hymenopter parasitoid emerging from a fly pupa

overall in closed spaces, causing extinction of their host: the carrion breeding flies. During the summer of 2003, the presence of the Enchyrtid *Tachinaephagus zaelandicus* was reported from dipters' puparia collected from a body in an advanced state of decomposition in northern Italy (Turchetto and Vanin 2004c; Turchetto et al. 2003). This species, probably native to Australia and New Zealand, has been introduced into various parts of the world in attempts to control pest species of synanthropic diptera (Brazil, United States, Africa, and New Caledonia), but no records were available for Europe or the northern regions.

The effect of global climate change has had direct effects not only in the community composition and the extinction or arrival of new species but also in the temporal succession of the carrion breeding species regarding their abundance and phenology.

It is commonly believed that the distinct seasonal pattern of the early colonizing blowfly species allows allocation of the time of death to a particular time interval, even in cases of long PMIs, if the empty puparia of these insects can be found. The order of succession of carrion invertebrates and the arrival and departure times of taxa involved are potentially predictable (Smith 1986).

However, climate change can induce annual variations in the arrival and departure times of carrion insects (Archer 2003). Succession patterns also differ greatly between seasons, as seasonal temperature differences affect decomposition rates, which can, in turn, affect succession rates. Some carrion taxa may also be seasonally active, whereas others are active all year (Archer and Elgar 2003). Carrion taxa may also show seasonal variations in their abundance. Climate change, however, can induce annual variations in arrival and departure times of carrion insects (Archer 2003) with evident consequences in the interpretation of the entomological evidence. In fact, complications may arise from variations between the same seasons in different years, regarding both climate and the population parameters of the carrion invertebrates. In southern Europe, the phenology of several species of Calliphoridae is changing; for example, *Lucilia* species are active also in early spring and in late autumn, with evident widening of their activity period, *Calliphora vicina* specimens were found active during the winter of 2006–2007 and in 2008 *Chrysomya albiceps* was collected at the beginning of November in Venetian region (Vanin, unpublished).

15.6 Discussion on the Validity of Traditional Methods to Calculate by Means of Insects

The debate on global climate change has largely failed to factor in the inherently chaotic, sensitively balanced, and threshold-laden nature of the Earth's climate system and the increased likelihood of abrupt climate change. Current speculation about the future climate and its impact have focused on the IPCC reports (1996a, b, 2001, 2007), which forecast gradual global warming of 1.4°–5.8°C over the current century. During the past century (1965–1995) the greatest extent of warming has been registered during winter in the Northern Hemisphere across Eurasia and inland northern America. At lower latitudes there has been extensive warming across the oceans and, with more seasonal variation, warming across the continents.

It is possible, however, for an abrupt cooling down to occur in selective areas of the world, such as in the North Atlantic regions, caused by the shutdown of the Gulf Stream. Intense cooling has occurred in the mid-latitudes around the central Pacific and in the area between Hudson Bay and Greenland. Environmental conditions are rapidly changing from recent years, induced both by the warming climate and by intense human interaction, such as deforestation, large monocultures, and industrialization. Often the small systematic trends are ignored by nonbiologists, but they may become important in the long term.

These local climate changes would occur quickly, even as the global climate continues to warm slowly. Hansen and colleagues (Hansen and Lebedeff 1987; Hansen et al. 1996) compared global and low-latitude temperature changes, showing that big environmental events, such as El Niño or major volcanic eruptions not only have a heavy impact on local temperatures but have also influenced global climatic fluctuations. Most short-term local changes are not caused by climate change but by land-use change and by natural fluctuations in abundance and distribution of species.

In this scenario it is difficult to predict the trend of animal dispersion. The Earth's warming induces species, especially the ectothermic invertebrates, to shift spontaneously toward lower latitudes and higher altitudes. The capability of adapting to changed conditions allows different native species to survive in their historic areas, but it is possible that they have to partially adapt to their ecological niche, their phenology, or their habitat. The current data allow us to foresee that some species will occupy new areas and others will become niche-species or extinct. Along with other insects, many fly species are saprophagous or preferentially necrophagous as adults and particularly as larvae, feeding on all decaying matter.

Most of these species are not only generalist in their feeding behavior but also in suffering temperature changes and thermal shocks by means of diapauses. In this way, for example, the tropical soldier fly, *Hermetia illucens*, has survived in the southern temperate regions of Europe, modifying its diapause length and its voltinism. In many years, thanks also to increasing temperatures, this alien species has become able to shift northward and to higher altitudes. Other saprophagous or necrophilous insects, strictly linked to humans and human food or waste, can live in the cities of cold lands when they could not survive in their surroundings. A strictly connected example may also be the presence of termites in the houses of Venice and Padova, which until now were only found in central and southern Italy. This finding demonstrates that medium to large cities are spots of constant warmth. Following Hansen and Lebedeff (1987), cities with more than 100,000 inhabitants have modified the regional temperature by 0.1°C or more. Karl and Williams (1987) estimated that global warming during the past century may be reduced by 0.1°C if these cities are excluded, and that the total urban effect on the current global climate is 0.1°–0.2°C on average.

The climatic variations, in terms of both temperatures and weather, directly modify biotopes and vegetation arrangement, affecting the composition of communities. The necrophilous arthropods, such as the other organisms, suffer a selective pressure to adapt to the new conditions or, if they are able, to shift toward new, more consistent areas. This last choice is evident for many species, which have rapidly expanded the cool margins of their geographical ranges.

This is, for example, the case with the blowfly *Chrysomya albiceps*, as reported by Grassberger et al. (2003). It was once abundant in the Austral Hemisphere and southern Europe but has now migrated northward to central Europe; or *Megaselia scalaris*, a species of tropical climates and of the Mediterranean Basin, found in Belgium by Dewaele and Leclercq (2002). The continuity of the increase in the population's ranges leads us to suppose a spontaneous migration. Often the presence of new species dramatically affects the native populations, being their predators and/or competitors, such as *C. albipes* versus *Lucilia sericata* (Grassberger et al. 2001) and *H. illucens* (Turchetto et al. 2001) or more resistant to pesticides (Turchetto 2000).

Taking into account all the considerations reported above, the classic forensic indications about the methods in forensic entomology must be critically revised. Without entering into the subject of the rate of the body's degradation and tissue decay, which can also in some way be affected and accelerated by the warming climate, we restrict our attention only to cadaveric fauna. The eight invasion waves of arthropods proposed by Mégnin in 1894 have been reevaluated many times with small variations, regarding overall the number of stages of carcass decomposition. Mégnin, however, lived at the end of the Little Ice Age, when the global temperature was 1°C lower than at present; moreover, the local temperatures were somewhat lower, and precipitation was more abundant. Few concentrated built-up areas constituted the landscape, as in most parts of Europe, whereas today the land and woods have been overtaken by the spread of cities and industries. The consequences of human activity such as energy emitted as heat and the transformation of natural environments in cultivated fields (often the largest monocultures) along with deforestation have added to the impact of natural events, such as desertification, the changed level of precipitation (much lesser in some regions or concentrated in a few days throughout the year), the alteration of monsoons, and the shift and higher force of hurricanes (shifted northward, arriving through the Atlantic in northwestern Europe and England), forcing the entire fauna to change. The trend of rising temperatures was confirmed by the most recent data of 2006, as shown in Fig. 15.13 for the past 30 years. Following the NOAA report in 2006 based on preliminary data, the globally averaged combined land and sea surface temperature was the warmest on record for December 2006 and the fifth warmest for January-December. Temperatures were above average in the United States, Europe, southern Asia, central Russia, and most of South America; and cooler-than-average conditions occurred in the Middle East region. Precipitation during December 2006 was above average in Scandinavia, England, Japan, central United States, south-eastern Africa, and most of South America; and drier than average conditions were observed in the eastern United States, central Europe, eastern Australia, and southern India. December 2006 was the warmest since global surface records began in 1880 for combined global land and ocean surface temperatures. December land surface temperatures were the fourth warmest on record, and ocean surface temperatures were the second warmest in the 127-year record, behind only 1997, during which the very strong 1997/1998 El Niño event was developing. ENSO conditions persisted in a warm phase (El Niño) during December.

In light of all these considerations, new tables of carrion decay and arthropod succession must be formulated for more precise evaluation of the PMI. We suggest tables on a regional scale that are revised moreover, at frequent time intervals, so environmental and climatic parameters can be considered constant. The astronomic division of the year into spring, summer, autumn, and winter are not given to precise data and would be better substituted with the calendar months including mean, maximum and minimum temperatures. This proposal, which takes into account the selective pressures of climate change on the necrophagous insects, could foster

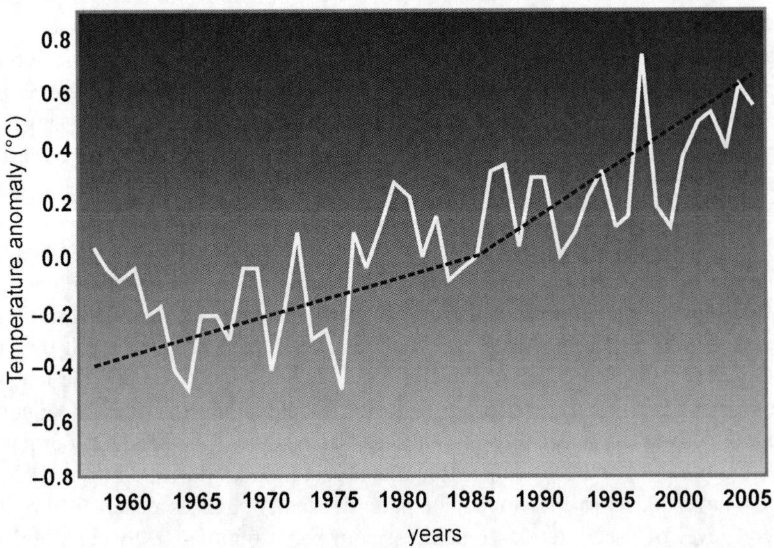

Fig. 15.13 Global mid-tropospheric temperature. Radiosonde measurements indicate that for the January–December period temperatures in the mid-troposphere (approximately 2–6 miles above the Earth's surface) were 0.56°C above average, creating the third warmest January–December since global measurements began in 1958 (data were obtained from http://www.ncdc.noaa.gov/oa/climate/research/2006/dec/global.html/)

agreement between the different methods of working of forensic entomologists strictly linked to the classic tables of faunal succession and those believing that every death find is a case apart.

References

Archer MS (2003) Annual variation in arrival and departure times of carrion insects from carcasses: implications for successive studies in forensic entomology. Aust J Zool 51:569–576

Archer MS, Elgar MA (2003) Yearly activity patterns in southern Victoria (Australia) of seasonally active carrion insects. Forensic Sci Int 132:173–176

Arnaldos I, Romera E, Garcia MD, Luna I (2001) An initial study on the succession of sarcosaprophagous Diptera (Insecta) on carrion in the southeastern Iberian peninsula. Int Legal Med 114:156–162

Bale JS (2002) Insects and low temperatures: from molecular biology to distributions and abundance. Philos Trans R Soc B Biol Sci 357:849–862

Bradshaw WE, Holzapfel CM (2001) Genetic shift in photoperiodic response correlated with global warming. Proc Natl Acad Sci USA 98:14509–14511

Brownstein JS, Holford TR, Fish D (2005) Effect of climate change on Lyme disease risk in North America. Ecohealth 2:38–46

Campobasso CP, Disney RHL, Introna F (2004) A case of *Megaselia scalaris* (Loew) (Dipt., Phoridae) breeding in a human corpse. Aggrawals Internet J Forensic Med Toxicol 5:3–5

IPCC (Intergovernmental Panel on Climate Change) (2001) IPCC: Third Assessment Report. http://www.ipcc.ch/ipccreports/assessments-reports.htm

IPCC (Intergovernmental Panel on Climate Change) (2007) IPCC: Fourth Assessment Report. http://www.ipcc.ch/ipccreports/assessments-reports.htm

Dewaele P, Leclercq M (2002) Les Phorides (Dipteres) sur cadavres humains en Europe occidentale. In: Proceedings of the first European forensic entomology seminar. Rosny Sous Bois, EAFE

Disney RHL (1994) Scuttle Flies: The Phoridae. Chapman & Hall, New York

Dyurgerov MB, Meier MF (2000) Twentieth century climate change: evidence from small glaciers. Proc Natl Acad Sci USA 97(4):1406–1411

Erzinclioglu Z (2000) Maggots, murder and men: memories and reflections of a forensic entomologist. Harley Books, Colchester, UK

Frank JH, McCoy ED (1994) Introduction to insect behavioral ecology: the good, the bad, and the beautiful; non-indigenous species in Florida-invasive adventive insects and other organisms in Florida. Fla Entomol 78:1–15

Frenot Y, Chown SL, Whinam J, Selkirk PM, Convey P, Skotnicki M, Bergstrom DM (2005) Biological invasions in the Antarctic: extent, impacts and implications. Biol Rev 80:45–72

Gagosian RB (2003) Abrupt climate change: should we be worried? World Economic Forum, Switzerland

Gobbi P, Toniolo M, Martínez-Sánchez A, Rojo S (2008) Life cycle of *Chrysomya megacephala* and *Protophormia terranovae* in SW Europe: introduced species and forensic entomology (Diptera: Calliphoridae). In: Proceedings of the sixth meeting of the European association for forensic entomology. Kolymbari, EAFE

Grassberger M, Frank C (2004) Initial study of arthropod succession on pig carrion in a central European urban habitat. J Med Entomol 41(3):511–523

Grassberger M, Friedrich E, Reiter C (2003) The blowfly *Chrysomya albiceps* (Wiedmann) (Diptera: Calliphoridae) as a new forensic indicator in central Europe. Int J Legal Med 117:75–81

Hall MJR, Smith KGV (1993) Diptera causing myiasis in man. In Lane RP, Crosskey RW Medical Insects and Arachnids, Chapman & Hall, London, p. 429–469

Hansen J, Lebedeff S (1987) Global trends of measured surface air temperature. J Geophys Res 92:13345–13372

Hansen J, Ruedy R, Sato M, Reynolds R (1996) Global surface air temperature in 1995: return to pre-Pinatubo level. Geophys Res Lett 23:1665–1668

Harrington R, Woiwod I, Sparks T (1999) Climate change and trophic interactions. Trends Ecol Evol 14:146–150

Hill JK, Thomas CD, Blakeley DS (1999a) Evolution of flight morphology in a butterfly that has recently expanded its geographic range. Oecologia 121:165–170

Hill JK, Thomas CD, Lewis O (1999b) Flight morphology in fragmented populations of rare British butterfly: *Hesperia comma*. Biol Coserv 87:277–284

Hill JK, Thomas CD, Fox R, Telfer MG, Willis SG, Asher J, Huntley B (2002) Responses of butterflies to twentieth century climate warming: implications for future ranges. Philos Trans R Soc B Biol Sci 269:2163–2171

Hughes C, Hill JK, Dytham C (2003) Evolutionary trade-offs between reproduction and dispersal in populations at expanding range boundaries. Philos Trans R Soc B Biol Sci 270(suppl 2):S147–S150

Introna F, Campobasso CP, Di-Fazio A (1998) Three case studies in forensic entomology from southern Italy. J Forensic Sci 43:210–214

IPCC (Intergovernmental Panel on Climate Change) (1996a) Climate change 1995: the science of climate change. Cambridge University Press, New York

IPCC (Intergovernmental Panel on Climate Change) (1996b) Impacts, adaptation, and migration of climate change: scientific-technical analyses. Cambridge University Press, New York

Kamata N, Esaki K, Kato K, Igeta Y, Wada K (2002) Potential impact of global warming on deciduous oak dieback caused by ambrosia fungus *Raffaelea* sp. carried by ambrosia beetle *Platypus quercivorus* (Coleoptera: Platypodidae) in Japan. Bull Entomol Res 92:119–126

Karl TR, Williams CN Jr (1987) An approach to adjusting climatological time series for discontinuous inhomogeneities. J Appl Meteorol 26:1744–1763

Lindgren E, Gustafson R (2001) Tick-borne encephalitis in Sweden and climate change. Lancet 358:16–18

Maistrello L, Lombroso L, Pedroni E, Reggiani A, Vanin S (2006) Summer raids of *Arocatus melanocephalus* (Heteroptera, Lygaeidae) in urban buildings in northern Italy: is climate change to blame? J Thermal Biol 31:594–598

Martínez-Sánchez A, Gobbi P, Velasquez Y, Rojo S (2007a) Biology of *Crysomya megacephala* (Fabricius, 1794) in Europe, new data and implications for forensic entomology research (Dipter: Calliphoridae). In: Proceedings of the fifth meeting of the European association for forensic entomology. Brussels, EAFE

Martínez-Sánchez A, Magaña C, Rojo S (2007b) First data about forensic importance of *Protophormia terranovae* (Robineau-Desvoidy, 1830) in Spain (Diptera: Calliphoridae). In: Proceedings of the fifth meeting of the European association for forensic entomology. Brussels, EAFE

Martìnez-Sánchez A, Rojo S, Marcos-Garcia MA (2000) Annual and spatial activity of dung flies and carrion in a Mediterranean holm-oak pasture ecosystem. Med Vet Entomol 14:56–63

McLaughlin JF, Hellmann JJ, Boggs CL, Ehrlich PR (2002) Climate change hastens population extinctions. Proc Natl Acad Sci USA 99:6070–6074

Megnin P (1894) La Faune des Cadavres: Applications de l'Entomologie à la Médecine Légale. Encyclopedie Scientifique des Aides-Memoires. Masson & Gauthier-Villars, Paris

Musolin DL, Numata H (2003) Timing of diapause induction and its life-history consequences in *Nezara viridula*: is it costly to expand the distribution range? Ecol Entomol 28:694–703

Musolin DL, Numata H (2004) Late-season induction of diapause in *Nezara viridula* and its effect on adult coloration and post-diapause reproductive performance. Entomol Exp Appl 111:1–6

Nicolussi K, Patzelt G (2006) Klimawandel und Veränderungen an der alpinen Waldgrenze-aktuelle Entwicklungen im Vergleich zur Nacheiszeit. Wien: BFW-Praxisinformation 10:3–5

Nimela J, Spence JR (1991) Distribution and abundance of an exotic ground-beetle (Carabidae): a test of community impact. Oikos 62:351–359

Parmesan C (2006) Ecological and evolutionary responses to recent climate change. Annu Rev Ecol Evol Syst 37:637–669

Rognes K (1997) Additions to the Swiss fauna of blowflies with an analysis of the systematic position of *Calliphora stylifera* (Pokorny, 1889) including a description of the female (Diptera, Calliphoridae). Mitteilschweiz Entomol Ges 70:63–76

Roy DB, Asher J (2003) Spatial trends in the sighting dates of British butterflies. Int J Biometeorol 47:188–192

Smith KGV (1986) A manual of forensic entomology. Trustees of the British Museum, London

Stenseth NC, Mysterud A, Ottersen G, Hurrell JW, Chan KS, Lima M (2002) Ecological effects of climate fluctuations. Science 297:1292–1296

Thomas CD, Lennon JJ (1999) Birds extend their ranges northwards. Nature 399:213

Thomas CD, Bodsworth EJ, Wilson RJ, Simmons AD, Davies ZG, Musch M, Conradt L (2001) Ecological and evolutionary processes at expanding range margins. Nature 411:577–581

Thomas CD, Cameron A, Green RE, Bakkenes M, Beaumont LJ, Collingham YC, Erasmus BF, De Siqueira MF, Grainger A, Hannah L, Hughes L, Huntley B, Van Jaarsveld AS, Midgley GF, Miles L, Ortega-Huerta MA, Peterson AT, Phillips OL, Williams SE (2004) Extinction risk from climate change. Nature 427:145–148

Turchetto M (2000) Implicazioni entomologico-forensi dell'introduzione in Italia della mosca neotropicale *Hermetia illucens* L. (Diptera: Stratiomyidae). Riv Ital Med Leg 22(6):1279–1290

Turchetto M, Vanin S (2004a) Forensic entomology and climatic change. Int Forensic Sci 46(suppl): 207–209

Turchetto M, Vanin S (2004b) Forensic entomology and globalisation. Parasitologia 46(1–2): 187–190

Turchetto M, Vanin S (2004c) Forensic evaluations of a crime case with monospecific necrophagous fly population infected by two parasitoid species. Aggrawals Internet J Forensic Med Toxicol 5(1):12–18

Turchetto M, Lafisca S, Costantini G (2001) Postmortem interval (PMI) determined by study sarcophagous biocenoses: three cases from the province of Venice (Italy). Forensic Sci Int 120:28–31

Turchetto M, Villemant C, Vanin S (2003) Two fly parasitoids collected during an entomo-forensic investigation: the widespread *Nasonia vitripennis* (Hymenoptera Pteromalidae) and the newly recorded *Tachinaephagus zealandicus* (Hymenoptera Encyrtidae). Boll Soc Entomol Ital 135(1):109–115

US National Academy of Sciences (2002) Abrupt climate change: inevitable surprises. National Academy of Sciences, National Research Council Committee on Abrupt Climate Change, National Academy Press, Washington, DC

Vanin S, Caenazzo L, Arseni A, Cecchetto G, Cattaneo C, Turchetto M (2009) Records of *Chrysomya albiceps* in Northern Italy: an ecological and forensic perspective. Mem Inst Osw Cruz 104(4):555–557

Vanin S, Uliana M, Bonato L, Maistrello L (2005) Nuove segnalazioni di *Leptoglossus occidentalis* (Heteroptera, Coreidae) nell'Italia nord-orientale. Lav Soc Ven Sc Nat 30:149

Vanin S, La Fisca A, Turchetto M (2007) Determination of the time of death of a brown bear *Ursus arctos arctos* L., by means of insects. Entomologia Mexicana 6(2):874–879

Velásquez Y, Martínez-Sánchez A, Rojo S (2008) Autumn colonization of pig carrion by blowflies (Diptera: Calliphoridae), in a mediterranean urban area (SE, Spain). In: Proceedings of the sixth meeting of the European association for forensic entomology. Kolymbari, EAFE

Venturi F (1956) Notulae Dipterologiche X. Specie nuove per l'Italia. Boll Soc Entomol Ital 3–4:56

Visser ME, Holleman LJ (2001) Warmer springs disrupt the synchrony of oak and winter moth phenology. Philos Trans R Soc B Biol Sci 268:289–294

Walther GR, Post E, Convey P, Menzel A, Parmesan C, Beebee TJ, Fromentin JM, Hoegh-Guldberg O, Bairlein F (2002) Ecological responses to recent climate change. Nature 416:389–395

Warren MS, Hill JK, Thomas JA, Asher J, Fox R, Huntley B, Roy DB, Telfer MG, Jeffcoate S, Harding P, Jeffcoate G, Willis SG, Greatorex-Davies JN, Moss D, Thomas CD (2001) Rapid responses of British butterflies to opposing forces of climate and habitat change. Nature 414:65–69

Work T, McCullough DG, Cavey JF, Komsa R (2005) Arrival rate of nonindigenous insect species into the United States through foreign trade. Biol Invas 7:323–332

Wuethrich B (2000) Ecology. How climate change alters rhythms of the wild. Science 287:793–795

Zera AJ, Denno RF (1997) Physiology and ecology of dispersal polymorphism in insects. Annu Rev Entomol 42:207–230

UNEP (United Nations Environment Programme) http://www.unep.ch/

Chapter 16
Future Trends in Forensic Entomology

Jens Amendt, Richard Zehner, Diana G. Johnson, and Jeffrey Wells

16.1 Introduction

The science of forensic entomology has had a staggered and interesting history (Nuorteva 1977; Smith 1986; Erzinçlioglu 1990; Marchenko 2001; Amendt et al. 2004). Its main application is the estimation of the postmortem interval (PMI), and Villet et al (this book, Chapter 7) highlight variables that affect insect development and in its consequence the calculation of this postmortem interval. Great strides have been made in basic and applied research, but there are many questions yet to be answered and there is still room for growth, as several other chapters in this book showed. While there is unquestionable the need for much more research to gather well-based data, there is also a need for quality assurance, standards and certification (Melbye and Jimenez 1997). In this chapter, we will discuss a selection of possible future trends in forensic entomology.

16.2 Research

16.2.1 Insect Succession

Insects colonize a corpse in a predictable sequence, therefore the set of species occurring on a corpse, both present and absent may indicate PMI (e.g. Hewadikaram and Goff 1991; Anderson 2001; Grassberger and Frank 2004; Tabor et al. 2004; De Jong and Hoback 2006). The results of the quoted papers are often site- and season-specific, even dealing with a more or less identical insect fauna. Factors that appear to influence succession pattern include the

J. Amendt (✉) and R. Zehner
Institute of Forensic Medicine, Frankfurt am Main, Germany

D.G. Johnson and J. Wells
Department of Biology, West Virginia University, Morgantown, USA

J. Amendt et al. (eds.), *Current Concepts in Forensic Entomology*,
DOI 10.1007/978-1-4020-9684-6_16,

illumination of the scene, size of the cadaver, time of the year, and type of habitat (Amendt et al. 2004; Hobischak et al. 2006; Joy et al. 2006).

Succession research is extremely labour intensive. This has made it difficult to produce replicated data suitable for a traditional type of statistical confidence interval about a PMI estimate (e.g. 20 or more experimental carcasses for one set of conditions, LaMotte and Wells 2000). Computer resampling methods (Schoenly 1992) may overcome this limitation. Therefore there is a great opportunity to increase such computer intensive applications in succession research.

It has long been observed that the adults of different carrion fly species differ in preference for habitat type (Norris 1965). Therefore, it may be possible to determine that a corpse was move following death, if the immature insects in a corpse are not typical of the site where the body occurs (Amendt et al. 2004). Two of the authors (J. A and R. Z) moved piglet carcasses between rural and urban sites in Germany, following one week of exposure to insect colonization, while leaving others in place. Although no carrion insect species was unique to a habitat, habitat-specific difference in the proportion of each species infesting a piglet could identify the habitat in which a carcass was first colonized. However, the experiments demonstrate that caution must be used in determining whether remains have been moved based solely on entomological evidence. Densely populated landscapes, especially in industrialised countries, exhibit no rigid separation of habitats, leading to a mixture of insect species communities.

Frost et al. (this book, chapter 6) discuss the importance of improving our knowledge about indoor arthropods and indoor scenarios: While the majority of cases with insect evidence occur indoors, nearly all insect succession studies take place outside.

16.2.2 Aquatic Entomology

Freshwater and marine fauna has received very little attention in forensic investigations (Nuorteva et al. 1974; Goff and Odom 1987; Haskell et al. 1989; Vance et al. 1995; Sorg et al. 1997; Davis and Goff 2000). Knowledge about the role of aquatic arthropods during decomposition is still scant (Keiper et al. 1997; Anderson 2001; Merrit and Wallace 2001; Hobischak and Anderson 2002). Compared to terrestrial habitats, decomposition in an aquatic environment is completely different. It occurs at roughly half the rate as on land, mainly due to the prevention of insect activity and cooler temperatures. Aquatic insects of forensic importance belong to the Ephemeroptera (mayflies), Trichoptera (caddisflies) and Diptera (true flies), the latter are mainly represented by Chironomidae (midges) and Simuliidae (black flies). But even these insects, unlike their terrestrial counterparts, are not obligatory sarcophages, but instead use the submerged carrion as a food source as well as a breeding site. The use of these insects for estimating time of death is more difficult and depends on the season and on other conditions of aquatic systems. Hobischak and Anderson (2002) evaluated whether or not arthropod succession in freshwater environments could be used to estimate the postmortem submersion interval. Although their data revealed a predictable succession of invertebrates, they stated that further studies are necessary to determine if the observed succession was carrion- dependent or season-dependent.

This research team also studied decomposition in a marine environment (Anderson and Hobischak 2004). Their results showed stages of decomposition different from patterns seen on land or in freshwater. Future avenues of research were suggested, such as allowing the carcass to float to the surface (see chapter 12, this book).

Limited work has been done regarding corpses that were allowed to decompose in a terrestrial environment and then transferred to an aquatic environment. Singh and Greenberg (1994) addressed this scenario by studying the effect of drowning on the survival of blowfly pupae. Data from this type of study would be valuable in aquatic cases where puparia (submerged, along with the corpse, after the larvae had pupariated) are discovered on a corpse.

16.2.3 Identification, Phylogenetics, and Population Genetics

Correctly identifying the insect specimen can be of crucial importance for a forensic entomological analysis (Smith 1986). One might think that this is self-evident for a forensic entomologist, but the situation is complicated. Many young immature stages of several forensically relevant taxa (like e.g. the Muscidae) are difficult to identify and would usually be reared to the adult stage for getting a safe ID. For some groups, like the sarcophagids, the situation is even worse, and morphological characters of the immature don't allow for identification.

Here, DNA methods may be a powerful tool. They have been used to identify specimens that are impossible or difficult to distinguish based on morphology. However, the work done so far has been very uneven in its coverage of forensically important insect taxa. The calliphorids are, by far, the most represented group in species identification (Harvey et al. 2003; Ratcliffe et al. 2003; Chen et al. 2004; Wells and Williams 2007). This reflects their importance as indicator species of the postmortem interval and, probably, the relative ease in identifying adult specimens to be used for DNA extraction. The sarcophagids, also important for postmortem interval estimation, require more specialized knowledge compared to the calliphorids, and fewer DNA data exist for these flies (Wells et al. 2001a; Ratcliffe et al. 2003; Zehner et al. 2004b). Scattered DNA records exist for a few species from other forensic insect groups, such as the phorid *Megaselia scalaris* (GenBank record AF217464, unpublished) and the muscid *Musca domestica* (AY818108, unpublished). Expanding the taxonomic breadth of DNA data for specimen identification will probably require that such studies be done more in cooperation with taxonomic experts who are not now involved in forensic entomology research.

In most published papers on DNA-based specimen identification, only a small number of individuals from each species were analyzed. The issue of replication when evaluating a species-diagnostic test has received little attention, even though the reliability of a procedure probably cannot be known until this issue is addressed (Wells and Williams 2007). In contrast to the DNA methods used to identify an individual (Evett and Weir 1998), there is no theoretical standard for estimating the replication needed to validate a species-diagnostic test. The development of such an analytical framework is clearly needed, not only for forensic entomology, but also for other forensic fields that involve species identification such as consumer fraud (Lenstra 2003), enforcement of conservation laws (Wan and Fang 2003), and bioterrorism (Jones et al. 2005).

Several loci have been used to make species determinations of forensically related fly species. The most popular mitochondrial DNA markers used for this purpose include the cytochrome oxidase subunit one (COI) gene (Wells and Sperling 2001; Zehner et al., 2004a, b; Wells and Williams 2007) and the cytochrome oxidase subunit two (COII) gene (Sperling et al. 1994; Wells and Sperling 2001; Wallman and Donnellan 2001). Other mitochondrial markers that have been used include the control region (Thyssen et al. 2005), subunit 5 of the NAD dehydrogenase (ND5) gene (Zehner et al., 2004a, b), and the tRNA-leucine gene (Wells and Sperling 2001). Nuclear markers have been utilized, as well, such as the internal transcribed spacer regions (Ratcliffe et al. 2003), and the gene for 28S ribosomal RNA (Stevens and Wall 2001). Although COI and COII are recommended by many authors as the best marker for identification of forensically important insects, some closely-related species cannot be differentiated in this manner (Wallman and Donnellan 2001; Wells et al. 2004). Future research should investigate other markers for closely-related species.

The methodologies vary, as well, and they include direct sequencing (Harvey et al. 2003; Chen et al. 2004; Zehner et al. 2004b; Ames et al. 2006), polymerase chain reaction-restriction fragment length polymorphism (PCR-RFLP) analysis (Sperling et al. 1994; Malgorn and Coquoz 1999; Thyssen et al. 2005), random amplified polymorphic DNA (RAPD) analysis (Benecke 1998), and allozyme electrophoresis (Wallman and Adams 2001). Although mtDNA sequencing is a particularly robust technique, the lesser-used methods (i.e., PCR-RFLP, RAPD, and allozyme electrophoresis) can be fast, inexpensive, and highly discriminating. Therefore, they deserve further evaluation as forensic tools.

Most of the genetic research in forensic entomology has focused on individual or species identification, as previously described. Investigations of larger phylogenetic patterns are limited (Stevens and Wall 1996; Stevens and Wall 1997; Wells and Sperling 1999; Bernasconi et al. 2000; Stevens and Wall 2001; Stevens et al. 2002; Wells et al. 2004, 2007). However, these basic studies can reveal forensically important information by uncovering a so-called "cryptic" species, one that could not be recognized by traditional means (Wallman et al. 2005). This kind of analysis may require a greater amount of genetic information than is normally used for simple identification. One approach is to use several mitochondrial genes (Wallman et al. 2005). In fact, it is possible to sequence the entire blowfly mitochondrial DNA molecule (Junqueira et al. 2004).

Perhaps more promising would be an analysis based on a combination of mitochondrial and nuclear loci. This has been done to a limited extent, for example using COI and 28S (Stevens et al. 2002). At present, however, it is much more difficult to generate useful nuclear DNA sequences compared to mitochondrial DNA when examining a previously unstudied taxonomic group. Fortunately, the phylogeny of the Order Diptera is an extremely active are of research (see www.inhs.uiuc.edu/cee/FLYTREE/). Forensic entomologists should pay close attention to new protocols developed for other fly taxa, because that technology might be applied to forensically important groups with little, or no, modification.

The population genetics of forensically important insects have been almost untouched. Yet such information is relevant to forensic entomologists for at least two reasons (Böhme 2006). The discovery of regional genetic differences would suggest that regional differences could exist in a forensically important characteristic of a species, such as development rate as a function of temperature. Regional genetic differences could also make it possible to infer the postmortem movement of a corpse if a larva from the body has a genotype characteristic of another location.

For these reasons, we expect to see an expansion in population genetic studies of carrion insects. Although STR genotypes are particularly valuable for this purpose, the development of STR methods for groups such as blowflies has been slow because such loci appear to be much less common in insect genomes compared to those of vertebrates (Ji et al. 2003). Nevertheless, some STR primers have been developed for some calliphorid species (Florin and Gyllenstrand 2002; Torres et al. 2004; Torres and Azeredo-Espin 2005). This demonstrates that it can be done, and it may be possible to apply these methods with little or no change to other, closely-related, species.

All of the aforementioned studies have generated valuable data and stand as the foundation of genetic research in forensic entomology. However, validation studies are essential before these methods are used for casework. That is, how confident can we be that a proposed method will accurately identify a specimen? In part, this depends on adequate replication, something that has not yet been determined (see above). However, this doesn't prevent validation efforts from being carried out. For example, Wells and Williams (2007) tested the accuracy of a published method for identifying individuals in the blowfly subfamily Chrysomyinae in North America (Wells and Sperling 2001). They obtained genetic data for hundreds of additional identified specimens, and analyzed them as if they were unknowns. All were correctly "identified" by the genetic method. More validation studies, such as this, are necessary.

16.2.4 Maggot Gut Content Studies

Identifying, and/or individualizing, the source of the gut content of forensically important insects could potentially identify a victim from a maggot left behind when a corpse was moved. A blood-feeding insect at a scene could connect a suspect to that scene if it were found to have fed on the suspect. Campobasso et al. (2005) reviewed these and other potential applications of this analysis.

The variety of taxa represented in gut content literature is limited. Mostly blowfly species have been examined (Linville et al. 2004; Linville and Wells 2002; Zehner et al. 2004b). The flesh flies and the muscids have received little attention in this area of research (Wells et al. 2001b). The remainder of the literature deals with mosquitoes (Culicidae) (Kreike and Kampfer 1999; Mukabana et al. 2002), lice (Pediculidae) (Replogle et al. 1994; Lord et al. 1998; Mumcuoglu et al. 2004), and a sap beetle (DiZinno et al. 2002). Future work in this area should incorporate a wider variety of species.

Short tandem repeat (STR) analysis is the standard for the determination of human nuclear DNA profiles. Countries around the world have adopted a somewhat overlapping set of standard STR loci for human identity testing (Butler 2005). Recent gut content studies have used STR loci as markers (Schiro 2001; Linville et al. 2004; Mumcuoglu et al. 2004; Zehner et al. 2004a).

Human forensic geneticists are constantly developing methods that are more discriminating and/or applicable to tissue samples that were previously unsuitable for analysis. Almost certainly these will be increasingly used for forensic insect gut content analysis. For example, Y-STR analysis is a new tool in forensic laboratories (Butler 2003). It has not yet become standard practice for most publicly-funded laboratories, but it will become more widely available in the future. Y-STR loci are so named because they reside on the Y chromosome, so they are useful for detecting human male DNA in a sample. One obvious application of this technology is the detection of semen on the remains of a victim of sexual assault-homicide. Limited work has been done to detect semen in the gut content of maggots (Clery 2001). This combination of STR profiling and gut content analysis could prove to be extremely useful in an investigation. Researchers working on gut content analyses should incorporate such new developments into their studies.

As pointed out by Campobasso et al. (2005), feeding stages usually are needed for the successful isolation of host DNA from the alimentary tract of the larvae. However, Carvalho et al (2005) reported fragments of 197 basepairs (bp) and 87 bp of host sheep DNA (satellite I region) in all stages of immatures, including 2-day-old pupae of *Calliphora dubia.*

16.2.5 Weather and Temperature Collection

It is usually essential to the success of a forensic entomological examination to obtain accurate weather data for the site of collection (Haskell and Williams 1990; Haskell et al. 2000; Amendt et al. 2007). Some research has been done to improve the accuracy of pre-discovery scene temperature estimates based on post-discovery measurements at the scene and at the nearest weather station (Archer 2004). Given the complicated and confounding effects of variation in microclimate when extrapolating from conditions at one site to another, we expect that forensic entomologists will further investigate techniques for accurately estimating past crime scene environmental conditions. Preliminary results indicate that satellite weather data is very strongly correlated to actual conditions (Hunt 2005). Therefore, we anticipate that forensic entomologists will increasingly rely on weather satellites for this purpose, and cooperate hand-in-hand with a meteorologist (Archer 2004; Scala and Wallace 2005; Amendt et al. 2007).

When collecting temperature data at the scene, perhaps the temperature of maggot masses should be recorded. As pointed out (Turner and Howard 1992; Ireland and Turner 2006), the temperature inside maggot masses can be much higher then the ambient temperature, creating a microclimate for potentially faster development of the maggots.

16.2.6 Parasitoids and PMI

Necrophagous taxa can get attacked by a special guild of insects called parasitoids. According to Godfray (1994), a parasitoid larvae feed exclusively on other arthropods, mainly insects, resulting in the death of parasitoid's host. They represent an extremely diverse group, mainly Hymenoptera, and constituting about 8.5% of all described insects (LaSalle and Gauld 1991, Godfray 1994). Fabritius and Klunker (1991) listed for Europe 83 parasitoid species, which attack the larval and pupal stages of synanthropic Diptera. The life-cycles of the common parasitoid species are known (e. g. Geden 1997) and even if the adults have already emerged and left the host, the pupal exuviae of the parasitic wasps can be identified a long time afterwards (Geden et al. 1998, Carlson et al. 1999). Despite their great number and our growing knowledge about their biology there are just few reports on the use of parasitoids in forensic entomology (Smith 1986, Haskell et al. 1997, Amendt et al. 2000, Anderson and Cervenka 2002, Grassberger and Frank 2003b, Voss et al. 2009). Especially pupal parasitoids of Diptera could play an especially important role in the estimation of the postmortem period as they hit the host stages in a well defined window of time at the beginning of their development (Anderson and Cervenka 2002). This means that the calculated developmental time of the parasitoid just has to add to the time of development of the host (e. g. Grassberger and Frank 2003b), therefore providing an extended PMI timeframe in cases where traditional forensic indicators have completed their development. However, when thinking about the potential influence, especially of larval parasitoids, it is important to take into account that they also could create significant problems for forensic entomology: Holdaway and Evans (1930) described e. g. the change of developmental times for Lucilia sericata after the attack of its parasitoid Alysia manducator which resulted in a premature pupariation. This clearly illustrates the need for further research in this field.

16.2.7 Multivariate Approaches

Although there is a strong tendency for forensic analysts to be very specialized, we think that forensic entomology would benefit from a mulivariate approach. That is, combining entomological and non-entomological quantitative measurements in the same analysis (Schoenly et al. 1991). Examples of non-entomological postmortem phenomena that can probably be examined along with entomological data include RNA decay rates (Anderson et al. 2005), changes in the volatile organic products of decay (Statheropoulos et al. 2005), and algal growth rate (applicable to cases in which decomposition occurs in an aquatic environment (Haefner et al. 2004)). Here, there seems to be no reason why the mathematical and computer models developed for PMI estimation based on insect data (e.g. Schoenly 1992; Wells and LaMotte 2001; Byrd and Allen 2001) can't be used for non-insect data, as well. A combined anthropological, botanical and entomological analysis may be especially useful when recovering buried bodies (Galloway et al. 2001), but should be proposed in general as more tools could refine or confirm the results.

16.2.8 Wildlife and Veterinary Forensic Entomology

Forensic entomology is commonly used in human death investigations. However, it can be equally applicable to wildlife crimes (Merck 2007). Anderson (1999) describes a case of illegal killing of two young bear cubs in Canada. Entomological data were used to determine the time of death. The time was consistent with the time that the defendants were seen at the scene and was used in their conviction. In addition to this published case study, research has been conducted using various wildlife species during the spring (Watson and Carlton 2003), fall (Watson and Carlton 2005), and winter (Watson and Carlton 2005) seasons.

16.3 Quality Assurance

16.3.1 Standardization of Methods

Forensic science laboratories must operate using prescribed methods and standards in order for their results to withstand scrutiny in the courtroom setting. The American Society of Crime Laboratory Directors/Laboratory Accreditation Board (ASCLD/LAB) has created voluntary programs "in which any crime laboratory may participate to demonstrate that its management, personnel, operational and technical procedure, equipment and physical facilities meet established standards" (Romano 2004). Accreditation is desirable because it demonstrates adherence to standard protocols which, in turn, confers confidence in the reliability of laboratory results. While an increasing number of laboratories, worldwide, are seeking accreditation by ASCLD/LAB International Standards of Organization (ISO) are applied world wide as well and many forensic institutes (including entomological labs), trying to implement ISO 17025.

This trend will also be seen in the sub-discipline of forensic entomology. In fact, Amendt et al. (2006) published the first set of standards and guidelines for use by entomologists, pathologists, investigators, and other professionals since the publication of Entomology and Death: A Procedural Guide (Catts and Haskell 1990). By following standards and guidelines, the results from entomological analyses conducted by independent analysts will be comparable. Furthermore, the quality of the work can be more easily assessed (Amendt et al. 2007). Ideally, these standards and guidelines will be adopted by practitioners in every country, universalizing the science.

The American Board of Forensic Entomology (ABFE) was formed in 1996 as a means of certifying forensic entomologists (http://www.research.missouri.edu/entomology). Being an ABFE Diplomate or Member reflects positively upon the forensic entomologist's experience and skill and bolsters their testimony as an expert witness In 2002 the European Association for Forensic Entomology was founded (www.eafe.org), which is just about to establish interlaboratory tests on an European level. The trend toward certification will continue as an increasing number of forensic entomologists are put on the witness stand to testify.

16.3.2 Central Repository

A centralized repository for reference specimens of forensic importance would facilitate research and create a dedicated location for type and voucher specimens. Practicing forensic entomologists, experts in a relevant group, or avid collectors throughout the world could contribute/borrow specimens to/from the collection. For example, a forensic entomologist, confronted with a vexing taxonomic dilemma, would not have to canvass other scientists in order to locate the appropriate voucher(s). Instead, they would be able to simply contact the central repository and access the specimen(s).

The forensic entomology collection could be housed in an existing museum, or exist in its own right. Visiting students could peruse the collection, expanding their own knowledge base or contributing valuable information to the collection. Additionally, a database of insect holdings, such as those maintained by leading museums and universities, could be created and posted on the internet. Furthermore, relationships with prominent collections could be cultivated for an even larger network of resources.

The collection could also serve as a repository for genetic material. Ethanol-preserved specimens, frozen tissue, samples of extracted DNA, or FTA cards (Harvey 2005), could be archived in this manner. Similar collections exist for other insect groups. The United States Department of Agriculture's Animal and Plant Health Inspection Service (USDA APHIS) created the Medfly Germplasm Repository (Otis, MA) for the curation of Mediterranean fruit fly genetic material. In Oberursel, Germany, the Institute für Bienenkunde houses a large collection of ethanol-preserved honeybees. Both of these repositories are valuable resources to researchers.

16.3.3 Specialized Professional Societies

Until recently, forensic entomologists typically conferred and presented their recent research results at large entomological or forensic science academic conferences. Participation in a large, multidisciplinary conference may aid in academic cross-fertilization, but the issues that are important to a subspecialty can be ignored. Also, large meetings can be confusing and intimidating to beginning students, and, therefore, are less-than-ideal for recruiting fresh talent.

The first such organization dedicated solely to forensic entomology was the European Association for Forensic Entomology (EAFE; www.eafe.org), formed in 2002. The establishment of the North American Forensic Entomology Association (NAFEA; www.nafea.net) followed shortly thereafter, in 2003. Both organizations now hold an annual academic conference for the presentation of research and more casual scientific exchange. Almost certainly, scientists from other geographic regions will form their own forensic entomology professional organizations.

16.3.4 Development of Full-Service

The establishment of laboratories designed for the mission of forensic entomology would benefit the science as a whole. The forensic entomology department of the Criminal Research Institute of the French Gendarmerie, created in 1992, is an excellent model (Gaudry et al. 2001). Their department is charged with three main duties: offering expert services in criminal investigations, educating crime scene investigators, and conducting research. Their first task involves the analysis of evidence collected at the scene (species identification and PMI estimation) and offering expert testimony at trial. Their second task, educating crime scene investigators, is extremely important, as well. Investigators appropriately trained in the collection and preservation of insect specimens at the scene will generate useable specimens. The third task, conducting research, not only provides information applicable to their casework, it also furthers the science of forensic entomology.

Their mission statements are well-designed, as is their organizational structure. The Criminal Institute of the French Gendarmerie is an interdisciplinary organization. Many sub-disciplines of forensic science can corroborate with ease and draw upon each other's expertise. For example, the toxicology department can be called upon in cases where the presence of drugs is suspected. Or, the biology department can generate sequence data for species determinations. Scanning electron micrographs can be generated by the microanalysis department. This is the ideal scenario. Often, even though a particular specialized analysis would be beneficial, the difficulty of obtaining these services is prohibitive. Within an interdisciplinary system, evidentiary samples can remain in-house instead of being out-sourced to experts in other locales. This not only increases the likelihood that samples will be subjected to further analysis, it also avoids potential chain of custody issues.

16.3.5 Public Relations

The popularity of forensic entomology has escalated in recent years due to television shows (Dirty Jobs, "Skull Cleaner", Season 3, Episode 9; CSI: Crime Scene Investigation, "Sex, Lies, and Larvae", Season 1, Episode 10; Forensic Files, "Insect Clues", Season 1, Episode 45), magazine articles (Gannon 1997; Sachs 1998), newspaper articles (Birch 1992; Staff 2005), fictional novels (Cornwell 2003), non-fiction books (Erzinçlioglu 2000; Goff 2000; Sachs 2001), museum exhibits ("CSI: Crime Scene Insects", M. Lee Goff, curator), and websites. Forensic entomology has been removed from obscurity and brought into the spotlight. Its popularity will surely continue to increase as more people learn about this fascinating science. Forensic entomologists need to take advantage of this period of enlightenment and tout the benefits of this unique science to laymen, crime scene investigators, and fellow scientists (e.g., pathologists, forensic scientists, entomologists). With more people involved, and more attention being given, resources will become available to continue research, investigative services, and outreach activities.

References

Amendt J, Krettek R, Zehner R (2004) Forensic entomology. Naturwissenschaften 91:51–65

Amendt J, Krettek R, Niess C, Zehner R, Bratzke H (2000) Forensic entomology in Germany. Forensic Sci Int 113:309–314

Amendt J, Campobasso CP, Gaudry E, Reiter C, LeBlanc HN, Hall MJR (2007) Best practice in forensic entomology–standards and guidelines. Int J Legal Med 121:90–104

Ames C, Turner B, Daniel B (2006) The use of mitochondrial cytochrome oxidase I gene (COI) to differentiate two UK blowfly species – *Calliphora vicina* and *Calliphora vomitoria*: Forensic Sci Int 164:179–182

Anderson GS (1999) Wildlife forensic entomology: determining time of death in two illegally killed black bear cubs. J Forensic Sci 44:856–859

Anderson GS (2001) Succession on carrion and its relationship to determining time of death. In: Byrd JH, Castner JL (eds) Forensic entomology – the utility of arthropods in legal investigations. CRC, Boca Raton, FL, pp 143–175

Anderson GS, Cervenka VJ (2002) Insects associated with the body: Their use and analyses. In: Haglund WD Sorg MH (eds) Advances in forensic taphonomy – method theory and archaeological perspectives. CRC, Boca Raton, London, pp 173–200

Anderson GS, Hobischak NR (2004) Decomposition of carrion in the marine environment in British Columbia Canada. Int J Legal Med 118:206–209

Anderson S, Howard B, Hobbs GR, Bishop CP (2005) A method for determining the age of a bloodstain. Forensic Sci Int 148:37–45

Archer MS (2004) The effect of time after body discovery on the accuracy of retrospective weather station ambient temperature corrections in forensic entomology. J Forensic Sci 49:553–559

Benecke M (1998) Random amplified polymorphic DNA (RAPD) typing of necrophageous insects (diptera coleoptera) in criminal forensic studies: validation and use in practice. Forensic Sci Int 98:157–168

Bernasconi MV, Valsangiacomo C, Piffaretti JC, Ward PI (2000) Phylogenetic relationships among Muscoidea (Diptera: Calyptratae) based on mitochondrial DNA sequences. Insect Mol Biol 9:67–74

Birch D (1992) Sought-after forensic entomologist digs deep for clues. The Sun, Baltimore, MD, pp 3A

Böhme P (2006) Population genetics of forensically important North American blow flies (Diptera: Calliphoridae) using the A+T-Rich region of mitochondrial DNA diploma. Thesis Üniversität Bonn

Butler JM (2003) Recent developments in Y-short tandem repeat and Y-single nucleotide polymorphism analysis. Forensic Sci Rev 15:91–111

Butler JM (2005). Forensic DNA Typing. The Biology & Technology Behind STR Markers. New York: Elsevier. 660 pp

Byrd JH, Allen JC (2001) Computer modeling of insect growth and its application to forensic entomology. In: Byrd JH, Castner JL (eds) Forensic entomology: the utility of arthropods in legal investigations, CRC, Boca Raton, FL, pp 303–330

Campobasso CP, Linville JG, Wells JD, Introna F (2005) Forensic genetic analysis on insect gut contents. Am J Forensic Med Pathol 26:161–165

Carlson DA, Geden CJ, Bernier UR (1999) Identification of pupal exuviae of Nasonia vitripennis and Muscidifurax raptorellus parasitoids using cuticular hydrocarbons. Biol. Control 15: 97–106

Carvalho F, Dadour IR, Groth DM, Harvey ML (2005) Isolation and detection of ingested DNA from the immature stages of *Calliphora dubia* (Diptera: Calliphoridae): a forensically important blowfly. For Sci Med Path 1:261–265

Catts EP, Haskell NH (1990) Entomology and death: a procedural guide. Joyce's Print Shop Inc, Clemson, SC

Chen W-Y, Hung T-H, Shiao S-F (2004) Molecular identification of forensically important blow fly species (Diptera: Calliphoridae) in Taiwan. J Med Entomol 41:47–57

Clery JM (2001) Stability of prostate specific antigen (PSA) and subsequent Y-STR typing of *Lucilia (Phaenicia) sericata* (Meigen) (Diptera: Calliphoridae) maggots reared from a simulated postmortem sexual assault. Forensic Sci Int 120:72–76

Cornwell P (2003) Blow fly. Penguin Group (USA)

Davis JB, Goff ML (2000) Decomposition patterns in terrestrial and intertidal habitats on Oahu Island and Coconut Island Hawaii. J Forensic Sci 45:836–842

De Jong GD, Hoback WW (2006) Effect of investigator disturbance in experimental forensic entomology: Succession and community composition. Med Vet Entomol 20:248–258

Disney RHL, Munk T (2004) Potential use of Braconidae (Hymenoptera) in forensic cases. Med Vet Entomol 18:442–444

DiZinno JA, Lord WD, Collins-Morton MB, Wilson MR, Goff ML (2002) Mitochondrial DNA sequencing of beetle larvae (Nitidulidae: Omosita) recovered from human bone. J Forensic Sci 47:1337–1339

Erzinçlioglu Z (1990) On the interpretation of maggot evidence in forensic cases. Med Sci Law 30:65–66

Erzinçlioglu Z (2000) Maggots, murder and men: memories and reflections of a forensic entomologist. Harley Books, Colchester, England

Evett IW, Weir BS (1998) Interpreting DNA evidence. Sinauer, Sunderland, MA

Fabritius K, Klunker R (1991) Die Larven- und Puparienparasitoide von synanthropen Fliegen. Angewandte Parasitologie 32: 1–20

Florin AB, Gyllenstrand N (2002) Isolation and characterization of polymorphic microsatellite markers in the blowflies *Lucilia illustris* and *Lucilia sericata*. Mol Ecol Notes 2:113–116

Galloway A, Walsh-Haney H, Byrd JH (2001) Recovering buried bodies and surface scatter: the associated anthropological botanical and entomological evidence In: Byrd JH Castner JL (eds) Forensic entomology: the utility of arthropods in legal investigations. CRC, Boca Raton, pp 223–262

Gannon R (1997) The Body farm. In: Popular Science (September) 76-82.

Gaudry E, Myskowiak J-B, Chauvet B, Pasquerault T, Lefebvre F, Malgorn Y (2001) Activity of the forensic entomology department of the French Gendarmerie. Forensic Sci Int 120:68–71

Geden CJ (1997) Development models for the filth fly parasitoids *Spalangia gemina*, *S cameroni*, and *Muscidifurax raptor* (Hymenoptera: Pteromalidae) under constant and variable temperatures. Biol Control 9:185–192

Goff ML (2000) A fly for the prosecution: how insect evidence helps solve crimes. Harvard University Press, Cambridge, MA

Goff ML, Odom CB (1987) Forensic entomology in the Hawaiian Islands: three case studies. Am J Foren Med Pathol 8:45–50

Godfray HCJ (1994) Parasitoids – behavioral and evolutionary ecology. Princeton University Press, Princeton New Jersey

Grassberger M, Frank C (2003) Temperature-related development of the parasitoid wasp *Nasonia vitripennis* as forensic indicator. Med Vet Entomol 17:257–262

Grassberger M, Frank C (2004) Initial study of arthropod succession on pig carrion in a central European urban habitat. J Med Entomol 41:511–523

Grassberger M, Reiter C (2002) Effect of temperature on development of the forensically important holarctic blowfly *Protophormia terraenovae* (Robineau-Desvoidy) (Diptera: Calliphoridae). Forensic Sci Int 128:177–182

Haefner JN, Wallace JR, Merritt RW (2004) Pig decomposition in lotic aquatic systems: the potential use of algal growth in establishing a postmortem submersion interval (PMSI). J Forensic Sci 49:1–7

Harvey ML (2005) An alternative for the extraction and storage of DNA from insects in forensic entomology. J Forensic Sci 50:1–3

Harvey ML, Dadour IR, Gaudieri S (2003) Mitochondrial DNA cytochrome oxidase I gene: potential for distinction between stages of some forensically important fly species (Diptera) in Western Australia. Forensic Sci Int 131:134–139

Haskell NH, Hall RD, Cervenka VJ, Clark MA (1997) On the body: insect's life stage presence, their postmortem artifacts. In: Haglund WD, Sorg MH (eds) Forensic Taphonomy: The postmortem fate of human remains. CRC Press, Boca Raton. pp 415–448

Haskell NH, Williams RE (1990) Collection of entomological evidence at the death scene. In: Catts EP, Haskell NH (eds) Entomology and death: a procedural guide. Joyce's Print Shop Inc, Clemson, SC

Haskell NH, Lord WD, Byrd JH (2000) Collection of entomological evidence during death investigations. In: Byrd JH, Castner JL (eds) Forensic entomology: the utility of arthropods in legal investigations. CRC, Boca Raton, FL

Haskell NH, McShaffrey DG, Hawley DA, Williams RE, Pless JE (1989) Use of aquatic insects in determining submersion interval. J Forens Sci 34:622–632

Hewadikaram KA, Goff ML (1991) Effect of carcass size on rate of decomposition and arthropod succession patterns. Am J Forensic Med Pathol 12:235–240

Hobischak NR, Anderson GS (2002) Time of submergence using aquatic invertebrate succession and decompositional changes. J Forensic Sci 47:142–151

Hobischak NR, VanLaerhoven SL, Anderson GS (2006) Successional patterns of diversity in insect fauna on carrion in sun and shade in the Boreal Forest Region of Canada near Edmonton, Alberta. Can Entomol 138:376–383

Holdaway FG, Evans AC (1930) Parasitism a stimulus to pupation: Alysia manducator in relation to the host Lucilia sericata. Nature 125: 598–599

Hunt P (2005) Third Annual North American Forensic Entomology Association Conference, Orlando, FL

Ireland S, Turner B (2006) The effects of larval crowding and food type on the size and development of the blowfly *Calliphora vomitoria*. Forensic Sci Int 159:175–181

Ji YJ, Zhang DX, Hewitt GM, Kang L, Li DM (2003) Polymorphic microsatellite loci for the cotton bollworm *Helicoverpa armigera* (Lepidoptera: Noctuidae) and some remarks on their isolation. Mol Ecol Notes 3:102–104

Jones SW, Dobson ME, Francesconi SC, Schoske R, Crawford R (2005) DNA assays for detection identification and individualization of select agent microorganisms. Croat Med J 46:522–529

Joy JE, Liette NL, Harrah HL (2006) Carrion fly (Diptera: Calliphoridae) larval colonization of sunlit and shaded pig carcasses in West Virginia, USA. Forensic Sci Int 164:183–192

Junqueira ACM, Lessinger AC, Torres TT, da Silva FR, Vettore AL, Arruda PA, Espin MLA (2004) The mitochondrial genome of the blowfly *Chrysomya chloropyga* (Diptera: Calliphoridae). Gene 339:7–15

Keiper JB, Chapman EG, Foote BA (1997) Midge larvae (Diptera: Chironomidae) as indicators of postmortem submersion interval of carcasses in a woodland stream: a preliminary report. J Forensic Sci 42:1074–1079

Kreike J, Kampfer S (1999) Isolation and characterization of human DNA from mosquitoes (Culicdae). Int J Legal Med 112:380–382

LaMotte LR, Wells JD (2000) p-Values for postmortem intervals from arthropod succession data. J Agric Biol Environ Stat 5:58–68

LaSalle J, Gauld ID (1991) Parasitic Hymenoptera and the biodiversity crisis. Redia 74: 315–334

Lenstra JA (2003) DNA methods for identifying plant and animal species in food. In: Lees M (ed) Food authenticity and traceability. CRC, Boca Raton

Linville JG, Wells JD (2002) Surface sterilization of a maggot using bleach does not interfere with mitochondrial DNA analysis of crop contents. J Forensic Sci 47:1–5

Linville JG, Hayes J, Wells JD (2004) Mitochondrial DNA and STR analyses of maggot crop contents: effect of specimen preservation technique. J Forensic Sci 49:1–4

Lord WD, DiZinno JA, Wilson MR, Budlowle B, Taplin D, Meinking TL (1998) Isolation amplification and sequencing of human mitochondrial DNA obtained from human crab louse *Pthirus pubis* (L) blood meals. J Forensic Sci 43:1097–1100

Malgorn Y, Coquoz R (1999) DNA typing for identification of some species of Calliphoridae: an interest in forensic entomology. Forensic Sci Int 102:111–119

Marchenko MJ (2001) Medicolegal relevance of cadaver entomofauna for the determination of time since death. Forensic Sci Int 120:89–109

Melbye J, Jimenez S (1997) Chain of custody from the field to the courtroom. In: Haglund WD, Sorg MH (eds) Forensic taphonomy: the post-mortem fate of human remains. CRC, Boca Raton, pp 65–75

Merck MD (2007) Veterinary forensics. Animal cruelty investigations. Blackell, Oxford 368

Merrit RW, Wallace JR (2001) The role of aquatic insects in forensic investigations In: Byrd JH Castner JL (eds) Forensic entomology – the utility of arthropods in legal investigations. CRC, Boca Raton, London, pp 177–222

Mukabana W, Takken RW, Knols BGJ (2002) Analysis of arthropod bloodmeals using molecular genetic markers. Trends Parasitol 18:505–509

Mumcuoglu K, Gallili YN, Reshef A, Brauner P (2004) Use of human lice in forensic entomology. J Med Entomol 41:803–806

Norris KR 1965 The bionomics of blow flies. Ann Rev Entomol 10: 47–68

Nuorteva P (1977) Sarcosaprophagous insects as forensic indicators. In: Tedeschi CG, Eckert WG, Tedeschi LG (eds) Forensic medicine: a study in trauma and environmental hazards. Saunders, Philadelphia, pp 1072–1095

Nuorteva P, Schumann H, Isokoski M, Laiho K (1974) Studies on the possibilities of using blowflies (Diptera: Calliphoridae) as medicolegal indicators in Finland. Ann Entomol Fenn 40:70–74

Ratcliffe ST, Webb DW, Weinzievr RA, Robertson HM (2003) PCR-RFLP identification of Diptera (Calliphoridae Muscidae and Sarcophagidae) – a generally applicable method. J Forensic Sci 48:1–3

Replogle J, Lord WD, Budlowle B, Meinking TL, Taplin D (1994) Identification of host DNA by amplified fragment length polymorphism analysis: preliminary analysis of human crab louse (Anoplura: Pediculidae) exreta. J Med Entomol 31:686–690

Reznik SY, Chernoguz DG, Zinovjeva KB (1992) Host searching, oviposition preferences and optimal synchronization in *Alysia manducator* (Hymenoptera: Braconidae), a parasitoid of the blowfly, *Calliphora vicina*. Oikos 65:81–88

Romano A (2004) American Society of Crime Laboratory Directors/Laboratory Accreditation Board (http://wwwascld-laborg/)

Sachs JS (1998) A maggot for the prosecution. In: Discover Magazine, pp 103–108
Sachs JS (2001) Corpse: nature forensics and the struggle to pinpoint time of death. Perseus Books Group, Cambridge, MA
Scala JR, Wallace JR (2005) Case study: the uncertainty of establishing a post-mortem (PMI) interval based on entomological evidence incorporating the influence of elevation on ambient temperature reconstruction. Proceedings of the American Academic Forensic Science 58th Annual Meeting. Seattle, WA
Schiro G J (2001) Extraction and quantification of human deoxyribonucleic acid and the amplification of human short tandem repeats and a sex identification marker from fly larvae found on decomposing tissue. MS Thesis University of Central Florida
Schoenly K (1992) A statistical analysis of successional patterns in carrion-arthropod assemblages: implications for forensic entomology and the determination of the postmortem interval. J Forensic Sci 37:1489–1513
Schoenly K, Griest K, Rhine S (1991) An experimental field protocol for investigating the postmortem interval using multidisciplinary indicators. J Forensic Sci 36:1395–1415
Singh D, Greenberg B (1994) Survival after submergence in the pupae of five species of blow flies (Diptera: Calliphoridae). J Med Entomol 31:757–759
Smith KGV (1986) A manual of forensic entomology. The Trustees British Museum, London
Sorg MH, Dearborn JH, Monahan EI, Ryan HF, Sweeney KG, David E (1997) Forensic taphonomy in marine contexts. In: Haglund WD, Sorg MH (eds) Forensic taphonomy: the postmortem fate of human remains. CRC, LLC, Boca Raton, pp 567–604
Sperling FAH, Anderson GS, Hickey DA (1994) A DNA-bases approach to the identification of insect species used for postmortem interval estimation. J Forensic Sci 39:418–427
Staff (2005) Love of bugs gives edge to death determinations. St Croix Source
Statheropoulos M, Spiliopoulou C, Agapiou A (2005) A study of volatile organic compounds evolved from the decaying human body. Forensic Sci Int 153:147–155
Stevens J, Wall R, Wells JD (2002) Paraphyly in Hawaiian hybrid blowfly populations and the evolutionary history of anthropophilic species. Insect Mol Biol 11:141–148
Stevens J, Wall R (1996) Species sub-species and hybrid populations of the blowflies *Lucilia cuprina* and *Lucilia sericata* (Diptera: Calliphoridae). Proc R Soc Lond B 263: 1335–1341
Stevens J, Wall R (1997) The evolution of ectoparasitism in the genus *Lucilia* (Diptera: Calliphoridae). Int J Parasitol 27:51–59
Stevens J, Wall R (2001) Genetic relationships between blowflies (Calliphoridae) of forensic importance. Forensic Sci Int 120:116–123
Tabor KL, Brewster CC, Fell RD (2004) Analysis of the successional patterns of insects on carrion in southwest Virginia. J Med Entomol 41:785–795
Thyssen PJ, Lessinger AJ, Azeredo-Espin AML, Linhares AX (2005) The value of PCR-RFLP molecular markers for the differentiation of immature stages of two necrophagous flies (Diptera: Calliphoridae) of potential forensic importance. Neotrop Entomol 34:777–783
Torres TT, Azeredo-Espin AML (2005) Development of new polymorphic microsatellite markers for the New World screw-worm *Cochliomyia hominivorax* (Diptera: Calliphoridae). Mol Ecol Notes 5:815–817
Torres TT, Brondani RPV, Garcia JE, Azeredo-Espin AML (2004) Isolation and characterization of microsatellite markers in the New World screw-worm *Cochliomyia hominivorax* (Diptera: Calliphoridae). Mol Ecol Notes 4:182–184
Turner B, Howard T (1992) Metabolic heat generation in dipteran larval aggregations: a consideration for forensic entomology. Med Vet Ent 6:179–181
Vance GM, VanDyk JK, Rowley WA (1995) A device for sampling aquatic insects associated with carrion in water. J Forens Sci 40:479–482
Voss SC, Spafford H, Dadour IR (2009) Annual and seasonal patterns of insect succession on decomposing remains at two locations in Western Australia, Forensic Sci. Int. doi:10.1016/j.forsciint.2009.08.014

Wallman JF, Adams M (2001) The forensic application of allozyme electrophoresis to the identification of blowfly larvae (Diptera: Calliphoridae) in Southern Australia. J Forensic Sci 46:681–684

Wallman JF, Donnellan SC (2001) The utility of mitochondrial DNA sequences for the identifcation of forensically important blowflies (Diptera: Calliphoridae) in Southeastern Australia. Forensic Sci Int 120:60–67

Wallman JF, Leys R, Hogendoorn K (2005) Molecular systematics of Australian carrion-breeding blowflies (Diptera: Calliphoridae) based on mitochondrial DNA. Invertebr Syst 19:1–15

Wan QH, Fang SG (2003) An extremely sensitive species-specific ARMS PCR test for the presence of tiger bone DNA. Forensic Sci Int 131:75–78

Watson EJ, Carlton CE (2003) Spring succession of necrophilous insects on wildlife carcasses in Louisiana. J Med Entomol 40:338–347

Watson EJ, Carlton CE (2005) Insect succession and decomposition of wildlife carcasses during fall and winter in Louisiana. J Med Entomol 42:193–203

Wells JD, Williams DW (2007) Validation of a DNA-based method for identifying Chrysomyinae (Diptera: Calliphoridae) used in a death investigation. Int J Legal Med 121:1–8

Wells JD, Sperling FAH (1999) Molecular phylogeny of *Chrysomya albiceps* and *Chrysomya rufifacies* (Diptera: Calliphoridae). J Med Entomol 36:222–226

Wells JD, Sperling FAH (2001) DNA-based identification of forensically important Chrysomyinae (Diptera: Calliphoridae). Forensic Sci Int 120:110–115

Wells JD, LaMotte LR (2001) Estimating the postmortem interval. In: Byrd JH, Castner JL (eds) Forensic entomology: the utility of arthropods in legal investigations. CRC, Boca Raton, FL, pp 263–286

Wells JD, Pape T, Sperling FAH (2001a) DNA-Based identification and molecular systematics of forensically important Sarcophagidae (Diptera). J Forensic Sci 46:1098–1102

Wells JD, Introna F, Di Vella G, Campobasso CP, Hayes J, Sperling FAH (2001b) Human and insect mitochondrial DNA analysis from maggots. J Forensic Sci 46:685–687

Wells JD, Lunt N, Villet MH (2004) Recent African derivation of *Chyrsomya putoria* from *Chrysomya chloropyga* and mitochondrial DNA paraphyly of cytochrome oxidase subunit one in blowflies of forensic importance. Med Vet Entomol 18:445–448

Wells JD, Wall R, Stevens JR (2007) Phylogenetic analysis of forensically important *Lucilia* flies based on cytochrome oxidase I sequence: a cautionary tale for forensic species determination. Int J Legal Med 121:229–233

Whiting AR (1967) The biology of the parasitic wasp Mormoniella vitripennis (Walker). Quarterly review of biology 42:333–406

Zehner R, Amendt J, Krettek R (2004a) STR typing of human DNA from fly larvae fed on decomposing bodies. J Forensic Sci 49:1–4

Zehner R, Amendt J, Schütt S, Sauer J, Krettek R, Povolný D (2004b) Genetic identification of forensically important flesh flies (Diptera: Sarcophagidae). Int J Legal Med 118:245–247

Index